W0018335

Interfacial Chemistry of Rocks and Soils

Surfactant Science Series

Founding Editor: Martin J. Schick (1918–1998)

Series Editor: Arthur T. Hubbard

Books in the Surfactant Science series emphasize surfaces and interfaces, including basic principles, major developments, and important applications. A substantial proportion of all physical phenomena involve interfaces in one way or another, and the practical and commercial applications of interface science are numerous. The series covers experimental phenomena, behavior and properties, major advances, experimental approaches, essential instrumental methods, theoretical strategies, and important applications. The level of presentation is intended for readers having a basic scientific training, such as advanced science students encountering the book topic for the first time, as well as scientific professionals refreshing their knowledge of engineering aspects of the topic and interface science.

Colloids in Drug Delivery
Edited by Monzer Fanun

Applied Surface Thermodynamics: Second Edition
Edited by A. W. Neumann, Robert David, and Yi Y. Zuo

Colloids in Biotechnology
Edited by Monzer Fanun

Electrokinetic Particle Transport in Micro/Nano-fluidics: Direct Numerical Simulation Analysis
Shizhi Qian and Ye Ai

Nuclear Magnetic Resonance Studies of Interfacial Phenomena
Vladimir M. Gun'ko and Vladimir V. Turov

The Science of Defoaming: Theory, Experiment and Applications
Peter R. Garrett

Soil Colloids: Properties and Ion Binding
Fernando V. Molina

Surface Tension and Related Thermodynamic Quantities of Aqueous Electrolyte Solutions
Norihiro Matubayasi

Electromagnetic, Mechanical, and Transport Properties of Composite Materials
Rajinder Pal

Silicone Dispersions
Edited by Yihan Liu

Wetting and Spreading Dynamics, Second Edition
Edited by Victor Starov, Manuel G. Velarde

Interfacial Chemistry of Rocks and Soils, Second Edition
Noémi Nagy, József Kónya

Surfactants from Renewable Raw Materials
Edited by Divya Bajpai Tripathy, Anjali Gupta, Arvind Kumar Jain, Anuradha Mishra

For more information about this series, please visit: https://www.crcpress.com/Surfactant-Science/book-series/CRCSURFACSCI

Interfacial Chemistry of Rocks and Soils
Second Edition

Noémi M. Nagy
József Kónya

CRC Press
Taylor & Francis Group
Boca Raton London New York

CRC Press is an imprint of the
Taylor & Francis Group, an **informa** business

The cover photo shows cation exchanged montmorillonites.

Second edition published 2022
by CRC Press
6000 Broken Sound Parkway NW, Suite 300, Boca Raton, FL 33487-2742

and by CRC Press
2 Park Square, Milton Park, Abingdon, Oxon, OX14 4RN

© 2022 Taylor & Francis Group, LLC

First edition published by CRC Press 2009
CRC Press is an imprint of Taylor & Francis Group, LLC

Reasonable efforts have been made to publish reliable data and information, but the author and publisher cannot assume responsibility for the validity of all materials or the consequences of their use. The authors and publishers have attempted to trace the copyright holders of all material reproduced in this publication and apologize to copyright holders if permission to publish in this form has not been obtained. If any copyright material has not been acknowledged please write and let us know so we may rectify in any future reprint.

Except as permitted under U.S. Copyright Law, no part of this book may be reprinted, reproduced, transmitted, or utilized in any form by any electronic, mechanical, or other means, now known or hereafter invented, including photocopying, microfilming, and recording, or in any information storage or retrieval system, without written permission from the publishers.

For permission to photocopy or use material electronically from this work, access www.copyright.com or contact the Copyright Clearance Center, Inc. (CCC), 222 Rosewood Drive, Danvers, MA 01923, 978-750-8400. For works that are not available on CCC please contact mpkbookspermissions@tandf.co.uk

Trademark notice: Product or corporate names may be trademarks or registered trademarks and are used only for identification and explanation without intent to infringe.

ISBN: 978-0-367-85682-3 (hbk)
ISBN: 978-1-032-11725-6 (pbk)
ISBN: 978-1-003-02008-0 (ebk)

DOI: 10.4324/9780429318726

Typeset in Times
by Deanta Global Publishing Services, Chennai, India

Contents

Preface to the First Edition ix

Preface to the Second Edition..xi

Authors.. xiii

Chapter 1   Components of Soil- and Rock-Solution Systems................................ 1

   1.1   Solid: Soil and Rock 1

       1.1.1   Mineral and Chemical Composition of Rocks
and Soils 2

           1.1.1.1   Silicates...3

           1.1.1.2   Oxides...9

           1.1.1.3   Other Minerals..9

       1.1.2   Organic Matter in Soil..9

       1.1.3   Particle and Pore Sizes, External and Internal
Surfaces 12

   1.2   Liquid: Soil and Groundwater Solutions 14

       1.2.1   Compositions of Natural Water................................. 14

       1.2.2   Parameters Affecting Chemical Speciation 15

       1.2.3   Dissolution and Precipitation of Solids 22

       1.2.4   Properties of Very Dilute Solutions 27

   1.3   Interface of Rock/Soil–Aqueous Solutions 29

       1.3.1   The Structure of the Interface 30

       1.3.2   Characterization of the Interface of Geological
System/Groundwater.. 33

           1.3.2.1   Characterization of External Surfaces........ 33

           1.3.2.2   Characterization of Internal Surfaces.........40

       1.3.3   Interfacial Processes Related to External and
Internal Surfaces 41

           1.3.3.1   Adsorption 41

           1.3.3.2   Ion Exchange 42

           1.3.3.3   Precipitation..45

           1.3.3.4   Evaluation and Separation of
Interfacial Processes 45

       1.3.4   Quantitative Treatment of Interfacial Processes 46

           1.3.4.1   Adsorption Isotherms 46

           1.3.4.2   Treatment of Ion Exchange Processes 51

           1.3.4.3   Treatment of Simultaneous Ion
Exchange and Adsorption Processes 65

           1.3.4.4   Treatment of Heterogeneous Isotope
Exchange..66

       1.3.5   Catalytic Effects of Clays...67

1.3.6   Kinetics of Interfacial Processes in Rocks
          and Soils 69
          1.3.6.1   Steps of Interfacial Reactions 70
          1.3.6.2   Kinetics of Heterogeneous Isotope
                    Exchange 75
          1.3.6.3   Migration of Water-Soluble Substances
                    in Rocks and Soils 75
          1.3.6.4   Disposal of Nuclear Waste in
                    Geological Formations 77
References 79

Chapter 2   Interfacial Processes in Geological Systems: Studies on
             Montmorillonite Model Substance 91

2.1   Use of Montmorillonite as a Model Substance: The
        Important Interfacial Processes on Montmorillonite 92
        2.1.1   Crystal Structure and Layer Charge of
                  Montmorillonite 92
        2.1.2   Clay–Water Interactions 95
        2.1.3   Edge Charges of Montmorillonite 96
        2.1.4   Montmorillonite as a Model Substance 97
2.2   Cation Exchange: Outer-Sphere Complexation 98
2.3   Synthesis and Characterization of Cation-Exchanged
        Montmorillonites 103
2.4   Surface Acid–Base Properties of Montmorillonite 107
        2.4.1   Formation of Edge Sites on Montmorillonite 107
        2.4.2   Effect of Permanent Charge on the Study of
                  Edge Charges 108
        2.4.3   Acid–Base Properties of Cation-Exchanged
                  Montmorillonites 109
2.5   Ion Adsorption on the External Surfaces 113
        2.5.1   Some Examples of Ion Sorption Processes on
                  Montmorillonite 114
                  2.5.1.1   Sorption of Zinc Ion on Montmorillonite 114
                  2.5.1.2   Sorption of Manganese Ion on
                            Montmorillonite 116
                  2.5.1.3   Sorption of Palladium Ion on
                            Montmorillonite 116
                  2.5.1.4   Sorption of Lead Ion on Montmorillonite 118
2.6   Separation of Interfacial Processes of Montmorillonite 118
2.7   Role of Hydrogen Ions in the Interfacial and Dissolution
        Processes of Montmorillonite 120
        2.7.1   Effect of Hydrogen Ion on the Cation Exchange
                  Processes 122
        2.7.2   Acidic Destruction of Montmorillonite 127

2.8   Effect of Complexation Agents 128
       2.8.1   Effect of Complex Formation in the EDTA/
               Ca-Montmorillonite System 132
       2.8.2   Effect of Complex Formation in the EDTA/
               Ca-Montmorillonite/Manganese(II) Ion System...... 136
       2.8.3   Effect of Stability Constants on the Cation
               Composition of the Interlayer Space 138
       2.8.4   Effect of Complex Formation in the EDTA/
               Ca-Montmorillonite/Lead(II) Ion System................ 140
2.9   Sorption of Organic Matter on Minerals.............................. 142
       2.9.1   Sorption of EDTA on Montmorillonite................... 143
       2.9.2   Sorption of Valine on Montmorillonite................... 144
               On the Layer Charges... 145
               On the Edge Sites 145
2.10   Transformations Initiated by Interfacial Processes of
       Montmorillonite... 149
       2.10.1 Oxidation of Mn(II) Ion and Formation
               of a Nanolayer in the Interlayer Space of
               Montmorillonite 150
       2.10.2 Formation of an Iron(III) Oxidhydroxide
               Nanolayer in the Interlayer Space of
               Montmorillonite 153
       2.10.3 Reduction of Ions 158
               2.10.3.1 Reduction of Silver and Palladium Ion..... 158
               2.10.3.2 Reduction of Fe(III) Ions........................ 160
       2.10.4 Heterogeneous Nucleation on Edge Site:
               Formation of Lead Oxide Fine Particles on the
               Edges of Montmorillonite 160
       2.10.5 Lanthanide Ion Exchange: Structural
               Modification Due to the Exchange with Light
               Lanthanide and Ittrium Ions.................................... 169
       2.10.6 Structural Change upon Heating Lanthanide-
               Bentonites... 171
2.11   Effect of Chemical Modification of Bentonite on
       Sorption Properties: Sorption of Anions.............................. 172
References 177

Chapter 3   Interfacial Reactions at Rock and Soil Interfaces 187

3.1   Relationship between the Interfacial Properties and the
       Geological Origin of Bentonite Clay.................................... 187
       3.1.1   Geological and Mineral Characteristics of
               Sajóbábony Bentonites 188
       3.1.2   Interfacial Properties of Bentonite Samples from
               Sajóbábony (HU).. 193

3.1.3   Relations of Geological Origin and Interfacial
        Properties 196
3.1.4   Applications of Bentonites of Different
        Interfacial Properties.. 197
3.2   Migration of Water-Soluble Substances in Rocks 198
3.2.1   Sorption and Migration of Carrier-Free
        Radioactive Isotopes in Rocks 198
        3.2.1.1   A Model Predicting Migration Rate on
                  the Basis of Mineral Composition 198
        3.2.1.2   The Application of the Linear Model
                  for the Sorption of Cs-137 and Sr-85
                  Ions.. 201
3.2.2   Effect of Sorption on the Migration of Ions in
        Bentonite 208
3.2.3   Effect of Precipitation in Migration 210
3.3   Interfacial Acid–Base Properties of Soils 212
3.4   Sorption of Cyanide Anion on Soil and Sediment 218
3.5   Sorption of Phosphate Anion on Soils.................................225
References 229

Chapter 4   Experimental Methods in Studying Interfacial Processes of
            Rocks and Soils 233

4.1   Analysis of the Solid Phase 233
      4.1.1   Methods of Chemical Analyses 233
      4.1.2   Study of Mineral Composition...............................235
      4.1.3   Study of Morphology 236
      4.1.4   Study of Soil Organic Matter 240
4.2   Analysis of the Liquid Phase..240
4.3   Study of Interfaces...241
References 247

Index..249

Preface to the First Edition

Geological formations, soils, and rocks are essential components of the environment. Their chemistry and chemical transformations play an important role in the quality of life for all living beings. Both natural and anthropogenic processes can result in effects that last for a long time. We have to comprehend the basic interactions that take place between geological materials and different substances in order to be able to drive the processes toward advantageous results. One important group of these interactions occurs on the interface of rocks and soil. In these so-called interfacial reactions, the structure of the soil and rocks does not change significantly; however, they are the first step toward any chemical and other transformations of the soils and rocks. Thus, it is very important to understand the interfacial processes of rocks and soils.

Soils and rocks can adsorb or absorb a wide variety of substances from the environment. The mechanism and strength of sorption, and the chemical transformations due to sorption, are influenced by mineral structure, interfacial forces, the geological environment, and the chemical species produced as a result of such factors. The interactions are complicated because of the large variety of chemical entities present. For example, the chemical structure of soils and rocks can vary, and there may be many different substances present, depending on the actual geochemical environment. Furthermore, the presence of groundwater, dissolved minerals, and other components as well as the pH, redox conditions, gaseous components (e.g., carbon dioxide), complex forming agents, organic matter, living organisms, etc., do have an important influence on the interactions.

In this book, the processes at solid/liquid interfaces of soil and rock, in most cases under environmental conditions, will be discussed. A scientifically correct description of interfacial processes requires the study of the properties of solid and liquid phases and the interface, as well as the interactions of these phases. Previous books typically focus on selected aspects of the subject, such as, for example, the properties of the solid phase or the interactions of selected substances such as heavy metal ions with solid/rock. We intend to present a comprehensive treatment of the solid–liquid-interface system, emphasizing the importance of the chemical species produced in a geological material/solution/interface interaction. We recommend the book to all chemists, geologists, and soil scientists working in interfacial, environmental, and soil sciences.

The book consists of four chapters. The first one deals with the individual components of the studied systems: the solid, the solution, and the interface. "Solid" means rocks and soils, namely, the main mineral and other solid components. In order that the solid/liquid interactions become possible, these must be located in the Earth's crust where the groundwater is present. The liquid phase refers to soil solution and groundwater, and also any solutions that are part of laboratory experiments studying interfacial properties with the objective of understanding the principles behind the reactions.

In Chapter 1, the characteristics and thermodynamics of the interface will be discussed, including the importance of chemical species. The kinetic aspects of interfacial reactions will be illustrated, too.

In Chapter 2 the interfacial processes are discussed in a model system. A clay mineral, montmorillonite, has been chosen to illustrate important interfacial processes of geological formations. Some transformations initiated by interfacial processes are demonstrated.

In Chapter 3 the interfacial processes of real rock and soil/solution systems are studied. It is shown how the basic phenomena of the interfacial processes can be applied for the treatment of practical geological, soil science, and environmental problems.

Finally, in Chapter 4 the most important analytical methods used in the study of the interfacial processes of rocks and soils will be discussed briefly.

The book summarizes the results and knowledge of the authors' research work in this field over several decades. The main part of the studies has been done by radioisotopic labeling in the Imre Lajos Isotope Laboratory, Department of Physical Chemistry, University of Debrecen, Hungary, so we had the opportunity to work together with many excellent colleagues and exceptional students. Many thanks for their contributions! We are grateful for their assistance in the experimental work and the useful discussions that helped to advance our understanding in this field. We thank Dr. Klára Kónya for the critical reading of the manuscript, for her remarks, and corrections.

Preface to the Second Edition

The first edition of *Interfacial Processes of Cocks and Soils*, published in 2009, has been improved and complemented in this second edition, especially by the addition of some novel results in the field.

In Chapter 1, the section on the comparison of ion exchange isotherms and the law of mass action is extended to heterovalent exchanges. The kinetic section is significantly improved. A new section is included on disposal of nuclear waste in geological formations.

In Chapter 2, new sections discuss the behavior of montmorillonite exchanged with trivalent cations, namely, the preparation of chromium (III) montmorillonite, the hydrolysis of iron (III) ions in the interlayer, and the structural transformation of montmorillonite lattice in some lanthanide montmorillonites. The mineral transformation upon heating of lanthanide bentonite is also shown. In addition, the chemical modification of bentonite for the sorption of anions (phosphate, arsenite, chloride, iodide, pertechnetate) discussed.

In Chapter 3, a new section deals with the sorption of phosphate ion on different soil types. The application of heterogeneous isotope exchange reaction for the determination of plant available phosphate species is illustrated.

In Chapter 4, two less known methods in the study of porous substances, the application of nuclear magnetic resonance and positron annihilation spectroscopy, are discussed; many thanks to Professor István Bányai and Dr. Mónika Kéri (Department of Physical Chemistry, University of Debrecen) for their assistance.

We thank Dr. Klára Kónya for the critical reading and improvement of the manuscript of the second edition.

The second edition was supported by the EU and co-financed by the European Regional Development Fund under the project GINOP-2.3.2-15-2016-00008 and the Hungarian National Research, Development, and Innovation Office (NKFIH K 120265).

Authors

Noémi M. Nagy is a professor of radiochemistry in the Imre Lajos Isotope Laboratory of Physical Chemistry, University of Debrecen, Hungary. She earned her M.Sc. and Ph.D. degrees in radiochemistry at that university. She has a D.Sc. degree in agrochemistry from the Hungarian Academy of Sciences. Dr Nagy has more than 35 years of experience in nuclear and radiochemistry teaching. Her research interest is the study of the interfacial processes of natural sorbents, including soils, rocks, clay minerals mainly by radioactive tracer methods. Recently, she has been dealing principally with studies of nuclear waste storage. She has written or co-written numerous peer-reviewed scientific papers and is the co-author four books in the fields of nuclear and radiochemistry as well as the interfacial chemistry of geological formations. She won the George Hevesy's award for nuclear safety in Hungary. Dr Nagy is the president of the Radiochemical Scientific Committee of Hungarian Academy of Sciences.

József Kónya is a professor of radiochemistry in the Imre Lajos Isotope Laboratory of Physical Chemistry, University of Debrecen, Hungary. He earned his M.Sc. and Ph.D. in physical chemistry from that same university. He holds a D.Sc. degree in Radiochemistry from the Hungarian Academy of Sciences. Dr Konya has more than 60 years of experience teaching in the fields of nuclear and radiochemistry. His research interest is the study of the interfacial processes of natural sorbents, including soils, rocks, and clay minerals mainly by radioactive tracer methods. Recently, he has been dealing principally with studies of nuclear waste storage. He has written or co-written numerous peer-reviewed articles and is the co-author of four books in the fields of nuclear and radiochemistry as well as the interfacial chemistry of geological formations. He won the George Hevesy's Award for nuclear safety in Hungary.

1 Components of Soil- and Rock-Solution Systems

1.1 SOLID: SOIL AND ROCK

The crust of the Earth is made up of rocks and soils, the latter produced by the weathering of rocks. Therefore, rocks and soils contain the same chemical elements, and their mineral composition is also similar. In addition, soils contain organic matter.

Rocks are defined as incoherent, naturally formed substances, mostly composed of minerals. They can be classified as igneous, sedimentary, and metamorphic, depending on their origin. Rocks are stable in the conditions under which they were formed. When environmental conditions are changed, they become thermodynamically unstable. This happens on the surface of the Earth when the rocks interact with the atmosphere, hydrosphere, and biosphere. The formation of soils is the result of the interactions between the rocks and the atmosphere, hydrosphere, or biosphere. The mineral transformations, including weathering, have been extensively studied and described as interactions of the lithosphere, hydrosphere, atmosphere, and biosphere (Kabata-Pendias and Pendias 1985, p.15). These processes are not limited to the solid phase; the liquid phase is also involved and plays an important role. As in all chemical processes, these interactions are driven toward achieving a chemical equilibrium, that is, the species formed must be thermodynamically more stable under the particular conditions than the original material was. The same can be said for any substances getting onto the rocks and soils regardless of its source (e.g., anthropogenic or other), which tend to transform if they are thermodynamically unstable. Obviously, the environment is not a closed system; therefore, the laws of thermodynamics will only determine the direction of the processes but not their final results. Furthermore, kinetic barriers also limit achievement of the equilibrium. The soil is an intermediate stage of this thermodynamically driven process during which rocks transform into other stable species under changed conditions.

Soil consists of an abiotic (as mentioned earlier) and a biotic (derived from living matter) part. This book deals with the abiotic part, the components of which are shown in Figure 1.1.

Abiotic soil components form a three-phase polydisperse system. About 50 v/v% of soil is solid, 50 v/v% is soil solution, and air, depending on humidity. As seen in Figure 1.1, the solid phase consists of mineral and organic components (Section 1.1.2). The main constituents of the gas phase (soil air) is composed of vapor and gases of the air (nitrogen, oxygen, carbon dioxide, etc.), and other gases such as hydrocarbons, radon, anthropogenic pollutions, etc., depending on local conditions. The soil solution is made up of the water-soluble components of solid and soil air. Its composition is determined by the geological conditions (composition of rocks and soils) and influenced by natural and anthropogenic factors.

DOI: 10.1201/9781003020080-1

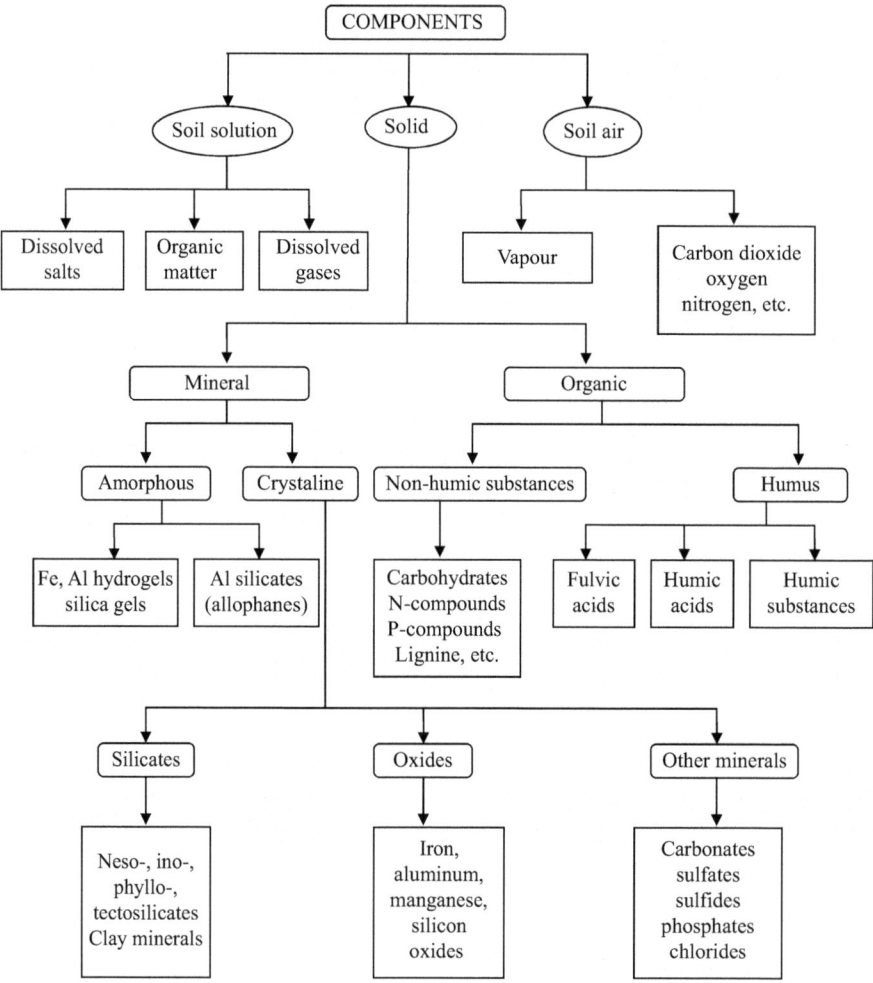

FIGURE 1.1 Components of the abiotic part of soil.

1.1.1 Mineral and Chemical Composition of Rocks and Soils

Minerals are natural, usually inorganic, substances with a definite and predictable chemical composition and physical properties (O' Donoghue 1990). Some organic substances (e.g., oxalates, amber, etc.) are also considered as minerals.

Minerals are arranged by the Dana and Strunz classification systems (Gaines et al. 1997; Strunz 1941). The groups are as follows:

Elements.
Sulfides.
Halogenides.
Oxides and hydroxides.

Nitrates, carbonates, and borates.
Sulfates, chromates, molybdates, and tungstates.
Phosphates, arsenates, and vanadates.
Silicates.
Organic compounds.

In the soil and rocks near to the surface, silicates, oxides, carbonates, sulfides, sulfates, and chlorides are the most important, and therefore, they are presented briefly here.

1.1.1.1 Silicates

The basic unit of silicates is the tetrahedral SiO_4^{4-}. Its central atom is silicon, surrounded by four oxygen atoms. In the literature, silicon is frequently presented as Si^{4+} ion. This representation simplifies the problem of the Si-O bond, which lies in the difference of electronegativities between silicon and oxygen (1.8), and it implies that the Si-O bond is partly covalent, partly ionic. The SiO_4^{4-} tetrahedron, therefore, can be considered as a complex anion (similarly, for example, to sulfate or carbonate), and the neutralizing cations form nearly pure ionic bond with it (Nemecz 1973).

The tetrahedrons can be present as isolated units, or they may form simple or double polymer chains or bands, layers, or three-dimensional network. When the SiO_4^{4-} tetrahedrons are bound in a two-dimensional layer, some of the oxygen atoms form the layer, with the rest of the oxygen atoms rising above the plane of the layer (these are the so-called peak oxygen atoms). The peak oxygen atoms then can form further chemical bonds. The negative charge of the tetrahedron is neutralized by cations, or with positively charged octahedral units or layers where the central atom of the octahedrons can be, for example, Al^{3+} or Mg^{2+}, surrounded by oxygen or hydroxide groups.

In the Strunz classification system these different silicate arrangements are classified in the following way:

Nesosilicate (isolated tetrahedron) structures: with $[SiO_4]^{4-}$ groups, cations of tetrahedral orientation.
Nesosubsilicate (isolated/semi-isolated tetrahedron) structures: with anions unfamiliar to tetrahedrons, cations of octahedral and tetrahedral orientation.
Sorosilicate (dimer) structures: with $[Si_2O_7]^{6-}$ gropus, without anions unfamiliar to tetrahedrons.
Unclassified silicate structures: mostly cations, of small size, such as Mg, Fe, Mn, and Cu.
Cyclosilicate (ring) structures: triplet rings $[Si_3O_9]^{6-}$.
Inosilicate (chain and band) structures: double chains $[Si_2O_6]^{4-}$.
Intermediate (layered/chain) structures.
Phyllosilicate (layered) structures: tetragonal or pseudotetragonal layered structures $[Si_4O_{10}]^{4-}$ and others.
Tectosilicate (network) structures: without anions unfamiliar to the tetrahedrons.

The most important types, characteristics, and representatives of silicates are summarized in Table 1.1.

As seen in Table 1.1, there are minerals for which the Si^{4+} of the tetrahedron is substituted by Al^{3+} through an isomorphic substitution in the tetrahedral layer. Since the substituting cation has a smaller charge than the substituted one, the result is a negative excess charge that is neutralized again by cations (feldspar) or positively charged octahedral layers (phyllosilicates).

Isomorphic substitution is especially significant in the case of clay minerals (hydrous aluminum silicates), which are formed by the weathering of the primary silicates of the igneous rocks. Most clay minerals belong to the phyllosilicates, but there are some of fibrous and amorphous structure.

The layered clay minerals can be classified on the basis of the number and quality of the layers:

1. Two-layer clays (1:1 clay, TO): consist of a tetrahedral silica and an octahedral alumina layer. The layers are fixed in a well-defined distance by the hydrogen bonds (H-OH) between the silica and alumina layers.
2. Three-layer clays (2:1 clays, TOT): an AlO(OH) octahedral sheet is between two tetrahedral SiO_4 layers. The distance of the layers can be varied (-O-O- bonds), so the layers can be expanded.
3. 1+1 or TOT+O clays: octahedral oxide layers (e.g., magnesium aluminum oxide) are encapsulated between the TOT layers.
4. Mixed structures.

Oxide and hydroxide ions around the central ions of the tetrahedral and octahedral layers for three anion sheets can be of the following types:

X-type: contains the oxide ions in hexagonal arrangements.
Y-type: a compact structure; every third oxide is substituted by hydroxide ions.
Z-type: similar to Y-type, but it contains only hydroxide ions.

The most important layered clay minerals are listed in Table 1.2 (Mackenzie and Mitchell 1966; Mackenzie 1975).

The formulas in Table 1.2 (and, of course, in Table 1.1) represent an idealized composition. The real structure and composition of the minerals varies from place to place, depending on the chemical composition and conditions under which they are formed. For example, trace elements also build into the crystal lattice and their concentration gives very useful petrogenic information (Charlet et al. 1994; McDonough and Sun 1995).

What can also be seen in Table 1.2 is that each group is usually further divided into two series: dioctahedral (D) and trioctahedral (T) according to the number of central positions of the octahedrons occupied by cations. In the trioctahedral minerals, every central position is filled, usually with Mg^{2+}, while in the dioctahedral minerals, two-thirds of the central position is filled with Al^{3+} ions (e.g., montmorillonite; Figure 2.1).

TABLE 1.1

Silicates

Type	Structure	General formula *	Representatives *
Nesosilicates	Isolated tetrahedrons with $[SiO_4]^{4-}$ groups. The size and physical parameters of the lattice depend on iron content.	M_2SiO_4 M:bivalent cation (Fe^{2+}, Mg^{2+}, Mn^{2+}, etc.)	Forsterite (Mg_2SiO_4) Fayalite (Fe_2SiO_4) Olivine (Forsterit:fayalit = 80:20)
Inosilicates	Chain and band structures joint with metal ions	$M_2Si_2O_6$ M:Ca^{2+}, Mg^{2+}, Fe^{2+}	Enstatite ($MgSiO_3$) Ferrosillite ($FeSiO_3$) Aaugite $[(Mg, Fe, Al) (Si, Al)_2O_6]$
	Double chains (amphiboles)	$M_7Si_8O_{22}(OH, F)_2$	Tremolite $[Ca_2Mg_5Si_8O_{22}(OH)_2]$ Aktinolite $[Ca_2(Fe^{III})_5Si_8O_{22}(OH)_2]$
Phyllosilicates	SiO_4^{4-} tetrahedrons form two-dimensional layers (T-layer), the excess charge is compensated by other two-dimensional octahedral layers (O-layer)	There are bivalent cations in the central positions of the octahedral layer (Mg^{2+}, Fe^{2+}). Every central position is filled (trioctahedral layer).	Talc $[Si^{IV}_8Mg^{VI}_6O_{20}(OH)_4]$ Biotite $K_2[(Al_2Si_6)^{IV}(Mg, Fe^{II})_6{}^{VI}O_{20}(OH)_4]$
		There are trivalent cations in the octahedral layer (Al^{3+}, Fe^{3+}). Two-thirds of the central positions is filled (dioctahedral layer).	Pyrophyllite $[Si^{IV}_8Al^{VI}_4O_{20}(OH)_4]$ Biotite $K_2[(Al_2Si_6)^{IV}(Al)_4{}^{VI}O_{20}(OH)_4]$

(Continued)

TABLE 1.1 (CONTINUED)
Silicates

Type	Structure	General formula *	Representatives *	
Tectosilicates	Three-dimensional structure. SiO_4^{4-} tetrahedrons are bonded by all four oxygens.	SiO_2	Quartz (SiO_2)	
	The ratio of Si-Al substitutions is significant (feldspars)	$M[(Si_{4-x}, Al_x)^{IV} O_8]$ M:Na^+, K^+, Ca^{2+} x:1 … 2	Orthoclase [$K(Al, Si_3)O_8$] Plagioclase feldspars (mixture of albite [$Na(Al, Si_3)O_8$] and anorthite [$K(Al_2, Si_2)O_8$])	
	The (Si, Al)O_4 tetrahedrons are connected so that one-, two- or three-dimensional holes are within the crystal network.		Zeolite group (natrolite)	Natrolite Analcime
			Zeolite group (heulandite)	Heulandite Clinoptilolite Mordenite
			Zeolite group (phillipsite)	Phillipsite Chabazite Gmelinite

* IV and VI in the upper indices show tetrahedral or octahedral configuration.

TABLE 1.2
Layered Clay Minerals

Type	Group	Series[a]	Mineral	Ideal formula	Charge/ unit cell	State of interlayer cations
TO	Kaolinite	D	Kaolinite Halloyzite	$Si_4{}^{IV}Al_4{}^{VI}O_{10}(OH)_8 \cdot nH_2O$	≈ 0	No
TOT	Illite	D	Illite	$(Si_{8-x}Al_x)^{IV}(Al_{4-y}Mg_y)^{VI}O_{20}(OH)_4$ $(K+M)_{(x+y)}$	≈ 2, or less	Dehydrated, fixed
		T	Hydrobiotite	$(Si_{8-x}Al_x)^{IV}(Mg_{6-y}Al_y)^{VI}O_{20}(OH)_4$ $(K+M)_{(x-y)}$		
	Vermicu- lite	D	Vermiculite	$(Si_{8-x}Al_x)^{IV}(Al_{4-y}Mg_y)^{VI}O_{20}(OH)_4 \cdot nH_2O$ $M^+{}_{(x+y)}$	$1.2 - 1.8$	Exchangeable, hydrated
		T	Vermiculite	$(Si_{8-x}Al_x)^{IV}(Mg_{6-y}Al_y)^{VI}O_{20}(OH)_4 \cdot nH_2O$ $\rightarrow M^+{}_{(x-y)}$		

$N = 0$
$n = 2$

M:hydrated cations
x and y: charge
of T and O
layers $(x > y)$

(Continued)

TABLE 1.2 (CONTINUED)
Layered Clay Minerals

Type	Group	Series[a]	Mineral	Ideal formula	Charge/unit cell	State of interlayer cations
	Smectite	D	Montmorillonite	$(Si_8)^{IV}(Al_{4-y}Mg_y)^{VI}O_{20}(OH)_4 \cdot nH_2O$ $\begin{array}{c} - \quad\quad - \\ \downarrow \\ M^+_y \end{array}$	0.5 – 1.2	Exchangeable, hydrated
			Beidellite	$(Si_{8-x}Al_x)^{IV}(Al_{4-y}Mg_y)^{VI}O_{20}(OH)_4 \cdot nH_2O$ $\begin{array}{c} - \quad\quad + \\ \downarrow \\ M^+_{(x-y)} \end{array}$		
			Nontronite	Great iron content		
		T	Hectorite	$(Si_8)^{IV}(Mg_{6-y}Li_y)^{VI}O_{20}(OH)_4 \cdot nH_2O$ $\begin{array}{c} - \quad\quad - \\ \downarrow \\ M^+_y \end{array}$		
			Saponite	$(Si_{8-x}Al_x)^{IV}(Mg_{6-y}Al_y)^{VI}O_{20}(OH)_4 \cdot nH_2O$ $\begin{array}{c} - \quad\quad + \\ \downarrow \\ M^+_{(x+y)} \end{array}$		
TOT+O	Chlorite	D-T	Sudoite			
		T	Chlinoclore	$(Mg_4Al_2)^{VI}(OH)_{12} \cdot (Si_6Al_2)^{IV}(Mg_5Al)^{VI}O_{20}(OH)_4$	≈ 2	Fixed

[a] D = dioctahedral; T = trioctahedral; D-T = di-trioctahedral.

Source: Mackenzie and Mitchell 1966; Mackenzie 1975.

In the case of layered clay minerals, isomorphic substitutions occur not only in the tetrahedral layers, but in the octahedral layers, too. When Al^{3+} is in the central positions of octahedrons (dioctahedral minerals), it may be substituted by Mg^{2+}, Fe^{2+}, or Fe^{3+}, producing a negative layer charge in the case of bivalent cation substitution, similarly to the Si^{4+}–Al^{3+} substitutions. However, when the central ion of the octahedrons is Mg^{2+} (trioctahedral minerals), and it is substituted by a trivalent Al^{3+}, a positively charged octahedral layer is formed. The net layer charge is given by the sum of the charges of the two types of layers. As seen in Table 1.2, a large variety of net layer charge (depending on the mineral) can exist; however, most of the time it is negative. When the net layer charge is negative, additional cations are needed for its neutralization. Depending on their size, the neutralizing cations can be dehydrated and be fixed in the lattice or interlayer space or may keep their hydrate water and remain exchangeable. Later in this chapter we will show that the exchange of these cations is one of the most important interfacial reactions of clays and soils (Chapter 1.3.3).

1.1.1.2 Oxides

Among the oxides, the silicon, aluminum, iron, and manganese oxides and hydroxides are the most important (Table 1.3).

Iron and manganese can have different oxidation states, depending on the redox conditions of the environment. Iron(II) compounds, however, are only stable under anaerobic conditions; they transform iron(III) compounds on contant with air and groundwaters (pH = 6–8), therefore, the interfacial processes of iron(II) oxides and hydroxide can play a smaller role under environmental conditions. (Note: The iron(II) of silicates can also transform into iron(III) during weathering.)

1.1.1.3 Other Minerals

Among the other minerals, the most important representatives of carbonates, sulfides, sulfates, and chlorides are summarized in Table 1.4.

Some of the minerals ($Na_2CO_3*10H_2O$, $MgSO_4*7H_2O$, $Na_2SO_4*10H_2O$, $NaCl$) in Table 1.4 have a good solubility in water, so the interfacial process of these minerals means dissolution and precipitation of these minerals. In the presence of groundwater, they take part in other interfacial processes as dissolved ions. In addition, carbonates ($CaCO_3$, $MgCO_3$, $CaMg(CO_3)_2$) are also dissolved as hydrocarbonates, depending on the partial pressure of carbon dioxide and the pH of the groundwater (Chapter 1.2.1).

1.1.2 Organic Matter in Soil

The redox conditions in the environment will determine which carbon species are thermodynamically stable: carbon dioxide under oxidative conditions, elementary carbon and hydrocarbons under reductive conditions. The oxidation state of organic substances is always between these two thermodynamically stable species. Since these organic substances are thermodynamically not stable, they are driven to transform into carbon dioxide or into elementary carbon/hydrocarbons; which processes,

TABLE 1.3
Oxides and Hydroxides

Type	General Formula	Representative
	Silicon (see Table 1.1)	
Crystalline	SiO_2	Quartz
Amorphous	$SiO_2.nH_2O$	Opal
	Aluminum	
Crystalline oxide	$\alpha-Al_2O_3$	Corundum
Crytalline hydroxide	$\gamma-Al(OH)_3$	Gypsite
Crystalline oxide hydroxide	$\gamma-AlO(OH)$	Boehmite
Amorphous hydroxide	$Al(OH)_3.nH_2O$	Amorphous aluminum hydroxide
	Iron	
Crystalline oxides	$\alpha-Fe_2O_3$	Hematite
	Fe_3O_4	Magnetite
Crystalline oxide hydroxide	$\alpha-FeO(OH)$	Goethite
	$\gamma-FeO(OH)$	Lepidokrokit
Crystalline hydroxides	$Fe(OH)_3.(n-x)H_2O$	Ferrihydrites
Amorphous hydroxide	$Fe(OH)_3.nH_2O$	Amorphous ferri hydroxide
	Manganese	
Crystalline oxide	$\beta-MnO_2$	Pyrolusite
Crystalline oxide hydroxide	$\gamma-MnO(OH)$	Manganite
Amorphous hydroxide	$Mn(OH)_3.nH_2O$	Amorphous manganese(III) hydroxide

however, being limited by reaction kinetic barriers. This means that the formation of thermodynamically stable species can be very slow and the life-time of the non-stable species is fairly long.

These organic substances are produced during the decomposition of plants and animals, so they are always present in the soil and also in the layers of rocks very close to the biosphere. For this reason, they are called soil organic matter (Sparks 2003; Stevenson 1982) and they play an important part in the interfacial processes in rocks and soils.

The soil organic matter can be divided into two parts:

1. Non-specific organic matter or non-humic substances: these are produced during the decomposition of plants and animals, and their structure is known and well-defined. Carbohydrates, organic N- and P-compounds, lignine, and organic acids are mentioned here.
2. Specific organic matter or humic substances: these are produced by the connection of the decomposition products. Their structure is not exactly known and depends on the production conditions. On the basis of solubility, humic substances are classified as fulvic acid, humic acid, humin, and hymatomelanic acid (Stevenson 1982).

TABLE 1.4
Other Important Minerals

Type	General formula	Representative
	Carbonates	
Carbonates with simple formula	$CaCO_3$	Calcite
	$CaCO_3$	Aragonite
	$MgCO_3$	Magnesite
	$Fe(II)CO_3$	Siderite
Hydrated carbonate	$Na_2CO_3*10H_2O$	Natron, native soda
Carbonate with compound formula	$CaMg(CO_3)_2$	Dolomite
	Sulfides and sulfates	
Anhydrous sulfate	$CaSO_4$	Anhydrite
Hydrated sulfates	$CaSO_4*2H_2O$	Gypsum
	$MgSO_4*7H_2O$	Boehmite
	$Na_2SO_4*10H_2O$	Mirabilite
Sulfide	FeS_2	Pyrite
	Phosphates	
Hydrated phosphates	$FePO_4*2H_2O$	Strengite
	$AlPO_4*2H_2O$	Variscite
	$Fe_3(PO_4)_4*8H_2O$	Vivianite
	Chlorides	
Sodium chloride	$NaCl$	Halite

For the purpose of studying interfacial processes, humic substances are considered as coiled polymer chains containing a relatively large number of functional groups (Ksiezopolska 2000). Since this book deals with the interfacial processes, the structure of organic matter itself is not discussed here; only the functional groups relevant in them are shown in Table 1.5.

Table 1.5 classifies the functional groups as simple and complex ones. The simple groups contain oxygen and nitrogen. The oxygen-containing groups are listed on the basis of the oxidation states of the connected carbon atoms, starting with the reduced states toward the oxidized states. The complex functional groups are formed in the reaction of the simple functional groups.

These functional groups are present in the non-humic and humic substances, too. Non-humic substances are usually small molecules with some functional groups, their protonation/deprotonation and complex-forming reactions with metal ions are relatively simple and can be described with thermodynamic dissociation and stability constants. Humic substances, however, contain many different functional groups that are in interaction with each other, so the processes can hardly be described quantitatively. Furthermore, metal ions can also be bound to two or more ligands of different molecules (ligand-metal ion-ligand bridges). As a result, the organic matter-metal ion interactions lead to a large variety of different chemical species, even in the case of one specific metal ion-functional group reactions. For example, a chelate formation of a metal ion with

TABLE 1.5

Characteristic Functional Groups of Soil Organic Matter

Functional group	Chemical structure	Character of the group
	Simple functional groups	
	Oxygen-containing groups	
Alcoholic OH	R—OH	Neutral
Phenolic OH	Ar—OH	Acidic
Enol (see ketone)[a]	R—C=C—OH	Acidic
Aldehyde	R—C(=O)H	Neutral
Ketone (see enol)[a]	R—C(=O)R'	Neutral
Quinone	O=⬡=O	Acidic
Carboxyl	R—C(=O)OH	Acidic
	Nitrogen-containing groups	
Amine	R—CH₂—NH₂	Basic
Amide	R—C(=O)—NH₂	Basic
	Complex functional groups	
Ether (by connection of alcoholic OHs)	R—CH₂—O—CH₂—R'	Neutral
Ester (by connection of alcoholic OH and carboxyle)	R—C(=O)—O—CH₂—R'	Neutral
Peptide (by connection of amine and carboxyle	R—C(=O)—NH—CH₂—R'	Amphoter

[a] Enol and keto groups transform into each other.

a small molecule gives a water-soluble substance, and this reaction with humic matter can result in a water-insoluble substance. These processes have an important role in the transportation of metal ion in the geological environment.

The functional groups of soil organic matter can take place in the usual organic chemical reactions with the constituents of soil, such as esterification with phosphate ion (Anderson, 1960).

1.1.3 PARTICLE AND PORE SIZES, EXTERNAL AND INTERNAL SURFACES

The interfacial processes take place on the surface of the solid (rock or soil), so the specific surface area (area per mass) is one of the most important parameters. The specific surface area depends on the particle size or the particle size distribution, the roughness of the surface. These factors determine the external specific surface area. In addition, the rocks and soils contain a network of pores with different sizes and shapes. Some of them are in a direct connection with the external surface (open pores). Therefore, depending on the relative sizes of the pores and the particles of the substance, the different substances of the surroundings can enter the pores. Two especially important types of the pores are:

1. The holes of humic substances.
2. The interlayer space of layered minerals, that is, the space between the parallel sheets.

These two types represent the internal surface area.

The components of the soils/rocks have different size, shape, and quality. The particle size of the organic components is usually in the colloid range (<500 nm); the mineral components have different dispersity. The first classification system of soils on the basis of particle sizes was constructed by Atterberg (1905). Practically, this classification has been used until now, though some countries have their own classifications, considering their widespread soil types. The size of soil particles will also determine how the soil fraction is named (e.g., clay, sand, silt, rock, etc.). Table 1.6 provides these names along with the standard diameter of the particles for the international classification system.

A value for porosity can be calculated from the bulk density (φ_{bulk}) and particle density ($\varphi_{particle}$):

$$\Phi = 1 - \frac{\varphi_{bulk}}{\varphi_{particle}} \tag{1.1}$$

The particle density or true density of a particulate solid or powder is the density of the particles that make up the powder, in contrast to the bulk density, which measures the average density of a large amount of the powder in a specific medium (usually air). The normal particle density of geological formations is assumed to be

TABLE 1.6
Classification of Soils by Particle Size

Particle	Diameter
Clay	<0.002 mm
Silt	0.002–0.02 mm
Fine sand	0.02–0.2 mm
Coarse sand	0.2–2.0 mm
Fine gravel	2.0–20 mm
Coarse gravel	20–200 mm
Stone	>200 mm

Source: Atterberg 1905.

approximately 2.65 g/cm³, although a better estimation can be obtained by examining the lithology of the particles.

Porosity is divided using the International Union of Pure and Applied Chemistry (IUPAC) system (Rouquerol et al. 1994), based on pore size, into the following groups: macropores (>50 nm), mesopores (2–50 nm), and micropores (<2 nm). Microporosity may then be subdivided into three subsequent categories: supermicropores (1.4–2.0 nm), micropores (0.5–1.4 nm), and ultramicropores (<0.5 nm). Both mineral and organic soil components have pores with different diameter. The holes and channels in the polymer chain of the humic substances as well as the interlayer space of the layered mineral have an important role in determining the specific surface area. The size of the interlayer space of layered minerals in dry state is some 10 nm, so they are considered to be micropores.

The total surface of the rock and soils is the sum of the external and the internal surfaces. The external surface area is usually called the surface area and it is measured by the adsorption of apolar nitrogen gas at temperatures near to the boiling point of liquid nitrogen (Brunauer et al. 1938). Since the diameter of nitrogen molecules is 0.38 nm (Zdravkov et al. 2007), they can penetrate into the macro- and mesopores, so their surface is included in the external surface area. Nitrogen molecules cannot penetrate into the micropores. When the rock or soil contains no humic substances and layered minerals, the total surface area is equal to the external surface area. To determine the total specific surface area when humic substances and expandable layered minerals are present in the sample, the adsorption of different dipolar molecules, such as water vapor, ethylene glycol (Fiedler, Wagner 1967), ethylene glycol monoethyl ether (EGME), methylene blue (MB), and para-nitrophenol (pNP) can be measured since these molecules expand the layers of the clay minerals, and thus, they are able to get in contact with and be adsorbed on the internal surfaces. However, in the case of some adsorbents, the nature of the exchangeable cations of swelling clay minerals (Chapter 1.3.3.2) significantly influences the results. For example, reliable surface areas can be obtained with MB when the layer charge is between 0.28–0.33 per half unit cell (Kahr and Madsen 1995). In the case of EGME, the exchangeable cation as well as the layer charge modifies its apparent cross-sectional area (Kellomaki et al. 1989; Chiou and Rutherford 1993, 1997; Michot and Villiéras 2008). Nevertheless, the method gives useful comparative internal surface area data for rocks and soils (Tiller and Smith 1990; Churchman et al. 1991).

To obtain the internal surface area, the external surface area is subtracted from the total surface area. In the case of humic substances, the ratio of the external surfaces is below 1%. For clay minerals, the internal surface area is about 80–95% of the total surface area.

1.2   LIQUID: SOIL AND GROUNDWATER SOLUTIONS

1.2.1   Compositions of Natural Water

The main source of groundwaters (vadose water) is the precipitation falling onto the Earth's surface. In smaller quantity, the groundwaters form from the so-called juvenile water, that is, water formed chemically within the Earth and brought to the surface in intrusive rocks. Natural groundwater may contain a large variety of dissolved species, depending on the composition of geological formations where they have been formed and with which it came into contact. Most of the time, its pH is close to neutral. It may be considered as a complex electrolyte solution containing hydrated cations and anions, ion associations, soluble organic compounds, complexes, and dissolved molecules (gases) (Figure 1.1). The most important cations are calcium, magnesium, sodium, and potassium ions; the most important anions are hydrocarbonate/carbonate, chloride, and sulfate ions. In some cases, ammonium, aluminum, and ferrous/ferric ions (cations) as well as nitrate, SiO_3^{2-}, and AlO_3^{3-} (anions) are present. In addition, there are trace elements in all groundwaters. The soluble organic compounds are fulvic acids with small molecular weight, carboxylic acids, and carbohydrates. The dissolved gases are mainly oxygen and carbon dioxide. The main component of most groundwaters is $Ca(HCO_3)_2$, the total quantity of the dissolved material is 300–500 mg/dm^3. The second group of waters contain $Mg(HCO_3)_2$ and the quantity of the dissolved substances is much greater (~1500 mg/dm^3). On dry areas, groundwaters contain soda ($NaHCO_3$), while on marshy areas waters frequently contain sulfate, also. Naturally, groundwaters can contain anthropogenic pollutants, too, strongly depending on human activity.

1.2.2 Parameters Affecting Chemical Speciation

We have seen that groundwater has several chemical components. These can be present as different chemical species, depending on the chemical and environmental conditions. There are many factors affecting the chemical speciation, including concentrations, pH, redox conditions, natural and artificial organic substances, complex-forming agents, living organisms, and substances produced in their metabolism (e.g., hydrogen sulfide, etc.). The chemical effects of these factors are discussed here.

The effect of pH is illustrated by the example of the chemical speciation of hydrocarbonate, which is abundant and usually the most important anion of groundwaters. The hydrocarbonate content of groundwaters is usually high. Hydrocarbonate can be present as different species depending on pH. It is shown in Figure 1.2, where the speciation of a typical Hungarian groundwater is shown at different pH values (467 mg/dm^3 ≈ 7.5 mmol/dm^3 hydrocarbonate, 48 mg/dm^3 ≈ 1.2 mmol/dm^3 calcium content). The chemical equilibria in a closed system are as follows.

The two steps of the protonation of carbonate ions (Equations 1.2 and 1.3):

$$2H^+ + CO_3^{2-} \rightleftarrows H_2CO_3 \qquad \lg K = 16.5 \qquad (1.2)$$

$$H^+ + CO_3^{2-} \rightleftarrows HCO_3^-, \qquad \lg K = 10.2 \qquad (1.3)$$

The hydrolysis of calcium ions (Equation 1.4):

$$Ca^{2+} + OH^- \rightleftarrows CaOH^+ \qquad \lg K = -12.2 \qquad (1.4)$$

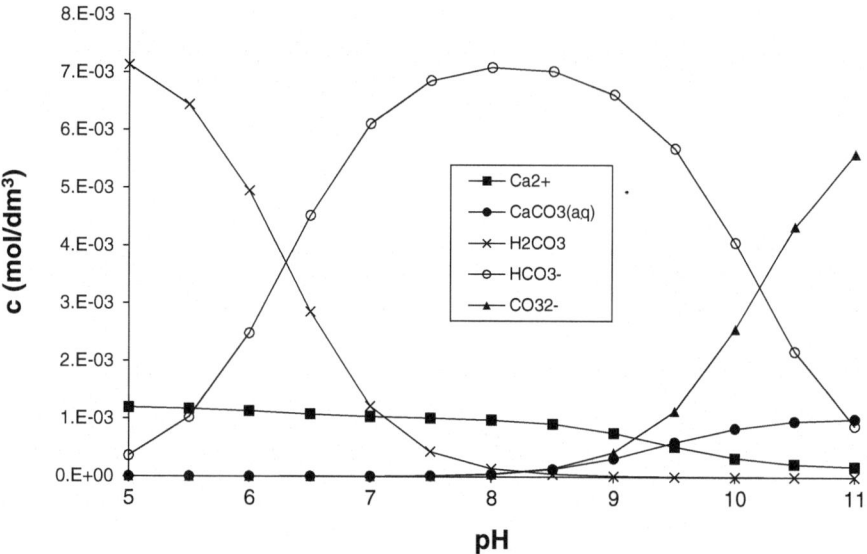

FIGURE 1.2   Chemical species of hydrocarbonate as a function of pH.

The reaction of calcium and carbonate ions, including the protonation of carbonate (Equations 1.5 and 1.6):

$$Ca^{2+} + H^+ + CO_3^{2-} \rightleftarrows CaHCO_3^+ \qquad \lg K = 11.6 \qquad (1.5)$$

$$Ca^{2+} + CO_3^{2-} \rightleftarrows CaCO_3 \qquad \lg K = 3.0 \qquad (1.6)$$

Equation 1.6 shows the precipitation of calcium carbonate, the quantity being calculated from the total calcium and dissolved calcium ions. For the sake of simplicity, it is considered as mol/dm³ units; however, in reality, it is a separated solid phase. The chemical species of hydrocarbonate as a function of pH are shown in Figure 1.2.

As seen in Figure 1.2, hydrocarbonate can be present as different species, depending on pH. At acidic pH values, H_2CO_3 is the dominant species in a closed system. In an open system, it is present as carbon dioxide, elaborating or dissolving, depending on its partial pressure. At neutral pH values, characteristic of natural waters, hydrocarbonate is dominant. As pH increases to the basic range, the concentration of carbonate increases. In the presence of cations forming insoluble carbonates (as calcium ion), precicipitates are formed (as calcium carbonate here). The groundwaters containing magnesium hydrocarbonates participate in similar reactions, only the equilibrium constant is different.

The oxygen content of groundwaters, or usually the geological environment, is also important. It determines the oxidation state of the different substances, that is, the redox equilibria. (Note: In the previous example of hydrocarbonate, it has no effect on the oxidation state.) A simplified picture of the redox equilibia is given by the equilibrium thermodynamic potential-pH diagrams (Pourbaix 1966).

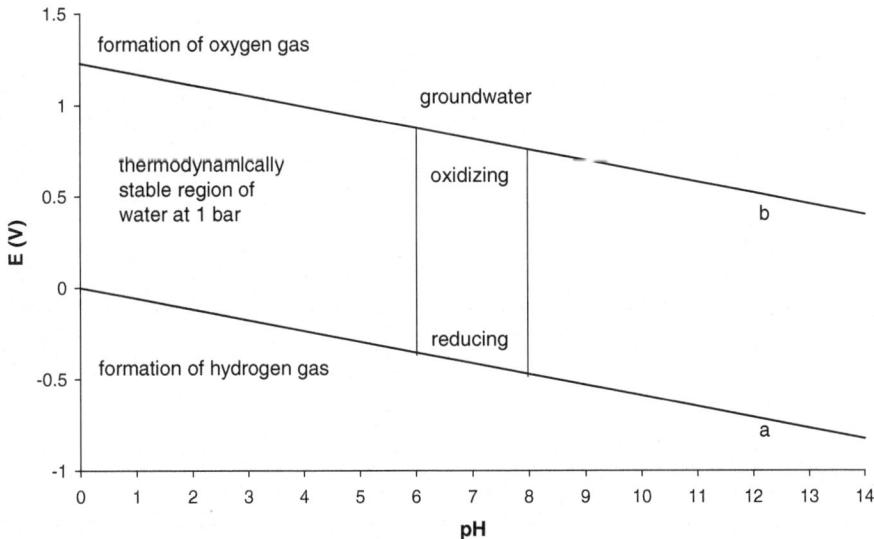

FIGURE 1.3 Potential-pH equilibrium diagram of water at 25°C.

The potential range of oxidation in groundwaters is determined by the pH-potential diagram of water, so the construction and evaluation of potential-pH diagrams are illustrated by the example of water (Figure 1.3).

The detailed explanations of the potential-pH diagrams can be found elsewhere (Pourbaix 1966); here a short discussion is given. The electrochemical potentials of the reactions between the two different oxidation states are calculated from the chemical potential (the free energy of the formation of the species) and a term indicating the presence of electrons. The calculated electrochemical potentials are plotted as a function of pH. In this way, the different pH and potential values can be determined where the different species are stable. In the case of water, its thermodynamically stable region of water is limited by two parallel lines labeled with a and b. Line a represents the equilibrium conditions of the reduction of water (or its hydrogen or oxonium ions) to hydrogen gas at 1 bar:

$$H_3O^+ + e^- \rightleftarrows \frac{1}{2}H_2 + H_2O \tag{1.7}$$

From the thermodynamic data, the redox potential of the Equation 1.7 is as follows:

$$E_{H^+/H_2} = E^0_{H^+/H_2} + \frac{RT}{F} \ln a_{H_3O^+} - \frac{RT}{F} \ln p_{H_2} = 0.000 - 0.059pH - 0.0295p_{H_2} \tag{1.8}$$

where $E^0_{H^+/H_2}$ is the standard potential (as redox potentials are expressed in the reference to standard hydrogen electrode [SHE], it is zero); R is the ideal gas constant, T is the temperature, F is the Faraday constant, $a_{H_3O^+}$ is the activity of hydrogen ions, and p_{H_2} is the partial pressure of hydrogen gas.

The line b shows the oxidation of water to oxygen gas:

$$6H_2O \rightleftarrows O_2 + 4H_3O^+ + 4e^- \tag{1.9}$$

$$E_{H_2O/O_2} = E^0_{H_2O/O_2} + \frac{RT}{F} \ln a_{H_3O^+} + \frac{RT}{4F} \ln p_{O_2} = 1.228 - 0.059pH + 0.015p_{O_2} \tag{1.10}$$

where p_{O_2} is the partial pressure of oxygen gas.

In groundwater, the range between the lines a and b is interesting. At typical pH values of groundwaters (pH = 6–8), oxidizing (aerobic; +400–+500 mV), anoxic, and reducing (anaerobic; below −100 mV) atmospheres are usually separated on the basis of redox potential.

To construct the pH-potential diagrams of the different elements, all their possible redox processes with water, oxygen, and hydrogen have to be taken into account, and the electrochemical potentials have to be calculated. In addition, the dissolution/precipitation equilibria (e.g., hydrolysis) have to be taken into consideration, as well. The main dissolved ions in groundwater (calcium, magnesium, sodium, and potassium cations; hydrocarbonate/carbonate, chloride, sulfate anions) have well-defined oxidation states; they have no characteristic redox processes under these conditions. The equilibrium thermodynamic potential-pH diagrams of two important redox elements of groundwaters and geological formations, iron and manganese, are shown in Figures 1.4 and 1.5. The figures show only the pH-potential range where water is stable (the range between a and b in Figure 1.3). The activities of dissolved iron and

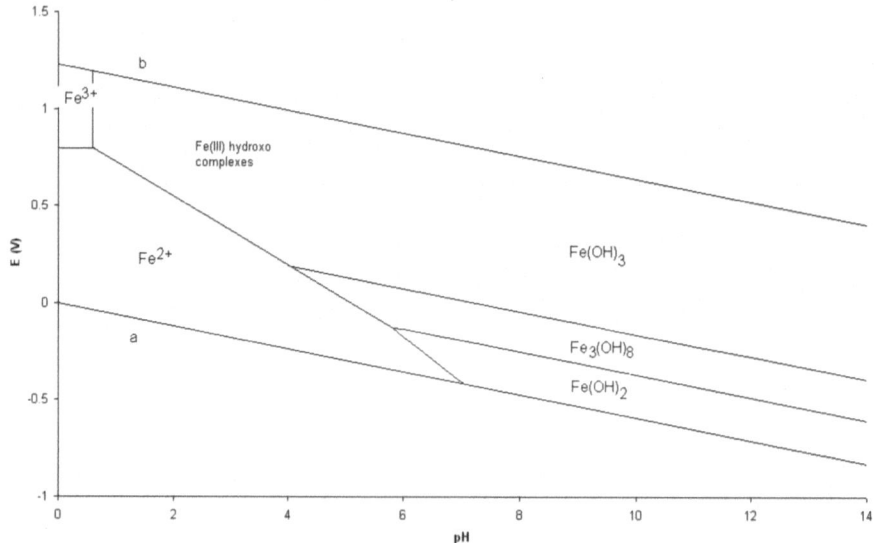

FIGURE 1.4   Potential-pH equilibrium diagram of iron at 25°C.

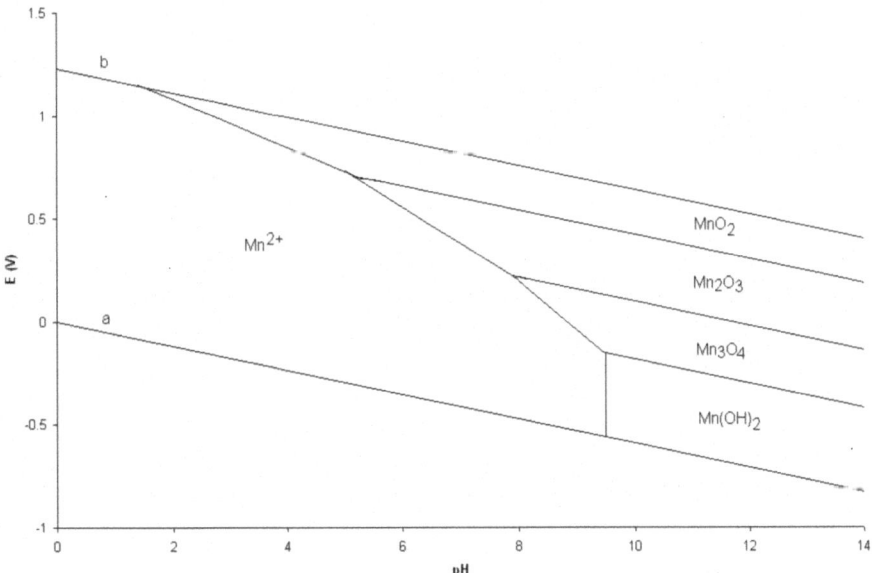

FIGURE 1.5   Potential- pH equilibrium diagram of manganese at 25°C.

manganese species are 10^{-3} mol dm^{-3}. To construct the diagrams the reactions are considered as follows.

For iron:

$$Fe^{2+} = Fe^{3+} + e^- \qquad E = 0.771 + 0.059 lg \frac{\left[Fe^{3+} \right]}{\left[Fe^{2+} \right]} \tag{1.11}$$

$$3Fe^{2+} + 4H_2O \rightleftarrows Fe_3O_4 + 8H^+ + 2e^- \qquad E = 0.980 - 0.2364pH - 0.0886 lg \left[Fe^{2+} \right] \tag{1.12}$$

$$2Fe^{2+} + 3H_2O \rightleftarrows Fe_2O_3 + 6H^+ + 2e^- \qquad E = 0.728 - 0.1773pH - 0.059 lg \left[Fe^{2+} \right] \tag{1.13}$$

$$2Fe^{3+} + 3H_2O \rightleftarrows Fe_2O_3 + 6H^+ \qquad lg \left[Fe^{3+} \right] = -0.72 - pH \tag{1.14}$$

$$2Fe_3O_4 + H_2O \rightleftarrows 3Fe_2O_3 + 2H^+ + 2e^- \qquad E = 0.221 - 0.059pH \tag{1.15}$$

$$Fe_3 \left(OH \right)_3 + H_2O \rightleftarrows 3Fe \left(OH \right)_3 + H^+ + e^- \qquad E = 0.429 - 0.059pH \tag{1.16}$$

For manganese:

$$3Mn^{2+} + 4H_2O \rightleftarrows Mn_3O_4 + 8H^+ + 2e^- \quad E = 1.824 - 0.2364pH - 0.0886lg\left[Mn^{2+}\right]$$

$$\text{(1.17)}$$

$$2Mn^{2+} + 3H_2O \rightleftarrows Mn_2O_3 + 6H^+ + 2e^- \quad E = 1.443 - 0.1773pH - 0.059lg\left[Mn^{2+}\right]$$

$$\text{(1.18)}$$

$$Mn^{2+} + 2H_2O \rightleftarrows MnO_2 + 4H^+ + 2e^- \quad E = 1.228 - 0.1182pH - 0.0295lg\left[Mn^{2+}\right]$$

$$\text{(1.19)}$$

$$3MnO + 2H_2O \rightleftarrows Mn_3O_4 + 2H^+ + 2e^- \quad E = 0.462 - 0.059pH \quad \text{(1.20)}$$

$$2Mn_3O_4 + H_2O \rightleftarrows 3Mn_2O_3 + 2H^+ + 2e^- \quad E = 0.689 - 0.059pH \quad \text{(1.21)}$$

$$Mn_2O_3 + H_2O \rightleftarrows 2MnO_2 + 2H^+ + 2e^- \quad E = 1.014 - 0.059pH \quad \text{(1.22)}$$

Figure 1.4 shows that, under oxidizing conditions, iron has +3 oxidation states and forms different solid hydroxide species. They transform into oxide hydroxides or oxides under dry conditions. Under reducing conditions, iron(II) can be dominant, for example, in reductive groundwaters. However, when at pH>7, iron(II) forms solid ferrous hydroxide or mixed ferric-ferrous hydroxide. Under dry conditions, it also transforms into oxide.

Figure 1.5 represents that manganese is stable as manganese(II) ions in broad pH-potential range, even under oxidizing conditions when pH is acidic. At high pH values, solid manganese hydroxides and oxides are formed, depending on the redox conditions. As the potential increases, the oxidation state of manganese increases.

In the previous examples, the concentration of the different species and the thermodynamically stable species were determined on the basis of equilibrium thermodynamic parameters, that is, chemical potential and, in practice, the stability constants of the substances (see, for example, Equations 1.2–1.6) and the electrochemical potential of the redox processes (Equations 1.7–1.10). The equilibrium calculations and potential-pH diagrams give very useful but limited information for two reasons. One is that the potential-pH diagrams use only two parameters (pH and potential) and do not take into account other effects (e.g., the effects of other ions, molecules). As previously shown, natural systems contain several other components (minerals, ions in groundwater, dissolved gases, organic matter, etc.), so the complete and correct thermodynamical treatment is rather problematic. The effect of other components can be calculated similarly to that of hydrocarbonate, but in this case the redox reactions are neglected. Thus, the overall effect of the many components and redox reactions is difficult to estimate. In addition, thermodynamic equilibria are valid exclusively in a closed system. Naturally, the geological environment cannot be considered as a closed system, so the equilibrium models show only the main tendencies of the processes.

The ions in the aqueous solutions are present as complexes. The most obvious complex is the aqua complex where the ion is surrounded by water molecules. The aqua complexes of ions are frequently called "free" ions. The presence of complexes is especially important in the interfacial reactions where the ions can keep the water molecules or lose them during the process. There are, however, complex-forming compounds in the environment that can form other complexes with the ions, especially with cations. These are inorganic anions (hydroxide, chloride, sulfate, phosphate, polyphosphates, carbonate, etc.) or organic compounds, including natural (carbohydrates, tartaric acid, citric acid, fulvic acid dissolved in soil solution, etc.) or anthropogenic (chelating agents of fertilizers, detergents, degradation products of plastics, etc.) complex-forming agents. It is worth emphasizing that the most important complex-forming agents of soils are the humic substances. They, however, are present mostly as colloid particles or solid phase, and therefore complex formation with their functional groups can be treated as interfacial reactions, when it causes the sorption of a dissolved species on a solid matter (Section 1.1.2).

Under geological conditions the substances dissolved in aqueous solutions are sorbed on the surface of the mineral and humic substances. The sorption processes are influenced by the composition of the solution, the stable chemical species. At pH $\approx$ 6–8, characteristic of the geological environment, some cations have hydrolytic tendencies, hydroxides and oxides can precipitate. When studying the interfacial reaction of rocks and soils, hydrolysis has to be avoided.

From the main cations of groundwaters, sodium and potassium ions do not form stable complexes, except aqua complexes; however, the other ions do. Complex formation plays an important role in the interactions of cations and rocks/soils because it significantly influences the species of cations in the groundwater. There are two other important factors in the formation of complexes and the interfacial processes: the stability and the charge of the complexes. The stability determines the ratio of the different species. The charge of the complex is important because several interfacial processes are determined by electrostatic interactions. Since the net surface charge of rocks and soils is usually negative under environmental conditions, interfacial processes (adsorption, ion exchange, etc.) occur with positive ions, including aqua complexes. The study of complex formation is important in the nutrient cycle of the ions in the soils, and it may have an important role in the decontamination of polluted soils.

The general principles of cation sorption and complex formation are discussed by Marcus and Kertes (1969) mainly for cation exchange resins. The anionic and neutral complexes are considered as non-adsorbed on cation exchangers (Bobleter and Bonn 1991). The anionic complexes of cations are sorbed on anion exchange resins. According to Schubert (Marcus and Kertes 1969), the cationic complexes are not sorbed on cation exchangers because they have smaller charge and greater size than hydrated cations (aqua complexes). According to Fronaeus (Marcus and Kertes 1969), the cationic complexes are also sorbed on cation exchangers. Both models consider pH to be constant.

Since the hydrated cations mainly take place in cation exchange reactions (Section 1.3.4.2), and their concentration is inversely proportional to the complex stability, the quantity of the sorbed cations on rocks and soils decreases as the stability of their complexes increases (Bonleter and Bonn 1991).

1.2.3   DISSOLUTION AND PRECIPITATION OF SOLIDS

As mentioned earlier, the composition of natural groundwaters depends on the composition of the geological formations where they originate from: they contain dissolved rock and soil components that were soluble under conditions (such as temperature and pressure) of their formation. Their dissolution is governed by the law of thermodynamics; that is, dissolution occurs when the solution is undersaturated with respect to components of rocks and soils. Provided that the solid components are present in sufficient quantity and there is no kinetic barrier, this process may lead to a thermodynamic equilibrium. The reversed process of dissolution is precipitation, that is, the formation of a solid phase from the dissolved components of a supersaturated solution. The composition of the precipitate is frequently different from that of the original dissolved mineral: a new solid phase can be formed (Stumm 1992).

The weathering of minerals can be understood as a continuous dissolution-precipitation process involving them. The process can be very complicated, leading to multi-phase equilibria, in which more than one solid phase, solution, and gas phases may be present. For example, primary silicates transform into secondary silicate minerals via such weathering reactions (Stumm and Wollast, 1990), as in the formation of kaolinite ($Al_2Si_2O_5(OH)_4$) from anorthite ($CaAl_2Si_2O_8$):

$$CaAl_2Si_2O_8 + 2CO_2 + 3H_2O \rightleftarrows Al_2Si_2O_5(OH)_4 + Ca^{2+} + 2HCO_3^- \quad (1.23)$$

or the formation of gibbsite ($Al(OH)_3$) and goethite ($FeOOH.H_2O$) from biotite ($KMgFe_2AlSi_3O_{10}(OH)_2$):

$$KMgFe_2AlSi_3O_{10}(OH)_2 + \frac{1}{2}O_2 + 3CO_2 + 11H_2O \rightleftarrows$$
$$Al(OH)_3 + 2FeOOH.H_2O + K^+ + Mg^{2+} + 3HCO_3^- + 3H_4SiO_4 \quad (1.24)$$

Equations 1.23 and 1.24 show the formation of some of the cations and anions mentioned previously in this chapter as the main dissolved ions present in groundwaters. It is to be noted, as well, that water and carbon dioxide are the main reactants in the weathering processes through the weakly acidic effect they produce together. Redox reactions may also take place; for example, iron(II) ion of biotite oxidizes to iron(III) after dissolution (Equation 1.24). When the weathering involves sulfide minerals, strong acids (such as sulfuric acid and sulfurous acid) form, further accelerating the weathering reactions. Other acids of anthropogenic origin (e.g., phosphoric acid, sulfuric acid, and nitric acid) have a similar effect.

The ions produced through the weathering of silicates can be classified into three groups:

1. Alkali metal ions (sodium and potassium ions) that are usually dissolved in groundwater.
2. Alkali earth metal ions (calcium and magnesium ions); their subsequent reactions are affected by carbon dioxide and its equivalent species (Figure 1.2). Iron(II) partly belongs here (under reducing conditions).
3. Silicon, aluminum, and iron(III) that form oxide, hydroxide colloids. The proportion of their aqua complexes is extremely low, and they transform to crystalline aluminosilicates.

The dissolution-precipitation equilibria are characterized by the thermodynamic equilibrium constants; and similar to potential-pH diagrams (Figures 1.4 and 1.5) stability diagrams can be constructed. The axes of the stability diagrams show the concentration of the relevant species, including pH (the concentration [activity] of hydrogen ions). Dissolution-precipitation reactions as well as redox reactions, if any, must also be taken into account when constructing such diagrams.

The stability diagram of oxide/hydroxide of aluminum and silicon minerals/aqueous solution systems are illustrated in Figures 1.6 and 1.7. In these diagrams, the following reactions were taken into account.

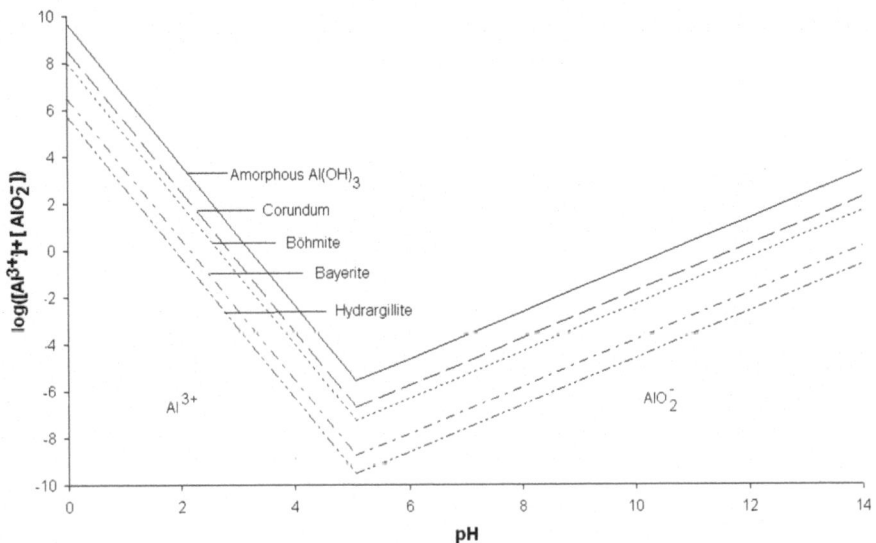

FIGURE 1.6 Stability diagram of aluminum oxides at 25°C.

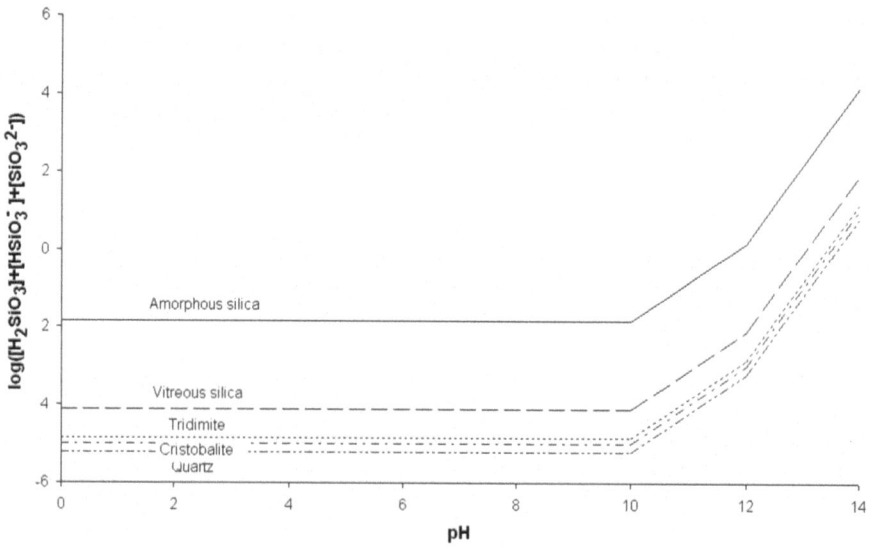

FIGURE 1.7 Stability diagram of silicon oxides at 25°C.

For aluminum:

$$2Al^{3+} + 3H_2O \rightleftharpoons Al_2O_3 + 6H^+ \qquad lg\left[Al^{3+}\right] = 5.70 - 3pH \qquad \text{hydrargillite} \quad (1.25)$$

$$lg\left[Al^{3+}\right] = 6.48 - 3pH \qquad \text{bayerit} \qquad (1.26)$$

$$lg\left[Al^{3+}\right] = 7.98 - 3pH \qquad \text{böhmite} \qquad (1.27)$$

$$lg\left[Al^{3+}\right] = 8.55 - 3pH \qquad \text{corundum} \qquad (1.28)$$

$$lg\left[Al^{3+}\right] = 9.66 - 3pH \qquad \text{amorphous} \qquad (1.29)$$
$$\text{hydroxide}$$

$$Al_2O_3 + H_2O \Leftrightarrow 2AlO_2^- + 2H^+ \qquad lg\left[AlO_2^-\right] = -14.6 + pH \quad \text{hydrargillite} \quad (1.30)$$

$$lg\left[AlO_2^-\right] = -13.82 + pH \quad \text{bayerit} \qquad (1.31)$$

$$lg\left[AlO_2^-\right] = -12.32 + pH \quad \text{böhmite} \qquad (1.32)$$

$$\lg\left[AlO_2^-\right] = -11.76 + pH \quad \text{corundum} \quad (1.33)$$

$$\lg\left[AlO_2^-\right] = -10.64 + pH \quad \text{amorphous} \quad (1.34)$$
$$\text{hydroxide}$$

$$Al^{3+} + 2H_2O \rightleftarrows AlO_2^- + 4H^+ \quad \lg\frac{\left[AlO_2^-\right]}{\left[Al^{3+}\right]} = -20.30 + 4pH \quad (1.35)$$

The concentration of Al^{3+} and AlO_2^- is equal at pH = 5.07.
   For silicon:

$$SiO_2 + H_2O \rightleftarrows H_2SiO_3 \quad \lg\left[H_2SiO_3\right] = -5.21 \quad \text{quartz} \quad (1.36)$$

$$\lg\left[H_2SiO_3\right] = -4.99 \quad \text{cristobalite} \quad (1.37)$$

$$\lg\left[H_2SiO_3\right] = -4.85 \quad \text{tridimite} \quad (1.38)$$

$$\lg\left[H_2SiO_3\right] = -4.11 \quad \text{vitreous silica}$$
$$(1.39)$$

$$\lg\left[H_2SiO_3\right] = -1.84 \quad \text{amorphous} \quad (1.40)$$
$$\text{silica}$$

$$SiO_2 + H_2O \rightleftarrows HSiO_3^- + H^+ \quad \lg\left[HSiO_3^-\right] = -15.21 + pH \quad \text{quartz} \quad (1.41)$$

$$\lg\left[HSiO_3^-\right] = -14.99 + pH \quad \text{cristobalite} \quad (1.42)$$

$$\lg\left[HSiO_3^-\right] = 14.85 \mid pH \quad \text{tridimite} \quad (1.43)$$

$$\lg\left[HSiO_3^-\right] = -14.11 + pH \quad \text{vitreous silica}$$
$$(1.44)$$

$$\lg\left[HSiO_3^-\right] = -11.84 + pH \quad \text{amorphous} \quad (1.45)$$
$$\text{silica}$$

$$SiO_2 + H_2O \rightleftarrows SiO_3^{2-} + 2H^+ \qquad lg\left[SiO_3^{2-}\right] = -27.21 + 2pH \qquad \text{quartz} \qquad (1.46)$$

$$lg\left[SiO_3^{2-}\right] = -26.99 + 2pH \qquad \text{cristobalite} \quad (1.47)$$

$$lg\left[SiO_3^{2-}\right] = -26.85 + 2pH \qquad \text{tridimite} \qquad (1.48)$$

$$lg\left[SiO_3^{2-}\right] = -26.11 + 2pH \qquad \text{vitreous silica}$$

$$(1.49)$$

$$lg\left[SiO_3^{2-}\right] = -23.84 + 2pH \qquad \text{amorphous} \quad (1.50)$$

$$\text{silica}$$

The protonation/deprotonation constant (pKs) of the $H_2SiO_3/HSiO_3^-$, and $HSiO_3^-/SiO_3^{2-}$ processes are 10.00 and 12.00, respectively.

Figures 1.6 and 1.7 show that aluminum as well as silicon do not participate in redox processes. They have a well-defined oxidation state under environmental conditions, namely, aluminum(III) and silicon(IV). Thus, the concentration of the dissolved species depends only on the pH.

In the case of aluminum, there are two soluble species: Al^{3+} cation at acidic pH values, and aluminate AlO_2^- anion at basic pH values. The latter is usually neglected in stability diagrams in the soil science literature (Lindsay 1979; Sparks 2003); however, it is an important process since, at pH>5.07, the concentration of aluminate is higher than the concentration of Al^{3+} ion. For example, at pH = 8, which is frequent under environmental conditions involving rocks and soils, the concentration of aluminate anion is 12 orders of magnitude higher than the concentration of Al^{3+} (Equation. 1.35). In addition to its relative importance compared to Al^{3+}, the absolute value of aluminate concentration can also be large, depending on the solid phase present.

As seen in Figure1.7, the metamonosilicic acid (H_2SiO_3) concentration of the solution is constant at a wide pH range (up to pH 10). It is not shown in Figure 1.7, but silica (SiO_2) has a large number of hydration products, for example, H_4SiO_4, $H_6Si_2O_7$, $H_8Si_3O_{10}$, and $H_{10}Si_4O_{13}$, that is, orthomono, orthodi, orthotri, and orthotetrasilicic acids, respectively (Pourbaix 1966; Latimer 1952). They also participate in partial dehydratation and deprotonation processes, and can be present simultaneously in an aqueous solution, resulting in a very complex system.

Of course, the stability diagrams can be constructed not only for the oxides but for other more complex minerals as well. From these diagrams, the stabilities of the minerals under a given condition can be predicted.

As seen previously in this chapter, geological systems have many components and different phases. Consequently, there are many processes and species to be taken into account when describing any equilibrium. Thus, the thermodynamic treatment of the systems demands a large number of mathematical calculations. Nowadays, several computer equilibrium models exist (Sparks 2003; Herbelin and Westall 1996) that take into account all chemical processes, including solutions, solids, and interfacial reactions (Section 1.3) if their thermodynamic equilibrium constants are known. These computer programs use the law of mass action, electroneutrality, mass balance, and redox conditions to construct a phase diagram. The concentration and distribution of the thermodynamically stable species are determined from the total concentrations of the components, the equilibrium constants, and redox potentials. In addition, when the concentrations of the species are determined experimentally, thermodynamic equilibrium constants are estimated. Thermodynamic equilibrium constants can be found in tables (e.g., Pourbaix 1966), or in some programs (e.g., MINTEQ, VisualMINTEQ) that have their own database. The program PHREEQC also considers the flowing param eters. It is important to note, again, that the systems are considered as closed systems in these thermodynamic calculations, which is an approximation, indicating only the possibility and direction of the processes.

1.2.4 Properties of Very Dilute Solutions

In nature, the concentration of some substances may be very low. Their chemical properties at these very low concentration ranges can be different from the usual properties seen in macroscopic concentrations. Thus, it is difficult to perform thermodynamic calculations in very diluted solutions, and the results are often only approximate. When confronted with such problems, experiments must be performed in order to describe the exact nature of such systems.

The phenomenon of very diluted solutions is well known in radiochemistry. Carrier-free radioactive isotopes could be mentioned as an example. The term denotes a radioisotope of an element in pure form, that is, essentially undiluted with a stable isotope. The chemical concentration of these radioisotopes is usually very low. For example, 1 kBq radioactivity (applied typically in a tracer experiment) is equivalent to cca. 2×10^{-12} mol in the case of ^{137}Cs or ^{90}Sr isotopes. In the case of such low concentrations, no chemical system can be considered homogeneous because all surfaces, the wall of the laboratory vessels, or any contaminants in the solution (such as air bubbles, small particles, great molecules, etc.) can initiate interfacial processes and the subsequent formation of heterogeneous phases (adsorption, colloid formation, precipitation, etc.). This is the result of the simple fact that the number of molecules on the surfaces is more than, or at least similar to, the number of particles in the solution. Even in a solution containing an alkali metal ion, Cs 137 ion in carrier-free concentration, colloid particles are formed under conditions that are not expected on the basis of thermodynamic parameters.

The following two examples show the effect of very low concentration on complex formation (Haissinsky 1964). In the first example, a mixture of three substances, A, B, and C is present. A reacts with either B or C as follows:

$$A + B \xrightarrow{k_1} X \qquad (1.51)$$

$$A + C \xrightarrow{k_2} Y \qquad (1.52)$$

where k_1 and k_2 are the rate constants of the reactions, and X and Y are the reaction products. Let $k_1 > k_2$, for example, $k_1/k_2 = 10^6$. Let us assume that X reacts with other components present in the studied system (M) in an equilibrium reaction:

$$X + M \rightleftarrows XM \qquad K = \frac{[XM]}{[X][M]} \qquad (1.53)$$

When B and C are present in similar concentration, the formation of X is strongly preferred because of the rate constants k_1/k_2, and the reaction in Equation 1.53 is shifted to the right. When, however, B is present in a very low concentration (e.g., 10^{-12} mol/dm^3), and the concentration of C is fairly high (e.g., 1 mol/dm^3), the formation of Y takes place. Under these conditions, XM, if any, will dissociate.

The second case is an example of the formation of different chemical species at very low concentration ranges. Let us see the complex formation of a cation in the presence of an excess of anions. With the increase of the ratio of anion, cation will obviously increase the stability of the formed complex. In the range of macroscopic concentration, if the concentration of anion and cation are similar, no complex formation can be detected. For example, in the case of lead and chloride ions, a slightly soluble compound, $PbCl_2$ is formed. However, when the concentration of lead ions is extremely low, using for example carrier-free Pb-212 isotope, and the chloride anions are present in macroscopic concentration, negatively charged $[PbCl_n]^{(n-2)-}$ are formed.

For understanding of the formation of heterogeneous phases at extremely low concentrations, two rules named Hahn's rules (Hahn 1926a, 1926b; Hahn and Imre 1929; Wahl and Bonner 1951) may serve as starting points:

1. The precipitation rule says that an element (isotope) in extremely low concentration coprecipitates with another element in high concentration if it fits into the crystal lattice of this other element.
2. The adsorption rule says that an ion in extremely low concentration adsorbs well on the surface of a solid crystal when the surface charge of solid and ion are opposite and the adsorbed compound is very insoluble. Thus, the ions in very low concentration can be precipitate even when the concentrations do not reach the solubility product (i.e., normally, they would stay in the solution) and/or radiocolloids may form.

When the inactive isotopes (carrier) of the radioactive isotope are added to the solution containing the isotope at low concentration, or the chemical/geological environment contains an inactive carrier, the usual chemical concentrations are present for chemical species in question, the thermodynamic relations can be used. For example, the calcium, iron, and hydrogen content of the rocks/soils and natural waters acts as the carrier of ^{45}Ca, ^{55}Fe, or ^{3}H isotopes, respectively. However, the chemical identity does not need to be the same; chemically similar substances may also behave as carriers. It is known, for example, that potassium or calcium ions are carriers of Cs-137 and Sr-90, respectively, in geological as well as biological systems. It should be mentioned though, that in such cases, the ratios of the isotopes and the carrier can be different in each phase of the heterogeneous system.

The problems connected to the low concentrations become more and more important with the prevalence of environmental pollution today. The pollutants can be present in very low concentrations in soil, water, and living organisms. Furthermore, the improvement of analytical method allows the detection and determination of smaller quantities than ever before, thus enabling us to take them into account in the description of geological systems.

1.3   INTERFACE OF ROCK/SOIL–AQUEOUS SOLUTIONS

In the previous paragraphs we dealt with the composition and thermodynamic equilibrium within two bulk phases, the solid and liquid. The homogeneous bulk phases are characterized by constant intensive properties throughout each phase. The reality, however, is that the particles (atoms, ions, and molecules) present in the dividing interface of the bulk phases have different properties from the particles inside the bulk phases. So, in a real system, there is a finite interface in which the properties (density, energy, entropy, etc.) change gradually going from one phase to the other one. When the interface contains only neutral particles (atoms or molecules), the thickness of this interfacial layer is a few molecules (cca. 1–2 nm), that is, the range of chemical forces. When there are ions in the interfacial layer, the thickness may be much larger because of the longer range of the electrostatic forces.

When talking about the importance of the interface in a given system, we have to compare the number of particles in the bulk phases and the interface. If the specific surface area, that is, the surface area/volume in mass ratio, is small, the number of particles in the interface can be neglected compared to the particles in the bulk phases. However, in the cases when the surface area is high, the number of the particles in the interface and in the bulk phases is similar. The interfacial properties become dominant in these systems. The colloid particles of soils and the porous rocks are good examples for systems in which interfacial properties are important. Furthermore, in the case of very diluted solutions, even a solid phase with a small surface area can have a significant effect on concentration. So, when evaluating the importance of interface, each reaction partner of the interfacial reactions has to be taken into account.

This book focuses on the reactions of the solid–liquid interface of geological formations and groundwaters, so the properties of these interfaces are discussed in detail.

1.3.1 The Structure of the Interface

The most important difference between particles inside the bulk and in the interfacial layer comes from the surrounding environment of the particles: the particles inside the bulk are in isotropic environment, while those in the interface are in anisotropic environment; thus, in the interlayer, the forces between the particles are unbalanced. To reduce the resulting surface pressure, some additional processes occur that must be taken into account. On clean surfaces (for example, on a solid surface in vacuum), these processes are the bond-length contraction or relaxation and reconstruction of the surface particles (Somorjai 1994). It results in significantly reduced spacing between the first and second layers compared to the bulk. The perturbation caused by this movement propagates a few layers into the bulk. The other effect is that the equilibrium position of the particles changes; that is the outermost layers can have different crystal structure than the bulk. This phenomenon is the reconstruction.

Adsorption of different substances on the surface is another way for the system to decrease the surface energy. Adsorption is an increase in the concentration of a gaseous or dissolved substance at the interface of a condensed and a gaseous or liquid phase, respectively, due to the operation of surface forces. The reversed process is called desorption.

The interface was treated mathematically at first by J.W. Gibbs, so the interface is also called Gibbs dividing plane (Everett 1972; Yildrim 2006) (Figure 1.8). Gibbs's concept allows applying thermodynamics to interfacial processes. It uses a reference system in which all extensive properties of the two bulk phases are unchanged divided by an imaginary Gibbs dividing layer. In a real interface, the number of moles of each component, charge, entropy, energy can be either positive or negative

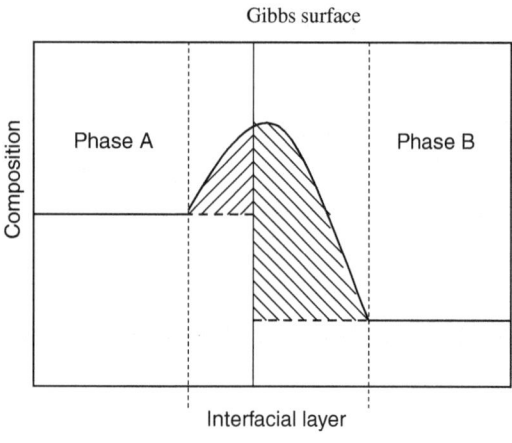

FIGURE 1.8 Schematic representation of the concentration profile (c_i) as a function of distance (spatial coordinate) normal to the phase boundary: full line (bold) in the real system; broken line in the reference system; chain-dotted line: boundaries of the interfacial layer.

compared to the bulk phases, illustrated by the lined area in Figure 1.8. These excess quantities are defined as surface excess of the various extensive properties that are used in thermodynamics.

Besides this, however, there is the effect of the excess charge in the interfacial layer, which results in the formation of an electric double layer. Therefore, the thermodynamic treatment of the interfacial reactions must be divided into two parts: chemical and electric contributions. The chemical reactions are taken into account in the usual way; the electric contribution means the electric work done by the particle when transferring through the electric double layer.

The structure of the electric double layers has been described by different models. All models consider the charges on solid phase to be compact, forming a plane layer; the charges in the liquid phase, however, are described by different distributions in each model. As a result, the variation of the potential as a function of the distance from the surface is different, depending on the model. The thickness of the double layer is the distance where the potential becomes equal to the bulk potential of the liquid phase.

The first double-layer model is the Helmholtz model (West 1965), which considers the electric double layer as a plane capacitor. In a solid/liquid interface, the electrode of the capacitor belongs to the two phases, and the charges are located in well-defined planes. The electric potential changes linearly between the two electrodes. In a real system, this model is often too simple, neglecting the diffusing effect of Brownian motion, however, it describes well those systems in which the background electrolyte has high concentration (great ionic strength).

Another model is the Gouy–Chapman model, which assumes that the charges are in a well-defined, compact plane in the solid phase, but the distribution of the counter-ions is diffuse in the liquid phase. Consequently, the potential changes exponentially in the liquid. However, capacities calculated by the model are significantly different from the experimentally measured capacities of the double layers.

The Stern model modifies the Gouy–Chapman model and divides ions present in the solution into two groups: a part of the ions is placed near the solid surface, forming the so-called Stern layer (similar to Helmholtz layer), and the other part has a diffuse distribution (Gouy–Chapman layer). It implies that the surface potential is linear in the Stern layer and exponential in the Gouy–Chapman layer.

Each double-layer model mentioned takes into account only the electrostatic interactions. If there are other interactions between the surface and the ions in the liquid phase, the plane layer (Helmholtz layer) must be divided into a further two parts: inner and outer Helmholtz layers. In the inner Helmholtz layer, the ions with the same charge as the surface are bonded by specific adsorption or chemical forces. This phenomenon is similar to the so-called inner-sphere complexation: the ions connected to the surface without their hydrate sphere, by chemical forces. Thus, when these ions are in the inner Helmholtz layer, they can be considered as such. The outer Helmholtz layer also contains the counter-ions, together with their hydrate sphere, in which case the interactions are mostly electrostatic. The variation of the potential in the liquid phase consists of two linear and an exponential range. This model was applied at first by Graham (1947, 1957) in

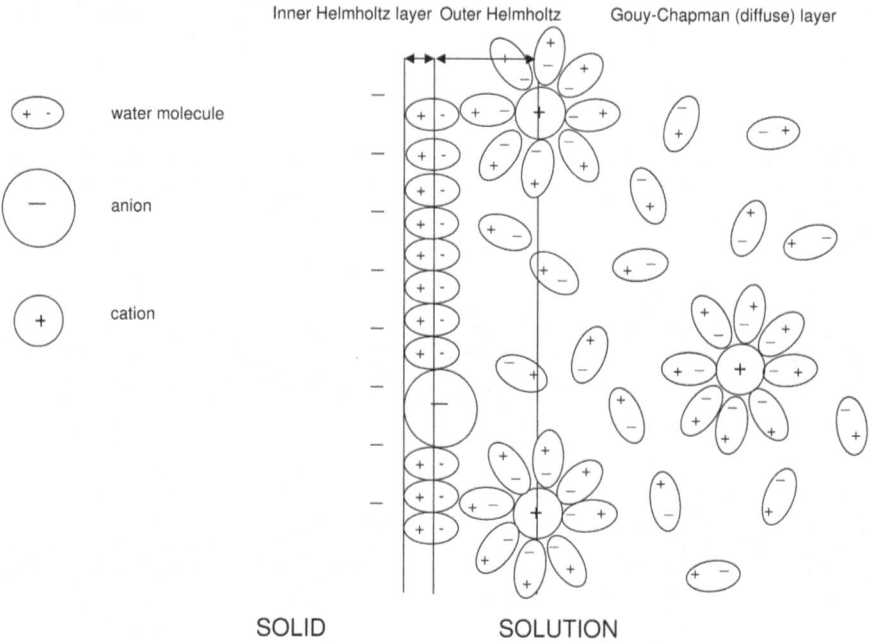

FIGURE 1.9   Model of electric double layer after Bockris et al. (1963).

electrochemistry. In surface chemistry, it is called the triple-layer model (Leroy and Revil 2003).

The Graham model was further developed by Bockris, Devanathan, and Müller (1963) who took into consideration that the polar molecules of the solvent (e.g., water) may also be adsorbed on the solid surface (Figure 1.9).

The solvent molecules form an oriented parallel, producing an electric potential that is added to the surface potential. This layer of solvent molecules can be protruded by the specifically adsorbed ions, or inner-sphere complexed ions. In this model, the solvent molecules together with the specifically adsorbed, inner-sphere complexed ions form the inner Helmholtz layer. Some authors divide the inner Helmholtz layer into two additional layers. For example, Grahame (1950) and Conway et al. (1951) assume that the relative permittivity of water varies along the double layer. In addition, the Stern variable surface charge-variable surface potential model (Bowden et al. 1977, 1980; Barrow et al. 1980, 1981) states that hydrogen and hydroxide ions, specifically adsorbed and inner-sphere complexed ions, the hydrated ions (outer sphere complexation) and diffuse layer give four layers. This model is known as the four-layer model.

In addition to the mentioned double-layer models, there are some others specifically used for the study of oxides (the modified triple-layer model, the generalized two-layer model, and the one-pK Stern model) (Sparks, 2003).

1.3.2 Characterization of the Interface of Geological System/Groundwater

As mentioned previously, the total surface of the rock and soils is the sum of external and internal surfaces. Both external and internal surfaces have charges, and so when the internal surface is significant (such as in the case of humic substances and expandable clay minerals), that extent of the interfacial layer is quite great. All systems have external surfaces, while internal surfaces are significant only in the case of certain minerals and organic matter; therefore, we will discuss first the properties of external surfaces.

1.3.2.1 Characterization of External Surfaces

The most important minerals in geological formations are silicates, aluminosilicates, and oxides. They have –OH sites on the edges of crystal lattices. These –OH sites are protonated or deprotonated depending on the pH of the solution. As a result, surface charges, so-called edge or pH-dependent charges, are formed, which has been described by the so-called surface complexation model. This is a thermodynamic equilibrium model that was developed for oxide surfaces (Davis et al. 1978; James and Parks 1982), but was later applied to different minerals, rocks, and soils for the evaluation of surface acid–base properties and ion sorption (e.g., Dodge et al. 2002; Giammar and Hering 2001; Liu and Tang 2002; Lomenech et al. 2003; Missana et al. 2003;, Nagy and Kónya 2007; Sperry and Peirce 1999; Stadler and Schindler; 1993a, 1993b; Tonkin et al. 2004; Wanner et al. 1994; Tournassat et al. 2004a, 2004b). In this model, the potential-determining ion is hydrogen, or hydroxide ion. Charged chemical species in the solution and the surface sites play an important role. When no specific adsorption occurs, the interfacial processes are directed by electrostatic forces, for example, a cation is sorbed on a negative surface site, and an anion is sorbed on a positively charged surface site.

1.3.2.1.1 Application of Surface Complexation Models for External Surfaces

The formation of surface charges in the surface complexation model is demonstrated on the example of aluminosilicates. Aluminosilicates have two types of surface sites, aluminol and silanol sites (van Olphen, 1977). These sites, depending on pH, may form both protonated and deprotonated surface complexes. From a thermodynamic equilibrium point of view, the protonated and deprotonated surface complexes can be characterized by the so-called intrinsic stability constants, considering the surface electric work. For aluminol sites:

$$-\text{AlOH} + \text{H}^+ \rightleftarrows -\text{AlOH}_2^+, \qquad K_{\text{AlOH}_2}^{\text{int}} = \frac{\left[\text{AlOH}_2^+\right]}{\left[\text{AlOH}\right]\left[\text{H}^+\right]\exp\left(-\dfrac{F\Psi}{RT}\right)} \qquad (1.54)$$

$$-\text{AlOH} \rightleftarrows -\text{AlO}^- + \text{H}^+, \qquad K_{\text{AlO}}^{\text{int}} = \frac{\left[\text{AlO}^-\right]\left[\text{H}^+\right]\exp\left(-\dfrac{F\Psi}{RT}\right)}{\left[\text{AlOH}\right]} \qquad (1.55)$$

where K^{int}'s are the intrinsic stability constants of the protonation and deprotonation processes, and ψ is the surface potential.

For silanol sites, due to the very low $\lg K^{int}_{SiOH_2}$, at pH>2 the protonation of silanol sites can be neglected (Hiemstra et al.1989), only deprotonation is taken into account:

$$-SiOH \rightleftarrows -SiO^- + H^+, \qquad K^{int}_{SiO} = \frac{\left[SiO^-\right]\left[H^+\right]\exp\left(-\dfrac{F\Psi}{RT}\right)}{\left[SiOH\right]} \qquad (1.56)$$

where K^{int} is the intrinsic stability constants of the deprotonation processes, ψ is the surface potential.

As seen from Equations 1.54–1.56, the intrinsic stability constants of surface reactions are dependent on two factors: a chemical and an electric contribution. The chemical contribution is taken into consideration by the mass balance; the electric contribution is treated by the charge balance. There are several surface complexation models that mainly differ in the description of the electric double layer that is used to calculate the surface potential, which is done by different double-layer models. These models have been mentioned previously in this chapter. Since, however, the terminology usually used in electrochemistry, colloid chemistry, and especially, in the discussions of surface complexation models is different, they are repeated again:

1. At high ionic strength, the electric double layer is considered to be plane; the so-called constant capacity model (Helmholtz model) is applied.
2. At low ionic strength, a diffuse double-layer model (Gouy–Chapman model) is used.
3. A more sophisticated model is the triple-layer model, allowing the surface reaction of the background electrolyte (Hayes et al. 1991). The potential-determining ions (hydrogen and hydroxide) are directly on the surface (inner Helmholtz layer), the other ions are at a certain distance from the surface (outer Helmholtz layer), and there is a diffuse layer, also.
4. Other models differentiate the inner-sphere complexation of hydrogen and hydroxide ions, other cations and anions, the outer sphere complexation of cations and anions and a diffuse layer. This demands the application of the four-layer model.

An excellent summary of the application of double-layer models in surface complexation modeling has been given by Sparks (2003). Here the permitted processes in the different models are summarized in Tables 1.7 and 1.8. There are examples in the literature where no electric work is considered, for example, the two-site protolysis non-electrostatic surface complexation and cation exchange model (Baeyens and Bradbury 1997, 2009).

TABLE 1.7
The Characteristics of Surface Complexation Models: reactions

Model	Adjustable parameters	Protolysis reaction	Surface complexation reaction	Surface mass balance
Diffuse-layer	Intrinsic stability constants; Number of surface sites	$-SOH + H^+ \rightleftarrows -SOH_2^+$ $-SOH - H^+ \rightleftarrows -SO^-$	$-SOH + M^+ \rightleftarrows -SOM^+ + H^+$ $-SOH + L^- \rightleftarrows -SL + OH^-$	$S_T =$ $[SOH]+[SOH_2^+]+[SO^-]$ $+[SL]+[SOMe^+]$
Constant capacitance	Intrinsic stability constants; Number of surface sites; Capacity of the plane (Helmholtz) layer			
Triple layer	Intrinsic stability constants; Number of surface sites; Capacity of the plane (inner and outer Helmholtz) layers		$-SOH + M^+ \rightleftarrows -SOM^+ + H^+$ $-SOH + L^- \rightleftarrows -SL + OH^-$ $-SOH + Ca^+ \rightleftarrows -SO^- - Ca^+ + H^+$ $-SOH + An^- + H^+ \rightleftarrows -SOH_2^+ - An^-$	$S_T =$ $[SOH]+[SOH_2^+]+[SO^-]$ $+[SOH_2^+-An]+[SO^-Cat^+]$ $+[SL]+[SOMe^+]$
Modified triple layer	Intrinsic stability constants; Number of surface sites; Capacity of the plane (inner and outer Helmholtz) layers		$-SOH + M^+ \rightleftarrows -SOM^+ + H^+$ $-SOH + L^- \rightleftarrows -SL + OH^-$ M^+ and L^- adsorbed as inner-sphere complexes, entering the inner Helmholtz layer $-SOH + Ca^+ \rightleftarrows -SO^- - Ca^+ + H^+$ $-SOH + An^- + H^+ \rightleftarrows -SOH_2^+ - An^-$	
Four-layer	Intrinsic stability constants; Number of surface sites ; Capacity of the three plane layers		$-SOH + M^+ \rightleftarrows -SOM^+ + H^+$ $-SOH + L^- \rightleftarrows -SL + OH^-$ M^+ and L^- adsorbed as inner-sphere complexes, the inner Helmholtz layer is divided into two additional layers $-SOH + Ca^+ \rightleftarrows -SO^- - Ca^+ + H^+$ $-SOH + An^- + H^+ \rightleftarrows -SOH_2^+ - An^-$	

(Continued)

TABLE 1.7 (CONTINUED)

The Characteristics of Surface Complexation Models: reactions

Model	Adjustable parameters	Protolysis reaction	Surface complexation reaction	Surface mass balance
One-pK Stern	Intrinsic stability constants; Number of surface sites ; Capacity of the plane (inner and outer Helmholtz) layers		$-SOH + M^+ \rightleftarrows -SOM^+ + H^+$ $-SOH + L \rightleftarrows -SL + OH^-$ $-SOH + Ca^+ \rightleftarrows -SO^- - Ca^+ + H^+$ $-SOH + An^- + H^+ \rightleftarrows -SOH_2^+ - An^-$ H^+ and OH^- ions adsorbed as inner-sphere complexes (inner Helmholtz layer) Other ions adsorbed as outer sphere complexes (outer Helmholtz layer)	
Generalized two-layer	Intrinsic stability constants; Number of surface sites; Capacity of the plane (Helmholtz) layer		$-SOH + M^+ \rightleftarrows -SOM^+ + H^+$ $-SOH + L^- \rightleftarrows -SL + OH^-$ Each ion is adsorbed in the surface plane and there is a diffuse layer	$S_T =$ [SOH]+[SOH$_2^+$]+[SO$^-$] +[SL]+[SOMe$^+$]

–SOH: surface site; M^+ the adsorbing cation; $L-$ ligand of adsorbed cation; Cat$^+$ and An$^-$ the ions of the background electrolyte.

Source: data from Hayes et al. 1991; Goldberg 1992; Sparks 2003.

TABLE 1.8
The Characteristics of Surface Complexation Models: Mathematical Relations

Model	Charge-potential relationships	Surface charge balance
Diffuse-layer	$-\sigma_0 = \sigma_d = -(8RTC\varepsilon_0\varepsilon_r)^{1/2}\sinh(zF\Psi_d/2RT)$ $= -0.1174\sqrt{I}\sinh(zF\Psi_d/2RT)$ $\Psi_0 = \Psi_d$	$\sigma_0 = B([SOH_2^+]-[SO^-])$ $\sigma_0 = \sigma_d$
Constant capacitance	$\sigma_0 = (CSa/F)\Psi_0$	
Triple layer	$\sigma_d = -(8RTC\varepsilon_0\varepsilon_r)^{1/2}\sinh(zF\Psi_d/2RT)$ $= -0.1174\sqrt{I}\sinh(zF\Psi_d/2RT)$ $\sigma_0 = (\Psi_0 - \Psi_\beta)C_1$ $\sigma_0 + \sigma_\beta = (\Psi_0 - \Psi_\beta)C_2 = -\sigma_d$	$\sigma_0 = B([SOH_2^+]+[SOH_2^+-An]+[SOMe^{2+}]-[SO^-]$ $-[SO-Cat^+])$ $\sigma_\beta = B([SO-Cat^+]-[SOH_2^+-An])$ $\sigma_0 = \sigma_H + \sigma_{is}$ $\sigma_0 = \sigma_H + \sigma_{os}$
Modified triple layer	$\sigma_0 = (\Psi_0 - \Psi_a)C_{oa}$ $\sigma_0 + \sigma_a = (\Psi_a - \Psi_\beta)C_{a\beta}$ $\sigma_d = -(\Psi_\beta - \Psi_d)C_{\beta d}$ $\sigma_d = -(8RTC\varepsilon_0\varepsilon_r)^{1/2}\sinh(zF\Psi_d/2RT)$ $= -0.1174\sqrt{I}\sinh(zF\Psi_d/2RT)$	$\sigma_0 = \sigma_H + \sigma_{is} + \sigma_{os}$
Four-layer		$\sigma_0 = \sigma_H + \sigma_{is} + \sigma_{os}$

(Continued)

TABLE 1.8 (CONTINUED)
The Characteristics of Surface Complexation Models: Mathematical Relations

Model	Charge-potential relationships	Surface charge balance
One-pK Stern	$\sigma_0 = (\Psi_0 - \Psi_d)C$	$\sigma_0 = \sigma_H$
	$\sigma_d = -(-8RTC\varepsilon_0\varepsilon_r)^{1/2}\sinh(zF\Psi_d / 2RT)$	
	$= -0.1174\sqrt{I}\sinh(zF\Psi_d / 2RT)$	
Generalized two-layer	$\sigma_0 = (CSa / F)\Psi_0$	$\sigma_0 = B([SOH_2^+]-(SO^-])$
		$\sigma_0 = \sigma_H + \sigma_{is}$
	$\sigma_d = -(-8RTC\varepsilon_0\varepsilon_r)^{1/2}\sinh(zF\Psi_d / 2RT)$	
	$= -0.1174\sqrt{I}\sinh(zF\Psi_d / 2RT)$	

Notes: σ_0 is the surface charge of the mineral, σ_d, σ_β, σ_H, σ_{is}, and σ_{os} are the charges in the different planes of double layer, namely σ_d is the charge of the diffuse layer, σ_β is the charge in the β plane, σ_H is the net proton charge, equal to the difference between the surface excess concentration of hydrogen and hydroxide ions, σ_{is} is the inner-sphere complex charge, and σ_{os} is the outer sphere complex charge.

Ψ's are the potential values at the different planes, C's are the capacitance density values of the plane layers.

R is the gas constant

T is the temperature

ε_0 is the permittivity of vacuum.

ε_r is the permittivity of the medium

z is the valence of the ion

F is the Faraday constant

I is the ionic strength

S is the specific surface area

a the concentration of the solid in the liquid

B is a constant and it converts surface charge from mol/dm^3 to C/m^2; B = F/aS.

The zero point of charge of the solid is an important parameter, meaning the pH value at which the number of the positive and negative surface sites is the same:

$$pH_{pzc} = \frac{pK_{SOH_2^+} - pK_{SO^-}}{2} \tag{1.57}$$

The thermodynamic equilibrium models, including surface complexation models, require the solution of a complex mathematical equation system. For this reason, many computer programs (e.g., CHEAQC, CHEMEQL, CHESS, EQ3/6, FITEQL, Geochemist's Workbench, HARPHRQ, JESS, MINTEQ and its versions, NETPATH, PHREEQC, PHRQPITZ, VisualMINTEQ, WHAM, etc.) have been developed to calculate the concentration and activity of chemical species, estimate the type and amount of minerals formed or dissolved, and to calculate the type and amount of sorbed complexes.

The number of surface sites, the surface area, and the structure of electric double layer, and its surface charge and potential have to be known in order to use these programs. The law of mass action, electroneutrality, and mass balances have to be taken into consideration.

These programs are able to model the geological systems of soil/rock–aqueous solution systems, that is the concentration and distribution of the thermodynamically stable species can be determined based on the total concentrations of the components and the parameters just mentioned. In addition, the programs can also be used to estimate thermodynamic equilibrium constants and/or surface parameters from the concentrations of the species determined through experiments. Thermodynamic equilibrium constants can be found in tables (Pourbaix 1966) or data bases (e.g., Common Thermodynamic Database Project, CHESS, MINTEQ, Visual MINTEQ, NEA Thermodynamical Data Base Project (TDB), JESS, Thermo-Calc Databases). Some programs (e.g., NETPATH, PHREEQC) consider the flowing parameters.

1.3.2.1.2 Deficiencies of Surface Complexation Models

Surface complexation models, similar to the other thermodynamic equilibrium models, have some deficiencies. As all models, the equilibrium models simplify the real system. Thus, in the case of geological systems that are very complicated (due to the fact that they are made up of many components, and these components can interact in a number of different ways), one of the deficiencies is that the resultant effect of the many components can only be estimated with difficulties. A further deficiency is that, although under given conditions (concentrations, pH, redox potential, etc.) the concentration of the different species can be calculated from the thermodynamic equilibrium constants, and these calculations are very useful for the planning of the migration and adsorption experiments and the evaluation of the results, thermodynamics gives only possibilities, and does not take into account the kinetic barriers. Kinetic barriers can be investigated experimentally. An additional hurdle comes from the adequate selection of the double layer model. It is known from the literature that equilibrium data can be well fitted with several double layer models. So, the correct description of a geological system requires other, independent data, for example, spectroscopic information on the surface complexes, in order to be able to select the right models. According to some authors (Kinniburgh et al.1983), even if it succeeds, the modeling of the surface electric properties may present unnecessary complications, particularly when many different surface sites have to be considered. In this case, the surface is energetically heterogeneous, and the surface heterogeneity must be taken into account (Sections 1.3.4.2.2 and 1.3.4.2.3).

As seen in Section 1.3.4.2, and especially in Section 1.3.4.2.1, surface complexation models can be applied to the processes on the internal surfaces; namely, for the ion exchange processes of the interlayer cations. Of course, the deficiencies mentioned there are valid in this case as well.

1.3.2.2   Characterization of Internal Surfaces

Interfacial processes also occur on the internal surface. A large internal surface area is characteristic of porous and layered materials, for example, humic substances and clay minerals. These have charges, of course, but the origin of the charges is different in the two cases. The origin of the charge of humic substances is the pH-dependent protonation and deprotonation of functional groups (Table 1.5) either on external or internal surfaces. In the case of layered minerals, however, the charge of the internal surface comes from the isomorphic substitutions of the central ions of the crystal lattice. The tetravalent silicon in the tetrahedral layer can be substituted by trivalent cation (e.g., aluminum ion), the trivalent cation of the octahedral layer is substituted by a bivalent cation (e.g., magnesium, iron(II) ions) (Table 1.2). The net charge of the layer is determined by these substitutions. It is usually negative and is found on the internal surfaces. The negative layer charge is neutralized by dehydrated, fixed (illite, chlorite) or hydrated exchangeable (smectite, vermiculite) cations in the interlayer space. Consequently, the layer charge of the internal surface is independent of pH (as opposed to external surface charge), so the layer charge is called permanent charge. The specific surface area of the internal surface is also independent of the particle size.

The ratio of the external and internal surfaces as well as the charges on the different surfaces is determined by the crystal structure and particle size. For clay minerals, the internal surface area is about 80–95% of the total surface area.

Layer charges can be calculated from the mineral and chemical composition. The mineral composition can be determined by the comparison of X-ray diffraction, thermo analytical, and surface area studies. The chemical composition is determined by a total chemical analysis of the sample. In the classical method, chemical analysis is made after acidic dissolution (Ross and Hendricks 1945). Nowadays, nondestructive analytical methods (e.g., electron microscopy, prompt gamma activation analysis, etc.) are also applied. Chemical composition is usually given as oxides (e.g., SiO_2, Al_2O_3, etc.). The cations are divided into three groups:

1. Silicon and aluminum, considered to be the central ion of the tetrahedrons or octahedrons, respectively.
2. The ions fitted into the octahedral sheet (Fe(II), Fe(III), Mg(II), and Li(I)).

3. The larger cations (potassium, sodium, and calcium ions), considered to be exchangeable cation in the interlayer space).

Then the real chemical composition is compared to the ideal formula of the mineral. For example, in the 2:1 clay minerals, the ideal unit cell contains eight atoms of silicon. When the real chemical composition shows less than eight atoms of silicon, the remaining atoms are considered as substituted by aluminum in the tetrahedral layer. The number of aluminum atoms in the tetrahedral layer is subtracted from the real chemical composition, and the difference is considered to be the number of aluminum ions in the octahedral layer. The ideal unit cell contains four atoms of aluminum, the lack is substituted by the cations in group 2. In this way, the cationic composition of the octahedral layer is given. The remaining magnesium ions, together with the ion in group 3, give the number of exchangeable cations. These can also be calculated by other standard methods (see Section 1.3.3.2).

1.3.3 INTERFACIAL PROCESSES RELATED TO EXTERNAL AND INTERNAL SURFACES

1.3.3.1 Adsorption

As seen previously, adsorption processes result in the decrease of surface energy. As mentioned in Section 1.3.1, adsorption is an increase in the concentration of a gaseous or dissolved substance at the interface of a condensed and a gaseous or liquid phase, respectively, due to the operation of surface forces. The main feature of the adsorption is that the adsorbed particles occupy the free sites of the interface. When two or more different substances are adsorbed, competitive adsorption occurs. In aqueous solution, competitive adsorption takes place in all cases because water molecules cover the total surface of the solid. The water concentration, however, is usually much greater than the dissolved substances, so the change in water concentration can be neglected. In the thermodynamic equations, the parameters characterizing water are included in other thermodynamic parameters.

A special case of adsorption is the neutralization of the charges on the internal and external surfaces. The hydrated cations neutralizing the permanent charge can be considered as adsorbed cations on the internal surfaces, or, in other words, in the interlayer space. Since the permanent charge is totally neutralized, the surface coverage will always be 1, and a monolayer of cations is present on the internal surfaces. The pH-dependent charges in water are neutralized by hydrogen or hydroxide ions, respectively.

Besides water and ions, other substances can also be adsorbed on both the external and internal surfaces. Since the surface of rocks and soils is strongly hydrophilic on a macroscopic scale, hydrophilic substances are more strongly adsorbed than hydrophobic substances. Some authors, however, say that the hydrophilic–hydrophobic character strongly varies on molecular scale. So, the oxygen atoms next to the isomorphic substitutions are hydrophilic, while the oxygen atoms far from them are hydrophobic. The hydrophobic oxygen atoms are, for example, responsible for the adsorption of hydrophobic pesticides (Laird 2004).

1.3.3.2 Ion Exchange

The other important interfacial process is ion exchange, during which the ions in the interlayer space or the external surfaces are changed to other ions. It happens in connection with the change of the composition of the liquid phase interacting on the solid. For example, the cation exchange of the interlayer space of a clay mineral is

$$z_2 Me_1^{z_1^+} + z_1 Me_2^{z_2^+} - S \rightleftarrows z_1 Me_2^{z_2^+} + z_2 Me_1^{z_1^+} - S \qquad (1.58)$$

where S means the surface, and Me_1 and Me_2 are the cations with z_1+ and z_2+ charges, respectively. When $z_1+ = z_2+$, the ion exchange is homovalent, and in other cases it is heterovalent.

During the exchange in the interlayer space, the cations keep their hydrate shell; this process is called outer-sphere complexation. It is because of electrostatic forces that the greater the charge and smaller the size of the hydrated cation, the more favorable the ion exchange is.

The ion exchange on the external surfaces can be found in Table 1.7, where the different complexation models are shown. The produced surface species are called outer- or inner-sphere surface complexes, depending on whether the ions remain their hydrate sphere or not, respectively. The inner-sphere complexation is frequently called *specific adsorption*.

The ion exchange on the external surfaces, especially the inner-sphere complexation, depends on the chemical properties of the ions. For example, the possibility of a covalent bond between the oxygen and the cation (having soft character; Pearson 1963, 1968; Buffle and Stumm 1994) promotes the dehydration of the cation.

Ion exchange processes are especially important in the geological environment. It is quantitatively characterized by the ion exchange capacity, that is, the total quantity of the exchangeable ions. It can be defined for cations (cation-exchange capacity, CEC) or anions (anion-exchange capacity, AEC). Mainly, cation exchange is significant because the permanent charges are generally negative and neutralized by cations. In addition, under geological conditions, the net external charge of the solid is also negative.

The CEC is the sum of the quantity of the cations neutralizing permanent and pH-dependent charges. For rocks and soils, it strongly depends on the mineral composition and humus content. The CEC practically originates from the clay and humus content of soils. The greater the clay mineral and humus content, the greater the CEC is. So, the CEC is connected to structural parameters and can be calculated from the mineral and elementary composition of a mineral fraction (Sections 1.3.2.1 and 1.3.2.2) and humus content. More often, however, cation exchange capacity is experimentally determined by standard methods, such as by cation exchange with ammonium (Richards 1957), alkali and alkali earth cations, organic cations (e.g., methylene blue), transition metal ions or organometal complexes (e.g., Cu-trien) (Bergaya et al. 2008b), or heterogeneous isotope exchange (Kónya and Nagy 1998; Nagy et al. 1998). It is important to note, however, that the methods operate at different pH values, so the competition of the cations with the hydrogen ions (Chapter 2, Section 2.7) can give different CEC values (Christidis 2008).

The cation exchange of layer silicates significantly influences some structural and colloid chemical properties. Depending on the charge of the cation, the interlayer space contains water in different quantities (Chapter 2, Section 2.1.2). So, the basal spacing (the distance between similar faces of adjacent layers) is different for monovalent, bivalent, or trivalent cations. For example, in monovalent montmorillonite, it is about 1.2 nm, and in bivalent and trivalent montmorillonite, it is about 1.5–1.6 nm.

When the size of the exchanged cation is similar to the pore sizes in the crystal lattice (e.g., potassium [Bouabid et al. 1991] and lithium ions [Komadel et al. 2003]), they can be built into the crystal lattice and decrease the layer charge. As an example, the illitization of smectites is mentioned.

As mentioned previously, because of the hydrophilic character of the clay minerals, water molecules can intrude into the interlayer spaces. The swelling of the layer leads to the extremely low water permeability of layer silicates. Water permeability, however, strongly depends on the cations in the interlayer spaces. The different cations coordinate different number of water molecules (Berend et al. 1995; Cases et al. 1997). Monovalent ions (e.g., sodium ions) decrease water permeability greater than bivalent ions (e.g., calcium ions). It can be seen well in saline soils where the water permeability is unfavorable.

Besides metal ions and hydrogen ions, other cationic substances (e.g., cationic organic molecules or cationic surfactants) can also be exchanged in the external and internal surfaces. The cationic surfactants can make the surface to hydrophobic (Lagaly and Dékány 2005). Furthermore, the distance between the layers is widened, enabling polymer chains to move in. This is an important condition of the preparation of nanocomposites containing layer silicates. Recently, the production and application of nanocomposites has become widespread, many papers are presented in this field (e.g., Paiva et al. 2008; Pavlidoua and Papaspyrides 2008).

The anion adsorption, then the exchange of the anions mainly takes place on the protonated surface sites of silicates and other oxides, hydroxides (e.g., iron, manganese oxides, and hydroxides), as well as on the positive functional group (e.g., protonated amino groups) of humic substances. It is directed by electrostatic forces. The degree of anion exchange of rocks and soils is usually much less than that of cation exchange.

Some anions (carbonate, phosphate, sulfate, borate, fluoride, and some organic anions) can be specifically adsorbed (Section 1.3.2.1) on the negative surface charges, changing the surface charge (Blesa et al. 1983; Peak et al. 2003; Wazne et al. 2003).

A special case of an ion-exchange process is the heterogeneous isotopic exchange, in which the two ions (Me_1 and Me_2) are the two different isotopes of the same element. Of the two isotopes, one is stable and present in macroscopic concentration, the other is radioactive with very low concentration. The isotopic exchange is controlled by the increase of the mixing entropy of the system. In equilibrium, the specific activity (activity/mass of the given element) is equal in the solid and liquid phases. The method can be applied well to interfacial studies because no other ion is introduced into the system. The exchange capacities can easily be determined by isotopic exchange. Similarly, the energy distribution of the surface can be studied (Section 1.3.4.4).

As mentioned in Sections 1.3.2.1 and 1.3.2.2, the CEC and specific surface area (both internal and external) are higher than those of clay minerals. The functional groups of soil organic matter (Table 1.5) can be deprotonated or protonated, depending on pH. It means that they have pH-dependent charges. The major functional groups can be deprotonated at pH values characteristic of soils (pH = 6–8) so they can sorb cations. It has been estimated that 20–70% if the CEC of soils originates from the soil organic matter (Stevenson 1982).

Obviously, the interaction between the functional groups of soil organic matter and cations plays an important role in the sorption of cation and, as a result in the nutrient cycle of soil and in the retention or transport of environmental and polluting cations. Though the different forms of soil organic matter contain molecules with different sizes, the functional groups, however, are the same, a given complexation reaction can have opposite effects, depending on the size and solubility of the reaction product. For example, the complex formation with carboxylic groups can keep the metal ions in solution in the case of acid molecules with a small size (citric acid, tartaric acid, etc., Chapter 2, Section 2.8.4), while the reaction with carboxyl groups of humic substances retard the metal ions, decreasing mobility.

The functional groups of the soil organic matter have different affinities to metal ions (Charberek and Martell 1959; Sparks 2003):

Enolate > amine > azo compounds > ring N > carboxylate > ether >carbonyl.

Since the soil organic matter contains more functional groups, metal ions can coordinate more than one functional group, forming rings, chelate complexes, and similar structures.

Besides the functional groups, the properties of cations also have an effect on the interactions. Similar to electrostatic interactions with the layer charge, the ionic potential (the ratio of charge to ionic radius), that is, the hard and soft character of the ions (Pearson 1963, 1968; Buffle and Stumm 1994) is important.

The affinity of the cations to the different functional groups can be characterized quantitatively with the stability constants of the complexes. The order of stability constants of metal cation-fulvic acid complexes has to be found as follows (Schnitzer and Hansen 1970):

$Fe^{3+} > Al^{3+} > Cu^{2+} > Ni^{2+} > Co^{2+} > Pb^{2+} > Ca^{2+} > Zn^{2+} > Mn^{2+} > Mg^{2+}$.

Besides the dissolved metal cations of the soil solution and groundwater, the soil organic matter forms complexes with clay minerals and metal oxides, for example, aluminum and iron oxides. Soil organic matter can coat the mineral surfaces partially or totally, depending on the concentration ratio between it and minerals. The clay-organic matter complexes can be formed by both physical (van der Waals, electrostatic) and chemical (hydrogen bonds, complexation, cation, and anion bridges) interactions. Some organic matter can be introduced into the interlayer space of clay minerals (Chapter 2, Section 2.9.2).

The formation of clay/oxide-soil organic matter complexes can modify the interfacial processes but causes no basic changes in the mechanism of interfacial reactions. The exchange sites of clay as well as surface organic matter can be reached by the cations (including hydrogen ions), but the cation exchange and the protonation/deprotonation processes can take place in the same way. At most, the rate of the reaction becomes slower because the ions have to penetrate through the pore system inside the soil organic matter (humic substances).

1.3.3.3 Precipitation

On the surfaces of geological formations, different precipitation processes can be observed. The first one occurs when the concentration of some components reaches the value of solubility product, the solution becomes oversaturated, and a new solid phase is precipitated (Section 1.2.3). The quantity of the precipitate depends only on the concentration of the solution. The precipitation takes place in a solution without the necessary presence of a solid surface. When, however, a solid phase, rocks, or soil is originally present, the precipitate is formed on it, and thus the total composition of the solid phase changes. When the precipitation forms colloid particles, especially in diluted solutions, they can be adsorbed on the solid, if it is present. This process is governed by the so-called theory of colloid adsorption (Derjaguin and Landau 1941; Verweij and G. Overbeek 1948).

The second form of precipitation cannot be understood from the thermodynamic properties of the solution: the solution is undersaturated, even so, precipitation takes place on the surface. This process is called surface precipitation. In this case there are three possibilities. One of them is when the precipitate is formed in a monomolecular layer. The second possibility is co-precipitation, when a component in low concentration coprecipitates with another component in high concentration if it can be built into the crystal lattice (Section 1.2.4). In this case, the thickness can be higher than monolayer. For example, cesium ions in very low concentration co-precipitate with iron and magnesium containing carbonates (Kónya et al. 2005; Chapter 3, Section 3.1.2). These types of surface precipitation can quantitatively be described by the adsorption equations (Section 1.3.4.1).

The third possibility of surface precipitation is that the cations sorbed on the edge sites can act as the initiator of a heterogeneous nucleation on the particle surface, followed by crystal growth. As a result, nano- and microparticles can be formed on these nuclei. The production of these particles is not expected from thermodynamic properties under the conditions of the solution and cannot be observed in the absence of a solid surface (Strawn and Sparks 1999; Nagy et al. 2003; Chapter 2, Section 2.10.4).

1.3.3.4 Evaluation and Separation of Interfacial Processes

In the geological and soil science literature, the ion exchange and precipitation are frequently considered as adsorption and thermodynamically described by adsorption equations, or isotherms. This is not correct because, as shown previously, the processes are principally different: adsorption is directed by the decrease of surface energy, and it takes place on the free surface sites; ion exchange is just a competitive process on an already covered surface, determined by the ionic composition of the liquid phase. Precipitation, including colloid formation, is governed by the composition of the liquid phase, the crystal structure (co-precipitation), or primary chemical forces.

Therefore, when studying interfacial reactions on rocks and soils, it must always be determined what the mechanism of the interfacial reaction is; what kind of processes take place. Also, the dominant processes responsible for the surface excess concentration must be identified. If this is not done, and the resultant process is evaluated without knowing it in conventional ways, incorrect thermodynamic data are obtained. The concepts of the adsorption, ion exchange, and surface precipitation have to be clearly differentiated, as done previously. When the character of the process can be neglected, only the surface accumulation is considered, and we can speak about sorption, including all of the surface processes (Nagy and Kónya 1988). In this case, only a phenomenological description can be given, and no thermodynamics can be applied.

1.3.4 Quantitative Treatment of Interfacial Processes

1.3.4.1 Adsorption Isotherms

Wang and Guo (2020a) have recently provided a general summary on adsorption isotherms.

1.3.4.1.1 Gibbs Isotherm

The correct thermodynamic treatment of the adsorption processes is possible only on liquid–gas, and liquid–liquid interfaces, where the surface energy or the surface tension of the liquid can be determined precisely. For these systems, the Gibbs adsorption isotherm can be applied. For example, on a liquid–liquid interface,

$$\Gamma = -\frac{c}{RT}\left(\frac{d\gamma}{dc}\right)_T \tag{1.59}$$

where Γ is the excess moles of the solute adsorbed per unit area at the interface, c is the molar concentration of the solute, γ is the surface tension, R is the gas constant, and T is the temperature.

On a liquid–gas interface, the partial pressure of the adsorbed gas is substituted in Equation 1.59.

On solid–gas and solid–liquid interfaces, only the excess surface concentration can be measured directly, and not the surface tension. The Gibbs adsorption isotherm is suitable for the calculation of the change of surface tension.

1.3.4.1.2 Langmuir Adsorption Isotherm

In geological surfaces, the solid–gas and solid–liquid interfaces are important, so the correct thermodynamic adsorption equation (Gibbs isotherm) cannot be used. Instead, other adsorption equations are applied, some of them containing thermodynamic approaches, others being empirical or semiempirical. One of the most widespread isotherms is the Langmuir equation, which was derived for the adsorption of gas molecules on planar surfaces (Langmuir 1918). It has four basic assumptions for the adsorption (Fowler 1935):

1. The adsorption takes place on the free sites of the adsorbent.
2. One adsorption site can adsorb one molecule; monolayer coverage is maximum.
3. The adsorption sites have the same energy (homogeneous surface), no interaction between the adsorbed molecules. It means that the adsorption energy is independent of coverage.
4. There is adsorption equilibrium between the phases.

These assumptions are very strict and usually not fulfilled in solid–gas systems. In practice, however, the Langmuir isotherm frequently describes the adsorption function quite well. This is the case for solid–liquid interfaces also, where the adsorption of the dissolved substance can be formally described by the Langmuir isotherm.

The Langmuir equation for gas adsorption is as follows:

$$a = z \frac{bp}{1+bp} \tag{1.60}$$

or for the adsorption of a dissolved substance:

$$a = z \frac{bc}{1+bc} \tag{1.61}$$

where a is the excess adsorbed amount per unit mass of the adsorbent, z is the number of surface sites (that is, the maximum adsorption capacity), p is the partial pressure of the gas in equilibrium, c is the concentration of the adsorbate in the equilibrium solution, and b is the parameter characterizing the adsorption energy. By statistical thermodynamics (Imre 1933a), it is shown that

$$b = \Delta V exp\left(-\frac{\Delta H_{ads}}{RT}\right) \tag{1.62}$$

where ΔV is the volume of the 1 mol adsorbed substance on the surface, ΔH_{ads} is the heat of adsorption.

By equivalent mathematical transformations, the Langmuir equation can be expressed in linear forms. From Equation 1.61, for example,

$$\frac{1}{a} = \frac{1}{z}\left(1+\frac{1}{cb}\right) \tag{1.63}$$

In this linear form, $1/a$ versus $1/c$ function is plotted, the slope is $1/b$, the intercept is $1/z$. This linear form is fairly suitable for the determination of the sorption parameter (b) since it is calculated from the slope. ·

The other linear form of the Langmuir isotherm is:

$$\frac{c}{a} = \frac{1}{z}\left(c + \frac{1}{b}\right)$$
(1.64)

The c/a versus c function is plotted here, the slope is $1/z$, the intercept is $1/(bz)$. This linear form is fairly suitable for the determination of the maximum adsorption capacity since it is calculated from the slope. Equation 1.64 is used more frequently than Equation 1.63.

In case of the adsorption of the dissolved substance (solid–liquid interface), the reciprocal amount of b can be used ($K=1/b$). So, Equation 1.64 can be rewritten as follows:

$$\frac{c}{a} = \frac{1}{z}\left(c + K\right)$$
(1.65)

The dimension of K is a concentration dimension, in an agreement with the dimension of c.

Among the four assumptions of the Langmuir adsorption equation, No. 3 is the most critical. It assumes that the surface is homogeneous and that there are no interactions between the adsorbed particles. On multicomponent adsorbents, such as rocks and soils, the surface is obviously heterogeneous, and so the adsorption energy may depend on the coverage. In addition, at great coverage values, the adsorbed particles may have interactions. In some lucky cases, the two effects are compensated, and the adsorption isotherm can be described by the Langmuir equation: Equation 1.65 gives a linear plot. In the other cases, the adsorption isotherm can be approached by a two-site Langmuir equation, assuming two sites with different affinities (low- and high-energy adsorption sites):

$$a = z_1 \frac{b_1 c}{1 + b_1 c} + z_2 \frac{b_2 c}{1 + b_2 c}$$
(1.66)

It is the first step in the construction of so-called composite isotherms describing the adsorption on surfaces containing k different adsorption sites. For example:

$$a = \sum_{i=1}^{k} z_i \frac{b_i c}{1 + b_i c}$$
(1.67)

where the total adsorbed quantity a is the sum of the adsorbed quantities on the sites with different adsorption energies. The values of b's contain the adsorption energy of each site (Equation 1.62; Kinniburgh et al. 1983). So, when the surface is heterogeneous or there is an interaction between the adsorbed particles, different composite isotherms can be constructed using the individual isotherms of the different sites and the site affinity distribution function. This will be discussed in detail for the ion exchange process (Section 1.3.4.2).

1.3.4.1.3 Special Isotherms for Heterogeneous Surfaces

When the energy distribution function shows that the adsorption energy is the logarithmic function of the coverage, the logarithm of the adsorbed quantity depends on the logarithm of the concentration of the liquid phase. Practically, it is expressed in the empirical Freundlich equation

$$a = k_F c^{1/n} \tag{1.68}$$

or in logarithmic form

$$ln\,a = ln\,k_F + \frac{1}{n} ln\,c \tag{1.69}$$

where k_F and $1/n$ are constants. The adsorption heat depends on the surface heterogeneity and the interactions of the adsorbed molecules. The constant $1/n$ characterizes the heterogeneity and the interactions of the adsorbed molecules together. When $1/n = 1$, the adsorbed quantity is proportional to the concentration (Henry's law). When $1/n < 1$, the surface is heterogeneous or the adsorbed particles repulse each other. When $1/n > 1$, the adsorbed molecules attract each other, the surface, however, can be heterogeneous, too.

As seen from the previous discussion, Langmuir and Freundlich, the mostly applied isotherms to describe the adsorption processes, assume that the surface sites have certain properties. The Langmuir isotherm assumes an ideal homogeneous surface with no interaction between adsorbed particles. The Freundlich isotherm considers a heterogeneous surface where the energy distribution (heat of adsorption) versus coverage function is logarithmic. Some scientists, however, simultaneously use Langmuir and Freundlich isotherms for the adsorption on rock and soil surfaces and found them satisfactory at the same time (e.g., El-Sofany et al. 2009; Freitas et al. 2009; Shah et al. 2009). When considering the energy of the surface sites, it is unsuitable because the two models contradict with each other. The Langmuir and Freundlich isotherms both seem to be equally suitable when a special coverage range is selected, but it cannot be correct for the total range of surface coverage.

When the adsorption energy is the linear function of surface coverage, the Temkin isotherm equation is used:

$$a = k_T ln\,c \tag{1.70}$$

where k_T is constant. It describes well the experimental isotherm at medium concentrations.

1.3.4.1.4 Competitive Langmuir Adsorption Isotherm

When two or more substances can be adsorbed on a surface, the so-called competitive isotherms are applied. The Langmuir isotherm was improved for the adsorption of two gases by Markham and Benton (1931). Of course, the four assumptions of

Langmuir isotherms are valid here, too. The surface adsorbed quantity of the component 1 is

$$a_1 = z \frac{b_1 p_1}{1 + b_1 p_1 + b_2 p_2} \tag{1.71}$$

A similar equation can be described for component 2. The ratio of the coverages of the two adsorbed components is

$$\frac{\Theta_1}{\Theta_2} = \frac{p_1}{p_2} \exp\left(\frac{\Delta H_2 - \Delta H_1}{RT}\right) = \frac{p_1 b_1}{p_2 b_2} \tag{1.72}$$

where Θs and ΔHs are the coverages and the adsorption enthalpies of the adsorbed gases at p's equilibrium pressures.

In the case of solid–liquid systems, a similar equation can be written for two adsorbed components using that $K_1 = 1/b_1$ and $K_2 = 1/b_2$ (Equation 1.65). In this case, the form of the competitive Langmuir isotherm analogous to Equation 1.64 for component 1 would be:

$$\frac{c_1}{a_1} = \frac{1}{z}\left(c_1 + K_1 + \frac{K_1}{K_2} c_2\right) \tag{1.73}$$

Similarly, for component 2:

$$\frac{c_2}{a_2} = \frac{1}{z}\left(c_2 + K_2 + \frac{K_2}{K_1} c_1\right) \tag{1.74}$$

As seen from Equations 1.73 and 1.74, the c_1/a_1 versus c_1 or c_2/a_2 versus c_2 functions are not linear because a composite amount containing the K's of both adsorbed components and the concentration of the other component is present, instead of the K constant in Equation 1.64.

1.3.4.1.5 Multilayer Adsorption: BET Isotherm

The isotherms discussed until now consider a monolayer coverage. Multi-layer adsorption, however, can also occur. This is the case for the nitrogen adsorption at the temperature of liquid nitrogen. The process was termed the BET (Brunauer–Emmett–Teller) method (Brunauer et al. 1938) for the determination of the external specific surface area (Section 1.1.3). The construction of the BET isotherm equation assumes that there is a difference between the first adsorbed layer and the additional layers. The first layer has a fixed number of adsorption sites with fixed adsorption enthalpy (ΔH_{ads}; similar to the Langmuir equation). The molecules in the second layer are condensed on the adsorbed molecules of the first layer, and so on. The second and other layers are considered as condensed layers, the characteristic enthalpy being equal to the condensation enthalpy (ΔH_{cond}). The adsorption and desorption rate constants are also different for the first and the other layers. Each layer is considered to be in equilibrium. The BET isotherm equation was originally derived by kinetic methods (Brunauer et al. 1938). Later Cassie (1945) and Hill (1946a, 1946b, 1949) obtained the same isotherm by statistic methods.

The linearized form of BET isotherms is:

$$\frac{p}{a(p_0 - p)} = \frac{1}{z}\left(\frac{1}{C} + \frac{C-1}{C}\right)\frac{p}{p_0} \tag{1.75}$$

where

$$C = exp\left(\frac{\Delta H_{ads} - \Delta H_{cond}}{RT}\right) \tag{1.76}$$

and p is the gas pressure in adsorption equilibrium, and p_0 is the gas pressure needed for saturation at the temperature of determination. The isotherm is linear when the left side of Equation 1.76 is plotted as a function of p/p_0. The external specific surface area is calculated from z (the maximum number of surface sites that is a monolayer) and the cross-sectional area of the adsorbate.

1.3.4.2 Treatment of Ion Exchange Processes

The other important surface reaction, the ion exchange (Equation 1.58), can also be treated by isotherm equations. Since the ion exchange is a competitive process on an already covered surface, at least two exchanging ions have to be taken into account. The simplest case is when the concentration or the molar fraction of the ions on the surface is plotted as a function of the concentration or molar fraction of the same ion in the equilibrium solution. When the function is diagonal, the two ions have the same surface preference. In most cases, however, the ions have different surface preferences. This is shown by the deviations from the diagonal. When the concentration or molar fractions of the exchanged ion (Me_1 in Equation 1.58) are plotted, the curves under the diagonal show surface preference for the exchanging ion (Me_2 in Equation 1.58); the curves above the diagonal show surface preference for the exchanged ion (Me_1 in Equation 1.58). The disadvantage of this isotherm is that it gives no quantitative information; only the relative preference of the ions can be determined from it.

In all quantitative treatments of the ion exchange, the quantities characterizing the process mostly depend on the molar fraction of the ions on the surface. This fact is treated in different ways in the models describing the ion exchange.

1. The models applying the law of mass actions (Gaines and Thomas 1953; Howery and Thomas 1965; Argersinger et al. 1950; Ekedahl et al. 1950; Högfeldt 1950) introduce the term of surface activity coefficient.
2. The models in which the ion-exchange processes are included into the surface complexation models take into account the surface electric work (Section 1.3.2.1; Tables 1.7 and 1.8).
3. Those that use competitive ion-exchange isotherm equations consider heterogeneous surface characterized by an energy distribution function (Sposito 1981; Kinniburgh 1983; de Wit et al. 1990; Borkovec and Koper 1994).

All models, however, have the same deficiency, that is, the concentrations in the solid and solution can be measured experimentally; however, the surface activity coefficients, the surface electric work, as well as the energy distribution function can only be estimated. It means that the models are adapted to the experimental data and the best-fitted model is used, and therefore the selected model has no thermodynamically significant meaning (Cernik et al. 1995). In this chapter, these problems will be illustrated and discussed.

1.3.4.2.1 Treatment of Ion Exchange by Law of Mass Action

As mentioned earlier, there are models describing the equilibrium of the exchange reactions that use the law of mass action. In a strict thermodynamic sense, the law of mass action is valid only for homogeneous systems. In order to be able to use it in solid–liquid systems, the layer containing the ions belonging to the solid is considered to be a part of the solution. (Surface complexation models use a similar approach.) So, the equilibrium constant (K) of the cation exchange reaction (Equation 1.58) can be expressed as follows:

$$K_{Me_1,Me_2} = \frac{a_{Me_2}^{z1} a_{Me_1-S}^{z2}}{a_{Me_1}^{z2} a_{Me_2-S}^{z1}} \tag{1.77}$$

where a's mean the activities of the ions on the solid ($Me - S$) and in the solution (Me). The activities in the solution can be calculated on the basis of the well-known Debye–Hückel theory from the concentrations (c) and activity coefficients (γ) at given ionic strength:

$$a = c\gamma \tag{1.78}$$

The activities of the ions on the solid surface cannot be calculated directly, so the so-called selectivity coefficients (K_{sel}) are described, and instead of ion activities on the solid surface, the surface concentration or molar fractions of the ions are used (Vanselow 1932):

$$K_{sel\,Me_1,Me_2} = \frac{X_{Me_1-S}^{z2} a_{Me_2}^{z1}}{X_{Me_2-S}^{z1} a_{Me_1}^{z2}} = \frac{X_{Me_1-S}^{z2} c_{Me_2}^{z1} \gamma_{Me_2}^{z1}}{X_{Me_2-S}^{z1} c_{Me_1}^{z2} \gamma_{Me_1}^{z2}} \tag{1.79}$$

The relationship between the equilibrium constant and selectivity coefficient is

$$K_{Me_1,Me_2} = K_{sel\,Me_1,Me_2} \frac{f_{Me_1-S}^{z2}}{f_{Me_2-S}^{z1}} \tag{1.80}$$

where fs are the surface activity coefficients of each ion, and $a_{Me-S} = X_{Me-S} f_{Me-S}$. Therefore, the selectivity coefficient depends on the surface activity coefficients, which vary at different surface compositions. The term surface activity coefficients implicitly means that two phenomena are present in real systems as opposed to ideal systems: surface heterogeneity and interactions of the ions on the surface.

The equilibrium constant (Equation 1.81) and the selectivity coefficients (Equations 1.82–1.83) have been calculated by the selectivity function, by the integration of the selectivity coefficient versus surface molar or equivalent fraction (Gaines and Thomas, 1953; Howery and Thomas 1965; Argersinger et al. 1950; Ekedahl et al. 1950; Högfeldt 1950):

$$lnK_{Me_1,Me_2} = (z_1 - z_2) + \int_0^1 lnK_{sel\,Me_1,Me_2}dX_{Me_1} \tag{1.81}$$

$$lnf_{Me_1^{z_2}} = (z_1 - z_2)X_{Me_2} - X_{Me_2}lnK_{sel\,Me_1,Me_2} - \int_{X_{Me_1}}^1 lnK_{sel\,Me_1,Me_2}dX_{Me_1} \tag{1.82}$$

$$lnf_{Me_2^{z_1}} = (z_2 - z_1)X_{Me_1} - X_{Me_1}lnK_{sel\,Me_1,Me_2} + \int_0^{X_{Me_1}} lnK_{sel\,Me_1,Me_2}dX_{Me_1} \tag{1.83}$$

The advantage of the application of the law of mass action is that the equilibrium exchange constant allows the calculation of other thermodynamic parameters, namely the Gibbs free energy (Equation 1.84):

$$\Delta G = -RTlnK \tag{1.84}$$

From the equilibrium constants at different temperatures, the enthalpy (ΔH) of the ion exchange can be determined by the van't Hoff equation:

$$lnK = -\frac{\Delta H}{RT} + Const \tag{1.85}$$

where *Const* is an integration constant. From the free energy and enthalpy, the entropy of the exchange process can be calculated:

$$\Delta G = \Delta H - T\Delta S \tag{1.86}$$

In the quantitative treatment of ion exchange processes, several authors used the law of mass action. The main difference among these approaches is how the activities and surface concentration of the ions are treated. The first such approach was the Kerr equation, which uses the concentration of the ions on the solid and liquid as well but totally neglected activity coefficients (Kerr 1928). The Vanselow (1932) equation applied activities in the solution and expressed the concentration of the ions on the solid phase in mole fraction, and in this way, it defined the selectivity coefficient (Equation 1.79).

When heterovalent exchange (e.g., exchange involving both mono- and bivalent ions) takes place, which is frequently seen in soil-solution systems, the quantitative treatment runs into further difficulties. In this case, the determination of mole fractions also becomes problematic. The Vanselow equation assumes that the monovalent and bivalent ions are equivalent when calculating mole fractions. Other empirical equations, however, simply introduce some factors for the ions with different valencies. For example, according to Krishnamoorthy and Overstreet (1949), this factor is 1 for monovalent ions, 1.5 for divalent ions, and 2 for trivalent ions, which is not in agreement in stoichiometry. Another model by Gapon (1933) uses the half power of the bivalent cation. These are empirical equations, their theoretical meaning is limited, and they do not deal with the problem of surface activities.

To estimate the surface activity coefficients, several different approaches have been suggested. For example, the surface can be considered heterogeneous, containing j kinds of homogeneous part with known ratios and equilibrium constants. In this case, the selectivity coefficient can be described by a power series (Barrer and Klinowski 1972):

$$lnK_{sel} = c_0 + 2c_1 X_{Me_1} + 3c_2 X_{Me_1}^2 + \ldots + (n-1)c_n X_{Me_1}^n \qquad (1.87)$$

where C's are constant coefficients. When the selectivity function is linear (there are two members on the right side of Equation 1.87), the surface can be considered homogeneous. When the selectivity function can be fitted by quadratic, cubic, etc. polinoms, the surface is heterogeneous. In practice, however, it is difficult to determine which polynomial expression describes more accurately the experimental data because the regressions of the fitted curves are close. Consequently, the number and quantity of the different sites cannot be determined, and only qualitative information can be obtained on heterogeneity.

The other interpretation of surface activity coefficients supposes that the interactions between the pairs of the exchanging and exchanged ions (Me_1–Me_1, Me_2–Me_2, and Me_1–Me_2, respectively) are not equivalent, so the surface activity is necessarily the function of surface composition (Högfeldt 1984, 1988; Kónya et al. 1989). This is the Högfeldt three-parameter model where the selectivity coefficient depends on the surface molar factions of the ions and their interactions, as follows:

$$lnK_{sel} = lnK_{sel}(1)X_{Me_1} + lnK_{sel}(0)X_{Me_2} + BX_{Me_1}X_{Me_2} \qquad (1.88)$$

where $lnK_{sel}(1)$ and $lnK_{sel}(0)$ are the selectivity coefficients measured at $X_{Mel} = 1$ or 0, respectively, and B is an empirical constant. The equilibrium constants obtained by this method are in a fairly good agreement with those obtained by the Gaines and Thomas method (Equation 1.81) (Högfeldt, 1984, 1988, 1990).

Both Equations 1.87 and 1.88 describe the selectivity coefficients by polynomial expressions. Equation 1.87 can be tested on ion exchangers mixed in well-defined ratios. In practice, however, we meet the reverse case: the selectivity coefficients are known at different surface concentrations, and the ratio of homogeneous part and their equilibrium constants should be estimated, the reality of which cannot be tested experimentally. This is the case for the Högfeldt model.

Using surface complexation models is another way to quantitatively describe the ion-exchange processes. Surface complexation models also apply the law of mass action, combined with the surface electric work (Tables 1.7 and 1.8). However, there are some theoretical problems with the calculations, namely, that the equations of intrinsic stability constants use concentrations instead of activities, without discussing the limit of their validity. In this quantitative treatment, the problem of surface activity coefficients is converted to electric problems. The surface complexation models require the capacity of the double layer. It, however, is not known; the problem is just obscured (Section 1.3.2.1.2).

The surface complexation models (Table 1.7 and 1.8) do not use selectivity coefficients but rather equilibrium constants, which include electric parameters for all processes. A special case for combining the selectivity coefficients and electric parameters was developed by Horst et al. (1990). According to it, the relationship between the selectivity coefficients, the equilibrium constant, and the parameters of the electric double layer is:

$$lnK_{sel} = lnK + \frac{F^2z}{RTCS}X_{Me_1} \qquad (1.89)$$

where C is the capacity of the electric double layer, S is the specific surface area. According to Equation 1.89, the lnK_{sel} versus X_{Me1} function must be linear, but this is not always the case. The authors choose to neglect these non-linear ranges of the selectivity function, and so, this model cannot be applied, for example, in very diluted solutions.

1.3.4.2.2   Treatment of Ion Exchange by Sorption Isotherms

Boyd et al. (1947) were the first to apply an isotherm equation to the quantitative treatment of ion-exchange reactions. They applied the competitive Langmuir isotherm to describe ion-exchange. It was correct, but they do not explicitly emphasize the basic difference between the adsorption and ion exchange. As mentioned previously, the adsorption and ion exchange are two principally different processes: adsorption is directed by the decrease of surface energy, and it takes place on the free surface sites; ion exchange is a competitive process on an already covered surface, determined by the ionic composition of the liquid phase.

In order to overcome the difficulties they met using a competitive Langmuir isotherm, Boyd et al. (1947) used Equation 1.71 and assumed that $1<<(b_1c_1 + b_2c_2)$, expressing that there are no unsaturated exchange sites (Cernik et al. 1996). They obtained the ion-exchange isotherm as follows:

$$a_1 = z\frac{b_1c_1}{b_1c_1 + b_2c_2} \qquad (1.90)$$

This equation can also be derived by taking into account the differences between the mechanism of adsorption and ion exchange (Nagy and Kónya 1992) through the following steps. Equation 1.72 shows the ratio of the surface coverages of the two particles. When we take into account that ion exchange occurs on the covered surface sites, we can assume that, for ion exchange, the surface is always covered by ions, mathematically:

$$\Theta_1 + \Theta_2 = 1 \tag{1.91}$$

From here,

$$\Theta_1 = 1 - \Theta_2 \tag{1.92}$$

Substituting into Equation 1.72 we obtain:

$$\Theta_1 = \frac{b_1 c_1}{b_1 c_1 + b_2 c_2} \quad \text{or} \quad \Theta_2 = \frac{b_2 c_2}{b_1 c_1 + b_2 c_2} \tag{1.93}$$

which is equivalent to Equation 1.90.

From these, we can arrive at the ion exchange isotherm equations analogous to Equations 1.73 and 1.74:

$$\frac{c_1}{a_1} = \frac{1}{\zeta}\left(c_1 + \frac{K_1}{K_2} c_2 \right) = \frac{m}{V}\frac{y_1}{x_1} = \frac{m}{V}\frac{1}{k_1} \tag{1.94}$$

$$\frac{c_2}{a_2} = \frac{1}{\zeta}\left(c_2 + \frac{K_2}{K_1} c_1 \right) = \frac{m}{V}\frac{y_2}{x_2} = \frac{m}{V}\frac{1}{k_2} \tag{1.95}$$

In Equation 1.94 we take into account that the equilibrium concentration of the solution can be expressed by the initial concentration of the liquid phase (c_{10}) and the relative quantity of the exchanging ion in the solution (y_1), that is,

$$c_1 = \frac{y_1 c_{10}}{V}$$

The adsorbed quantity per unit mass of the solid phase is:

$$a_1 = \frac{x_1 c_{10}}{m}$$

where m is the mass of the solid phase, V is the volume of the solution, x_1 is the relative quantity of the exchanging ion on the solid, assuming that $x_1 + y_1 = 1$, and k_1 is the distribution coefficient of the ion Me_1 ($k_1 = x_1/y_1$). A similar equation can be written for the exchanged cation (Me_2; Equation 1.95). In Equations 1.94 and 1.95, the number of exchange sites is signed by ζ, just to differentiate the adsorption and ion exchange sites. Comparing the adsorption (Equations 1.73 and 1.74) and ion exchange (Equations 1.94 and 1.95) isotherms, we can see that one parameter (K_1) is absent on the right side of the ion exchange isotherms. Equations 1.94 and 1.95 can also be derived using statistical thermodynamics (Sposito 1981).

It is in general, in the literature, that ion-exchange (and adsorption) processes are evaluated by a linearized isotherm. The number of active sites and the K isotherm parameter characteristic of the sorption affinities are determined from the slope and the intercept of the plot. When the isotherm is not linear, heterogeneous surface is supposed. The isotherms are divided into linear portions, and different ζ and K values are determined for the linear portions. However, for competitive processes, including competitive adsorption and ion-exchange, it is not correct because the so-called intercept is a complex quantity, containing the concentration of the competing substance.

A better version to take into account of the surface heterogeneity is the construction of so-called composite isotherms. With regard to Equations 1.94 and 1.95, isotherms refer to homogeneous surface (similar to the Langmuir isotherm of adsorption with the four basic assumptions mentioned earlier). When the surface is heterogeneous, having a range of nonequivalent sites, the so-called composite isotherms (Θ^* versus c functions) are constructed, which are the suitable weighted sums of the individual exchange isotherms (e.g., Equation 1.93) of each site (Kinniburgh et al. 1983; Riemsdijk et al. 1986):

$$\Theta_2^*(c) = \int_0^\infty w(b)\Theta_2(b,c)dc \qquad (1.96)$$

where $w(K)$ is the fraction of sites having a given b between $b + db$ ($b = b_2/b_1$), and $w(b)$ is a site affinity distribution function. The composite isotherm can experimentally be measured, and the site affinity distribution function and local isotherms can be determined. Kinniburgh et al. (1983) show four possible composite isotherms. Here, using the formalism of this book we obtain

Discrete-site Langmuir for ion exchange:

$$\Theta_2^*(c) = \sum_{i=1}^{k} w_i \frac{b_2 c_2}{b_1 c_1 + b_2 c_2} \qquad (1.97)$$

where w_i is the relative occupancy of sites of type I, and the sum of w_is is 1.

Generalized Freundlich:

$$\Theta_2^*(c) = \left(\frac{b_2 c_2}{b_1 c_1 + b_2 c_2}\right)^{1/n} \qquad (1.98)$$

Langmuir–Freundlich:

$$\Theta_2^*(c) = \frac{(b_2 c_2)^{1/n}}{(b_1 c_1)^{1/n} + (b_2 c_2)^{1/n}} \qquad (1.99)$$

Tóth:

$$\Theta_2^*(c) = \frac{b_2 c_2}{\left((b_1 c_1)^{1/n} + (b_2 c_2)^{1/n}\right)^n} \qquad (1.100)$$

In the different composite isotherms, the site affinity distribution function is expressed in the value of w or $1/n$.

The application of the composite isotherms enables us to model the ion exchange on heterogeneous surfaces, such as rocks and soils. When the structure and composition of the sorbent is well known, we can choose the most probable site affinity distribution function. If not, it is desirable to fit the composite isotherm using different models. The just-described four isotherms provide an opportunity for this. In addition, when adsorption and ion exchange can take place simultaneously, adsorption and ion-exchange isotherms and site distribution functions can be combined (Cernik et al. 1996).

1.3.4.2.3 Comparison of Law of Mass Action and Sorption Isotherms

The two treatments of ion exchange reactions are equivalent: exchange isotherm equations can be transformed into the ones that used the law of mass action. To understand it, we have to take into consideration that quantitative treatments using the law of mass action and isotherm equations apply different symbols and approaches. The treatments using law of mass action applies the activities signed by a. The letter a is used for the adsorbed and exchanged quantities in the isotherms, where concentrations are used. The surface molar or equivalent fraction is signed by X or Θ in the law of mass action or in the ion-exchange isotherm equations, respectively ($X = \Theta = a/\zeta$). When transforming the ion-exchange isotherm into the equation of the law of mass action, these differences have to be taken into consideration. For example, when writing Equation 1.94 by the indexes used in the law of mass action,

$$\frac{c_{Me_1}}{a_{Me_1}} = \frac{1}{\zeta}\left(c_{Me_1} + \frac{K_1}{K_2}c_{Me_2}\right) \tag{1.101}$$

for homovalent ion exchange (the exchange of two monovalent, bivalent, or trivalent cations), we assume that

$$\zeta = a_{Me_1} + a_{Me_2} \tag{1.102}$$

from Equation 1.101 we obtained:

$$\frac{c_{Me_1}}{a_{Me_1}}\zeta - c_{Me_1} = c_{Me_1}\left(\frac{\zeta}{a_{Me_1}} - 1\right) = \frac{K_1}{K_2}c_{Me_2} \tag{1.103}$$

$$\frac{c_{Me_1}}{c_{Me_2}}\left(\frac{\zeta}{a_{Me_1}} - 1\right) = \frac{K_1}{K_2} = \frac{1}{K_{Me_1, Me_2}} \tag{1.104}$$

By substituting Equation 1.102:

$$\frac{c_{Me_1}}{c_{Me_2}}\left(\frac{a_{Me_1}+a_{Me_2}-a_{Me_1}}{a_{Mc_1}}\right)=\frac{c_{Me_1}a_{Me_2}}{c_{Me_2}a_{Me_1}}=\frac{c_{Me_1}\dfrac{a_{Me_2}}{\zeta}}{c_{Me_2}\dfrac{a_{Me_1}}{\zeta}}=\frac{1}{K_{Me_1,Me_2}} \quad (1.105)$$

This equation is reciprocal to Equation 1.79, except that concentrations are used here instead of activities ($\gamma \approx 1$) (Vanselow [1932] equation for homovalent exchange), that is, $c_{Me1} = a_{Me1}$ and $c_{Me2} = a_{Me2}$ are the concentration or the activities of the ions in the solution.

The transformation of the equilibrium (or selectivity) constants and the ion-exchange isotherms can be made easily only for homovalent ion exchange because the ion-exchange isotherms usually do not take into consideration the heterovalent character of the ion exchange. This causes additional serious problems in the evaluation of isotherm parameters. It is shown for the exchange of monovalent and divalent cations (Nagy et al. 2016). Firstly, we assume that there is a monovalent ion exchanger and an exchange from the monovalent ions to bivalent ones:

$$2Me_1 - S + Me_2{}^{2+} = Me_2 - S + 2Me_1{}^+ \quad (1.106)$$

The indexes 1 and 2 mean the valances of the ions. The equilibrium constant of the process (Equation 1.106) is:

$$K_{1,2} = \frac{a_2 c_1^2}{a_1^2 c_2} \quad (1.107)$$

The number of exchange sites (ζ) can be expressed both for the monovalent and bivalent ions. For monovalent ions (ζ_{mono}):

$$\zeta_{mono} = a_1 + 2a_2 \quad (1.108)$$

and from here:

$$a_1 = \zeta_{mono} - 2a_2 \quad (1.109)$$

By substituting Equation 1.109 into Equation 1.107, we obtain:

$$K_{1,2} = \frac{a_2 c_1^2}{(\zeta_{mono} - 2a_2)^2 c_2} \quad (1.110)$$

and from here:

$$1 = \frac{1}{K_{1,2}} * \frac{a_2 c_1^2}{(\zeta_{mono} - 2a_2)^2 c_2} \quad (1.111)$$

After some equivalent mathematical transformations (Equations 1.112–1.114):

$$\frac{\zeta_{mono} - 2a_2}{a_2} = \frac{\zeta_{mono}}{a_2} - 2 = \frac{1}{K_{1,2}} * \frac{c_1^2}{(\zeta_{mono} - 2a_2)c_2} \tag{1.112}$$

$$c_2\left(\frac{\zeta_{mono}}{a_2} - 2\right) = \frac{1}{K_{1,2}} \frac{c_1^2}{(\zeta_{mono} - 2a_2)} \tag{1.113}$$

$$\zeta_{mono}\frac{c_2}{a_2} = 2c_2 + \frac{1}{K_{1,2}} * \frac{c_1^2}{(\zeta_{mono} - 2a_2)} \tag{1.114}$$

Finally, we obtain a c_2/a_2 versus c_2 relation:

$$\frac{c_2}{a_2} = \frac{1}{\zeta_{mono}}\left(2c_2 + \frac{1}{K_{1,2}} \frac{c_1^2}{(\zeta_{mono} - 2a_2)}\right) \tag{1.115}$$

Equation.1.115 is a sorption isotherm which describes the c/a versus c function for the monovalent ion in the ion exchange process of monovalent ions to bivalent ions.

A similar equation can also be obtained for the reverse exchange, that is, in the case of the exchange of bivalent ions for monovalent ions. The reverse reaction of Equation 1.106 is

$$Me_2 - S + 2Me_1^+ = 2Me_1 - S + Me_2^{2+} \tag{1.116}$$

The equilibrium constant is the reciprocal of Equation 1.107:

$$K_{2,1} = \frac{c_2 a_1^2}{c_1^2 a_2} = \frac{1}{K_{1,2}} \tag{1.117}$$

From Equation 1.108, we can express a_2 such as:

$$a_2 = \frac{\zeta_{mono} - a_1}{2} \tag{1.118}$$

By substituting Equation 1.118 into Equation 1.117, we obtain:

$$K_{2,1} = \frac{c_2 a_1^2}{c_1^2 \dfrac{\zeta_{mono} - a_1}{2}} \tag{1.119}$$

Equation 1.119 can be transformed in the same way as Equation 1.110 and we obtain the formal isotherm equation:

$$\frac{c_1}{a_1} = \frac{1}{\zeta_{mono}}\left(c_1 + \frac{2}{K_{2,1}} \frac{c_2 a_1}{c_1}\right) \tag{1.120}$$

As mentioned previously in this section, the number of exchange sites can be expressed to bivalent ions (ζ_{bi}) as follows:

$$\zeta_{bi} = \frac{a_1}{2} + a_2 \tag{1.121}$$

From here, the concentration of the monovalent and bivalent ions, respectively, in the solid ion exchanger is expressed as:

$$a_1 = 2\left(\zeta_{bi} - a_2\right) \tag{1.122}$$

and:

$$a_2 = \zeta_{bi} - \frac{a_1}{2} \tag{1.123}$$

By substituting Equations 1.122 and 1.123 into Equations 1.107 and 1.117, respectively, and after similar mathematical transformations, we obtain the c_2/a_2 versus c_2 functions:

$$\frac{c_2}{a_2} = \frac{1}{\zeta_{bi}} \left(c_2 + \frac{1}{K_{1,2}} \frac{c_1^2}{2^2 * \left(\zeta_{bi} - a_2\right)} \right) \tag{1.124}$$

and:

$$\frac{c_1}{a_1} = \frac{1}{\zeta_{bi}} \left(\frac{1}{2} c_1 + \frac{1}{K_{2,1}} \frac{a_1 c_2}{c_1} \right) \tag{1.125}$$

Equations 1.115, 1.120, 1.124, and 1.125 can be plotted as a c_{Me}/a_{Me} versus c_{Me}, or by the usual signs of isotherms equations c/a versus c function. As seen, the second member in the parenthesis on the right side of Equations 1.115, 1.120, 1.124, and 1.125 is a more complex quantity than in the case of homovalent exchange, containing variable quantities, namely, the concentrations of both ions in the solution and the sorbed quantity of the exchanging ion, as well. So, when evaluating c/a versus c function by the ion exchange isotherm (e.g., by Equation 1.94), the "intercept" depends on a lot of parameters, and so provides no information on the energy distribution of the surface. The evaluation of the experimental data in a hydrogen ion-calcium-montmorillonite system will be shown in Chapter 2, Section 2.7.1.

Similarly, the reaction equations and equilibrium constants of monovalent and trivalent, bivalent, and trivalent ions, respectively, can be defined. The number of exchange sites can be expressed for all of the monovalent, bivalent, and trivalent cations; and from here the c/a versus c functions can be derived. These functions are summarized in Table 1.9. Some examples of how these equations apply to bentonite are discussed in Nagy et al. 2016. The main conclusion is that the number of exchange sites could be determined with high accuracy; the standard deviation of the equilibrium constants of the cation exchange is quite large, thus only the relative preference of the ions can be determined.

TABLE 1.9
c/a vs c Functions of Homo and Heterovalent Ion Exchanges Derived from the Law of Mass Action

$$\frac{c_A}{a_A} = \frac{1}{\zeta}\left(c_A + \frac{c_B}{K_{B,A}}\right) \quad \text{(Equation 101)}$$

Homovalent ion exchange

$$\frac{c_B}{a_B} = \frac{1}{\zeta}(c_B + K_{B,A}c_A) = \frac{1}{\zeta}\left(c_B + \frac{c_A}{K_{A,B}}\right) \quad \text{(Equation 1 101)}$$

Heterovalent ion exchange

Mono and bivalent ions		
Number of exchange sites for monovalent ion	$\dfrac{c_1}{a_1} = \dfrac{1}{\zeta_{mono}}\left(c_1 + \dfrac{2}{K_{2,1}}\dfrac{a_1 c_2}{c_1}\right)$ (Equation 1.120)	$\dfrac{c_2}{a_2} = \dfrac{1}{\zeta_{mono}}\left(2c_2 + \dfrac{1}{K_{1,2}}\left(\zeta_{mono}-2a_2\right)c_1^2\right)$ (Equation 1.115)
Number of exchange sites expressed for bivalent ion	$\dfrac{c_1}{a_1} = \dfrac{1}{\zeta_{bi}}\left(\dfrac{1}{2}c_1 + \dfrac{1}{K_{2,1}}\dfrac{a_1 c_2}{c_1}\right)$ (Equation 1.125)	$\dfrac{c_2}{a_2} = \dfrac{1}{\zeta_{bi}}\left(c_2 + \dfrac{1}{K_{1,2}}\,2^2\,*\left(\zeta_{bi}-a_2\right)c_1^2\right)$ (Equation 1.124)
Mono- and trivalent ions		
Number of exchange sites for monovalent ion	$\dfrac{c_1}{a_1} = \dfrac{1}{\zeta_{mono}}\left(c_1 + \dfrac{3}{K_{3,1}}\dfrac{a_1^2 c_3}{c_1^2}\right)$	$\dfrac{c_3}{a_3} = \dfrac{1}{\zeta_{mono}}\left(3c_3 + \dfrac{1}{K_{1,3}}\,3^3\left[\zeta_{mono}-3a_3\right]^2 c_1^3\right)$
Number of exchange sites expressed for trivalent ion	$\dfrac{c_1}{a_1} = \dfrac{1}{\zeta_{tri}}\left(\dfrac{1}{3}c_1 + \dfrac{1}{K_{3,1}}\dfrac{a_1^2 c_3}{c_1^2}\right)$	$\dfrac{c_3}{a_3} = \dfrac{1}{\zeta_{tri}}\left(c_3 + \dfrac{1}{K_{1,3}}\,3^3\left[\zeta_{tri}-a_3\right]^2 c_1^3\right)$
Bi- and trivalent ions		
Number of exchange sites for monovalent ion	$\dfrac{c_2}{a_2} = \dfrac{1}{\zeta_{mono}}\left(2c_2 + \dfrac{3^2}{K_{3,2}}*\left[\zeta_{mono}-2a_2\right]\dfrac{a_2^2 c_3^2}{c_2^2}\right)$	$\dfrac{c_3}{a_3} = \dfrac{1}{\zeta_{mono}}\left(3c_3 + \dfrac{2^3}{K_{2,3}}\left[\zeta_{mono}-3a_3\right]^2 \dfrac{a_3 c_2^3}{c_3}\right)$
Number of exchange sites expressed for bivalent ion	$\dfrac{c_2}{a_2} = \dfrac{1}{\zeta_{bi}}\left(c_2 + \dfrac{1.5^2}{K_{3,2}}*\left[\zeta_{bi}-a_2\right]\dfrac{a_2^2 c_3^2}{c_2^2}\right)$	$\dfrac{c_3}{a_3} = \dfrac{1}{\zeta_{bi}}\left(1.5c_3 + \dfrac{1}{K_{2,3}}\left[\zeta_{bi}-1.5a_3\right]^2 \dfrac{a_3 c_2^3}{c_3}\right)$

(Continued)

TABLE 1.9 (CONTINUED)
c/a vs c Functions of Homo and Heterovalent Ion Exchanges Derived from the Law of Mass Action

Number of exchange sites expressed for trivalent ion		

$$\frac{c_2}{a_2} = \frac{1}{\zeta_{sri}} \left(\frac{2}{3} c_2 + \frac{1}{K_{3,2}} * \frac{a_2^2 c_3^2}{\left[\zeta_{sri} - \frac{2}{3} a_2 \right] c_2^2} \right) \quad 41$$

$$\frac{c_3}{a_3} = \frac{1}{\zeta_{sri}} \left(c_3 + \frac{1}{K_{2,3}} \frac{\frac{2}{3}^3 a_3 c_2^3}{\left[\zeta_{sri} - a_3 \right]^2 c_3} \right) \quad 42$$

Note: For the homovalent ion exchange, the concentrations are expressed in moles, the two columns contain the distribution of the A and B ions, as a function of the solution concentration of A and B, respectively, and $K_{B,A} = K_B/K_A = 1/K_{A,B}$. For the heterovalent ion exchanges, the concentrations can be expressed for the ions with different valencies (second column). The third and fourth columns contain the distribution of the A and B ions, as a function of the solution concentration of A and B, respectively.

1.3.4.2.4 Pitfall of Applications of Adsorption and Ion Exchange Isotherms

As a summary of the literature, we can say that the treatments of the ion-exchange and adsorption isotherms tend to use a linearized form of the isotherm equation. However, the isotherms are frequently not linear. The divergence from the linearity is evaluated by surface heterogeneity and the interactions between the particles of the sorbed substances. When we carefully study the isotherms, we can see that this way of treatment is not suitable for competitive interfacial processes, including competitive adsorption and ion exchange. As mentioned earlier, ion exchange is always competitive. In this case, Equations 1.73 and 1.74, or 1.94 and 1.95, have to be applied. As it seems from these equations, isotherms cannot be linear because the concentration of competing substances is also in the equation, which is a variable quantity. For this reason, isotherms have to be evaluated by a parameter estimation that takes into account the concentration of the competing substance. Fortunately, today it is not problematic because there are many parameter-estimating computer programs (PLOT, SCIENTIST, etc.) in which optional functions can be defined.

To illustrate the effect of the competing substance, the results of model calculations are shown in Figure 1.10 in the case of ion exchange (Kónya and Nagy 2009, 2013). Consider a system with different parameters: the mass of the sorbent is 1 g, the volume of the solution is 1 dm^3, and the number of ion-exchange and adsorption sites is 10^{-3} mol/g. Let the ratio of $K_1{:}K_2$ be 0.5, 1, 2, respectively. As shown in Equation 1.98, in the case of ion-exchange, the ratio of $K_1/K_2 = 1/K_{Me1,Me2}$, that is, the ratio of K's of the individual ions is equal to the reciprocal of the equilibrium constant or the selectivity coefficient. These data are real in ion exchange, that is, for montmorillonite (Chapter 2, Table 2.2). The concentrations of the competing ion (c_2) in Equations

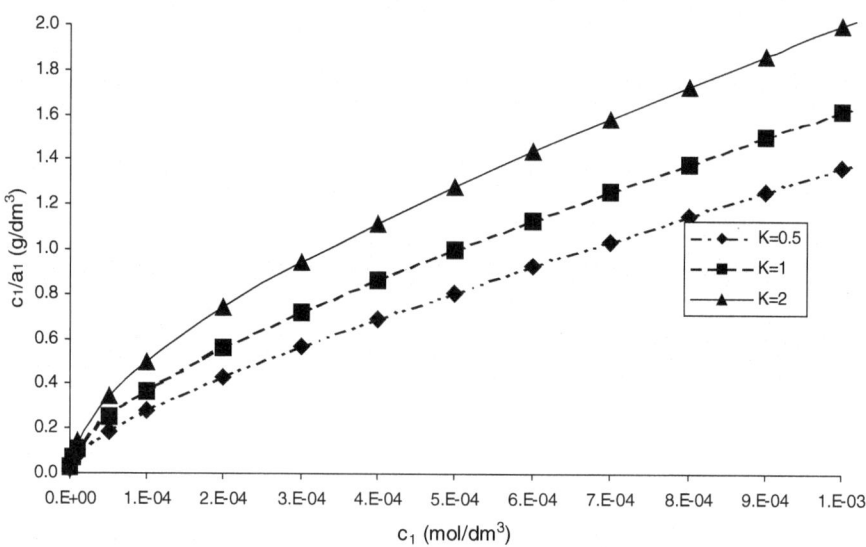

FIGURE 1.10 Model calculations of ion exchange and competitive adsorption isotherm.

1.73 or 1.94 are calculated from the sorbed quantity (a_1). Since 1 g and 1 dm^3 are used in the model calculation, the numerical value of c_2 and a_1 are equal.

As seen in Figure 1.10, the plot of the isotherms is usually not linear. The shape of the plot is determined by the ratios of K_1/K_2 and c_2. No linear portion of the isotherm can be expected.

In the case of competitive adsorption, similar calculations cannot be done exactly, because the value of c_2 is unknown, although it increases as a_1 increases. So, similar plots should be obtained. The value of K_1 only shifts the plot up. Linear plots can be obtained when K_1 or c_2 is great, that is, the sorption affinity of Me_1 ion is small compared to the sorption affinity of Me_2. Both cases mean that practically no competition occurs.

During the model calculations, surface heterogeneity and interactions of the ions are neglected; after all, the shape of the isotherm is curved. If there is any surface heterogeneity or interaction between the sorbed ions, it cannot be concluded from the isotherm. (The evaluation of an ion exchange isotherm will be shown in Chapter 2, Section 2.2.) In order to study the energy distribution of surface sites, thermodynamic equations, including isotherms, have to be selected according to the sorption mechanism. All variable quantities have to be determined that are involved in the applied model (e.g., concentration of the competing substances in both phases). These quantities have to be treated as independent variables during the estimation of the sorption parameters. This treatment can be combined with methods for the determination of energy distribution functions.

1.3.4.3 Treatment of Simultaneous Ion Exchange and Adsorption Processes

The situation is more complicated in the case of simultaneous ion-exchange and adsorption processes. In this case, the total sorbed quantity of the Me_1 cation (a_1) is the sum of the quantities sorbed by ion exchange $(a_{1\,ie})$ and adsorption (a_{1ads}):

$$a_1 = a_{1ie} + a_{1ads} \tag{1.126}$$

Let us consider that the competitive processes can be different in the two processes. For example, during ion exchange, two cations, during the adsorption on the edge sites, a cation and hydrogen ions, even water, can compete. The sorption affinity of the individual ions can also be different in the following two ways: ion exchange and adsorption. As a result, the K values of the ions can be different, that is $K_{1ie} \neq K_{1ads}$, or $K_{2ie} \neq K_{2ads}$. The equality of the pairs of K's is also permitted.

From Equations 1.94 and 1.73 we obtain:

$$a_{1ie} = \frac{\zeta c_1}{c_1 + \dfrac{K_{1ie}}{K_{2ie}} c_2} \tag{1.127}$$

Or:

$$a_{1ads} = \frac{z c_1}{c_1 + K_{1ads} + \dfrac{K_{1ads}}{K_{2ads}} c_2} \tag{1.128}$$

Equations 1.127 and 1.128 can be substituted into Equation 1.126:

$$a_1 = \frac{\zeta c_1}{c_1 + \dfrac{K_{1ie}}{K_{2ie}} c_2} + \frac{z c_1}{c_1 + K_{1ads} + \dfrac{K_{1ads}}{K_{2ads}} c_2} \qquad (1.129)$$

For the evaluation of the isotherm, Equation 1.129 should be used. Since there are several parameters (up to 7), the experimental separation of ion exchange and adsorption can be simpler (Chapter 2, Section 2.6).

On the basis of those discussed here, we can say that the thermodynamic parameters of ion exchange determined by linear isotherm equations are not correct. We determine the mechanism of sorption (ion exchange and/or adsorption) and then choose the isotherm equation. When simultaneous interfacial processes take place, the dominant processes, if any, have to be selected and experimentally separated, or Equation 1.129 should be used.

1.3.4.4   Treatment of Heterogeneous Isotope Exchange

A special case of ion exchange, or even molecule exchange, reactions is isotope exchange reaction. In this case, the reactants and products are chemically identical but have different isotopic composition. As is well known, the thermodynamic drive of a chemical reaction is the change in Gibbs free energy:

$$\Delta G = \Delta H - T \Delta S \qquad (1.130)$$

where G is the free energy, H is the enthalpy, and S is the entropy. During isotope exchange, the enthalpy of the reaction is zero because there are no chemical changes. Therefore, the reaction is driven by the change of entropy (ΔS) which is called mixing entropy:

$$\Delta G = -T \Delta S \qquad (1.131)$$

The isotope exchange processes are classified on the basis of the number of phases present as homogeneous and heterogeneous isotope exchange. Because of the scope of this book, only the heterogeneous isotope exchange is discussed.

The equations of heterogeneous isotope exchange in equilibrium are simpler than the ion-exchange equations because the two ions are chemically the same. In the treatment by the law of mass action, it means that the equilibrium constant is equal to 1. The selectivity coefficients at $X = 0$ and $X = 1$ can be determined by measuring the heterogeneous isotope exchange in which the concentration of the radioactive isotopes is very low and approaches zero (carrier-free radioactive isotope).

Heterogeneous isotope exchange can be used in the study of surface heterogeneity. To do so, the solid surface is equilibrated with a solution containing the radioactive isotopes diluted with an inactive carrier. After equilibrium is established, the relative activity of the radioactive isotope (y) in the solution is multiplied with the original solution concentration (c_0), and the obtained value is considered as the equilibrium concentration (c_2) in the isotherm equations. Similarly, the quantity sorbed on the solid surface (a_2) can be obtained: the relative activity of the radioactive isotope (x) on the solid is multiplied with the original solution concentration (c_0). (Note that $x + y = 1$.) Obviously, these are not real concentrations (which remain unchanged) but rather a way to construct ion-exchange isotherms or the selectivity functions, which will provide information on surface heterogeneity in the given system.

In the case of heterogeneous isotopic exchange, the ratio of $K_1/K_2 = 1$. From Equation 1.95 we can conclude that:

$$\frac{c_2}{a_2} = \frac{ym}{xV} = \frac{1}{\zeta}(c_2 + c_1)$$

(1.132)

where m is the mass of the solid phase, V is the volume of the solution, c_1 is the concentration exchanged in the process, and ζ is the number of active sites. The real concentration of the solution is $c_0 = c_1 + c_2$. When the surface is homogeneous, the c_2/a_2 versus c_2 function is linear. This is the basis of the Paneth's method for the determination of specific surface area (Paneth 1922). When the function is not linear, different ζ's can be obtained. Thus, information on site distribution can be gained.

Another application of heterogeneous isotope exchange is the determination of the ratios of differently sorbed ions or molecules taken up by geological formations, such as soil. An important example is the phosphate ion which can be sorbed in different ways; in the literature, weakly and tightly sorbed phosphate species are mentioned. The ratio of these species is essential information in plant nutrition. Using heterogeneous isotope exchange studies, we can determine the ratio of weakly and tightly sorbed phosphate, and consequently the phosphate available for plants. It is due to the fact that only weakly sorbed species can take part in heterogeneous reactions in the relatively short time-scale laboratory studies. These studies will be discussed in Chapter 3, Section 3.5.

The kinetics of heterogeneous isotope exchange will be discussed in Section 1.3.6.2.

1.3.5 CATALYTIC EFFECTS OF CLAYS

As a result of the interfacial processes on rocks and soils, the structure and chemical bonds of the sorbed compounds can be changed. For this reason, different chemical reactions can be initiated in which the components of rocks or soils act as catalysts. The most important mineral catalysts are the zeolites and clay minerals. Naturally, the different oxides also have catalytic effects, and nowadays some of them are being artificially produced for catalytic purposes such as framework silicates (zeolites), the most effective and selective catalysts in organic syntheses. The catalytic applications of zeolites are too wide to summarize in this book, so we deal with the catalytic effects of clay minerals.

Clay minerals are effective catalysts of many organic reactions, frequently showing product, region, or shape selectivity. The other important advantage of clay minerals as catalysts is that the reaction conditions are frequently milder than in traditional procedures. The catalyst can be eliminated by filtration, and the use of hazardous solvents can be avoided, which reduces the quantity of the waste. Consequently, clay minerals are often employed as catalysts in the so-called green chemistry.

Clay minerals or rocks are used as catalysts in both natural and chemically modified forms. Their very first application dates back to the 1960s when acid-treated clays were used for cracking in the oil industry (Franz et al. 1959). In the last decades, several reviews have been published on the catalytic effects of clay minerals (Theng 1974, 1982; Solomon and Hawthorne 1983; Laszlo 1986a, 1986b; Adams 1987; McCabe 1996; Adams and McCabe 2008).

The most important aspect of the catalytic activity of clay minerals is that they have Brönsted as well as Lewis acidity. Brönsted acidity comes from the terminal hydroxyl groups and from the bridging oxygen atoms. High acidity is due especially to the latter, and it is influenced by the exchangeable cations present in the clay since strong Brönsted acidity derives from the dissociation of water coordinated to the interlayer cations. The acid strength is increased with the increase of ionic potential, which is the ratio of charge and size of the ion. Lewis acidity originates from the incompletely coordinated Al^{3+} and Fe^{3+} ions, exposed at the edges. The two types of acidity can transform into each other, for example at high temperature, the number of sites with Lewis acidity increases because of the dehydration of the interlayer cations. The types of surface acidity can be determined experimentally (Heller-Kallai 2002).

As seen previously, the acidity depends on the interlayer cation, and so the catalytic activity can be affected by cation exchange processes. In addition, when the interlayer cations, or even the cations of the octahedral sheet (e.g., Fe^{2+} or Fe^{3+}), have different oxidation states, clay minerals can catalyze redox reactions, too.

Geometric factors play a role in the catalytic processes on clays. Smectites, for example, consist of regular, parallel sheets. The size and swelling properties of the interlayer space determine the size and shape of the molecules intercalating between the layers, which will have an effect on the selectivity of the reaction.

Clays can be used as catalysts in natural form or after purification, but in order to increase the catalytic activity, they can also be chemically modified. One of the possibilities for modifying the catalytic activity is the cation exchange, which has already been discussed. The other possibility is the acidic treatment of the clay, which also increases the acidity. The acidic treatment is made by concentrated hydrochloric, sulfuric, or phosphoric acid, and consequently, the crystal structure of the produced clay is strongly destroyed (Jang et al. 1999). However, from the catalytic point of view, this may be desirable.

The third possibility for the chemical modification is through the synthesis of so-called pillared clays. The term *pillaring* means the production of catalytically active mesoporous materials (Bergaya et al. 2008a). The pillared clays have several interesting properties: large surface area, high pore volume, and tunable pore size (from micropore to mesopores), high thermal stability, and strong surface acidity and catalytic activity (Adams and McCabe 2008). Pillaring has two main steps: the first step is the intercalation of different compounds (inorganic and organic cations or molecules); the second is heat treatment. Successful pillaring requires that the intercalated species must not fill the entire interlayer space, so that it provides the accessible porosity for the product.

Many cation-exchanged clays are suitable for the production of metal-oxide pillared clays. The hydrolysis of the cation helps the pillaring step, so, at first, aluminum ion was applied as a pillaring agent. Later, other elements were also used, for example, zirconia, chromium, iron, transition metal elements; and some lantanoids, organometallic complexes, surfactants, and polymers.

Clay minerals are frequently applied as supports, simply impregnated with a solution containing real catalysts.

As mentioned earlier, clay catalysts are heavily used in organic syntheses (Laszlo 1986; Adams and McCabe 2008), some examples being Brönsted acid catalysts in hydrogen isotope exchange reactions; ether, ester, lacton, amid, and peptide productions; formation of protective groups on carbonyl groups, cyclic anhydride and heterocyclics; rearrangements reactions and electrophylic aromatic substitutions. A number of these reactions have a carbocationic mechanism (Laszlo 1986). Clays are used as Lewis acid catalysts in the formation of C-C bonds, that is, alkylation and acylation, including Friedel–Crafts reactions; cracking and isomerization; rearrangements and electrophylic aromatic substitutions. Both cation-exchanged and pillared clays are used for oxidation and reduction reactions, including hydrogenation.

Since the wide variety of both organic substances and clays are present in nature, it is safe to assume that there are organic reactions initiated by clays under geological conditions, too. The mechanism of these reactions is probably carbocationic. In addition, the cationic species can be adsorbed on the negative charges of clays or exchanged in the interlayer space. These adsorption and cation exchange reactions, resulting in the rearrangements of the electron structure, can be the starting points of organic synthesis in nature. In the literature, this field is treated as the "origin of life" (prebiotic synthesis of protein precursors (e.g., Bujdak et al. 1994, 1995; Bujdak and Rode 1997; Ferris et al. 1996; Ferris 2002; Porter et al. 2000; Brack 2008). In our opinion, these organic reactions can also play an important role in the formation of humic substances. This is proved by the fact that the concentration of clay and soil organic matter of soils often shows some correlation (Sparks 2003). To our best knowledge, such studies have not been done until now.

1.3.6 KINETICS OF INTERFACIAL PROCESSES IN ROCKS AND SOILS

The interfacial processes in rocks and soils have been studied mostly thermodynamically; the kinetic investigations are limited by two important factors. One of them is that the time scale of geological processes can vary significantly: there are very fast (millisecond) and very long (thousands to millions of years processes). Very fast and very slow processes are challenging to study experimentally. Therefore, kinetic investigations can only involve processes with a medium reaction rate. The other limiting factor is the complexity of the systems. Many parameters affect the processes, for example, the composition of the phases, the heterogeneity of the surfaces, interactions and transformations of the sorbed particles, etc. Heterogeneous reactions may result in a wide variety of simultaneous and consecutive steps, each taking place according to the laws of classical reaction kinetics, with the overall kinetics being the result of the different combination of these processes. The relative importance of the processes is not always known. Thus, it is difficult to achieve the most important objective kinetic studies, namely to pinpoint the rate-determining processes. Another complicating factor is that the laboratory and field experiments can hardly be performed under the same conditions. These difficulties are frequently handled by the application of empirical rate equations. The disadvantage of this method is that it does not allow gaining information about the mechanism of the processes. Theoretical kinetic studies, on the other hand, may provide the limits of the application of the kinetic parameters. Combining the two approaches, the physical meanings to kinetic parameters can be given, providing a way to interpret the results of empirical kinetic equations.

1.3.6.1 Steps of Interfacial Reactions

To study any interfacial reactions of soils and rocks from kinetic points of view, as in thermodynamic studies, we have to take into consideration processes that occur both in the bulk phases and on their interface. The processes in bulk phases are important as they transport the reactive species to and from the interface. Accordingly, the literature of heterogeneous reactions divides the kinetics of the reactions into three basic steps. The first is the transport of the substances from the bulk to the interface. This means the convection, mixing, and diffusion of the dissolved substance from the solution phase to the interface, that is, the reaction zone. The transport in the solution phase has two steps: the movement of the dissolved substances in the bulk solution and through a so-called adhesion layer. Convection and mixing influences only the transport in the bulk phases; in the adhesion layer, only diffusion, called film diffusion, is possible, determined by the concentration gradient through the adhesion layer. Mixing decreases the thickness of the adhesion layer. Without any mixing, the thickness of the adhesion layer can be as thick as some tenth to a millimeter. Under intense mixing, the thickness of the adhesion layer can be decreased by 2–3 orders of magnitude. Therefore, comparing to the molecular dimensions (cca 10^{-10} m), the thickness of adhesion layer (at least 10^{-5} m or thicker) is large. For this reason, diffusion plays an important role in interfacial processes. After diffusion through the adhesion layer, the substance may participate in interfacial (physical or chemical) reactions, for example, adsorption, ion exchange, precipitation, chemical reactions, etc. Finally, the reaction products can leave the interface and return to the bulk phases. This can occur through intra-particular, inter-particular, surface diffusion, or recrystallization.

The interfacial processes can be greatly affected by the size of the pores of the solid phase. In the macropores, the size of the pores is greater than the thickness of the adhesion layer, and the diffusion is similar to the diffusion through the adhesion layer, that is, film diffusion. In meso- and micropores, the diffusion is called pore diffusion which is a type of inter-particular diffusion.

In a steady state, the rate of the overall interfacial reaction is determined by the slowest step. In laboratory experiments, the transport in the bulk solution can usually be fast, so it is not a rate-determining step. In many cases, diffusion through the adhesion layer is considered to be the rate-determining step. In this case, the kinetics may be treated as a diffusion problem. However, the chemical and physical interfacial reactions of rocks and soils can also be slow and may be rate-determining steps.

In the first approach described in the following text, the kinetics of interfacial reactions of rocks and soils will be treated as a diffusion problem. Then, we will discuss kinetic models, in which the concentration of free sorption sites is considered to be a rate-determining factor, that includes the rate of the interfacial reaction, even if it is not expressed explicitly. There are kinetic models which take into consideration the equilibrium state of the sorption reaction. They can also be treated as models including the rate-determining role of interfacial reaction.

As is well known in physical chemistry, diffusion is mathematically described by two Fick's laws. When the concentration gradient of the adhesion layer is constant, Fick's first law can be applied for film diffusion, as well. In geological systems, however, the concentration gradient of a diffusing species simultaneously changes in time and place, so the second law has to be applied. Fick's second law cannot be solved generally; only partial solutions exist under well-determined boundary and initial conditions (Crank 1975). In the case of natural rocks and soils, there is a complicated pore network, so well-defined boundary and initial conditions can be difficult to define. As a consequence, a general kinetic diffusion equation cannot be derived. In addition, as discussed previously, the surface of rocks and soils can have sites with different sorption energy, which also complicates the kinetic description. For these reasons, the investigators choose different ways to deal with the diffusion problem such as looking for an empirical equation that is fitting with the experimental data or constructing different kinetic models taking into account different factors, including diffusion processes and interfacial chemical and physical reactions.

Two examples of empirical equations are shown here. One of them is the Elovich equation:

$$a_t = \left(\frac{1}{\beta}\right)\ln(\alpha\beta) + \left(\frac{1}{\beta}\right)\ln t \tag{1.133}$$

where a_t is the amount of the sorbed substance per unit mass of the solid at time t, and α and β are constants in an experiment (Chien and Clayton 1980; Sparks, 1989).

The other example is the power function equation, or, by another name, fractional power or the modified Freundlich equation (Wahba and Zaghloul 2007):

$$a_t = kt^v \tag{1.134}$$

where a_t is the amount of the sorbed substance per unit mass of the solid phase at time t, k, and v are constants, and v is positive and <1 (Kuo and Lotse 1974; Havlin and Wesfall 1985).

Some authors describe the kinetics of adsorption process as an exponential function or by a linear combination of exponential terms analogous to a series of simultaneous first-order reactions. Neuman and Neuman (1958) described the kinetics of the adsorption process using the following equation:

$$x_t = x_\infty - A_1 \exp(-k_1 t) - A_2 \exp(-k_2 t) - \ldots - A_n \exp(-k_n t) \tag{1.135}$$

where t is the reaction time; x_t the relative sorption quantity at time t; $k_1, k_2, \ldots k_n$ are the rate constants; and $A_1, A_2, \ldots A_n$ are constants such that the sum is equal to x_∞, the relative sorbed quantity in equilibrium. Imre (1933b) already used this equation to describe an adsorption process on ionic crystals.

When only one exponential member is considered, a pseudo-first-order reaction is supposed as firstly supposed by Lagergren (1898).

The advantage of the application of Equation 1.135 is that the kinetics of all of the possible steps (film diffusion, intra-, inter-particle diffusion, chemical reaction, surface diffusion, and recrystallization) can be described by exponential terms (Sparks 1985, Table IV). It means that each member of the Equation 1.135 has a given physical meaning. In addition, under laboratory conditions, experiments can be conducted so that the number of the rate-controlling steps is less. For example, Liu et al. (1995) studied the short time kinetics of strontium ion uptake in natural samples and determined the rate coefficients of the steps. The rate coefficients of the steps were determined by changing the stirring speed. They suggested that the rate-limiting step is film diffusion in static condition (without stirring), particle diffusion in the stirred condition (at 600 rpm), and chemical reaction in the vortex condition (stirred at 2240 rpm), and therefore they applied one-term exponential equations (first-order kinetics).

When the kinetics of an interfacial reaction can be described by Equation 1.135, the processes can be differentiated and the numerical value of the rate constants (k_n's) can be determined provided that the difference between the values is at least one order of magnitude. In addition to this, the number of rate-controlling processes is influenced by the structure of the soil or rock. For example, the ion exchange kinetics of strontium ion on natural clay samples (Nemes et al. 2005) was described by forms of first-rate kinetic equation with different terms. The ion exchange process was reduced to two steps, such as film diffusion and inter-participle diffusion. However, when clay samples contained a significant amount of cristobalite, which can form a gel phase, three steps such as gel diffusion, film diffusion, and particle diffusion were needed to correctly describe the kinetics.

When the inter-particular diffusion is considered as the rate-determining step, the parabolic diffusion equation can be applied (Zimens 1945; Boyd et al. 1947; Crank 1975; Chute and Quirk 1967; Wollast 1967; Jardine and Sparks 1984; Sparks 1999).

$$\frac{a_t}{a_\infty} = \frac{4}{\pi^{1/2}} \frac{Dt^{1/2}}{r^2} - \frac{Dt}{r^2} \tag{1.136}$$

where r is the average radius of the soil particles, a_t is defined in Equation 1.134, a_∞ is the quantity of the sorbed substance at equilibrium, and D is the diffusion coefficient. By equivalent mathematical transformation,

$$\frac{1}{t} \frac{a_t}{a_\infty} = \frac{4}{\pi^{1/2}} \frac{D}{t^{1/2} r^2} - \frac{D}{r^2} \tag{1.137}$$

that is $\dfrac{1}{t} \dfrac{a_t}{a_\infty}$ versus $1/t^{1/2}$ function is linear. The diffusion coefficient can be determined from the slope as well as the intercept. This equation is sometimes expressed and applied in the soil science literature as follows (Sparks 1985, 2003; Stucki and Lee 1999):

$$\frac{a_t}{a_\infty} = R_d t^{1/2} + const \qquad (1.138)$$

When comparing these equations, we can see that Equation 1.138 is not mathematically equivalent to Equations 1.136 and 1.137. Equation 1.138 can be obtained by mathematical approximations, and it provides a good fit to the experimental points of the kinetic curves obtained in experimental studies (Gülen and Aslan 2020).

It happens quite frequently that the experimental data can be equally well described by different kinetic models. This is the consequence of the solution to the kinetic equations, being independent on the rate-determining steps, results in exponential equations. So, the validity of the rate constants always has to be evaluated from physico-chemical, structural points of view. For example, the activation energy of the processes should be calculated from the kinetic curves at different temperatures. The typical activation energy of film diffusion is 17–21 kJ/mol, that of inter-particle diffusion is 21–42 kJ/mol (Sparks 1985). Other processes (such as intra-particular diffusion and recrystallization) have greater activation energies. Thus, potential processes can be chosen based on activation energies. The structural effects (Nemes et al. 2005) were discussed previously.

In the last decades, the kinetic models have been combined with the equilibrium data of the interfacial processes, taking into account that soils and rocks are heterogeneous and consequently have different sites. These models are called non-equilibrium models (Wu and Gschwend 1986; Miller and Pedit 1992; Pedit and Miller 1995; Fuller et al. 1993; Sparks 2003). These models describe processes when a fast reaction (physical or chemical) is followed by one or more slower reaction(s). In these cases, Fick's second law is expressed in that the diffusion coefficient is corrected by an equilibrium thermodynamic parameter of the fast reaction (e.g., by a distribution coefficient), that is, the fast reaction is always assumed in equilibrium. In this way, the net processes are characterized by apparent diffusion coefficients. However, such reactions can be equally well described using Equation 1.135.

Similarly, to equilibrium studies, experimental kinetic curves are frequently fitted using different models and the plot with the smallest regression is considered to be the model describing the interfacial process. In addition to the models mentioned previously in this section (empirical, first-order kinetics, parabolic diffusion models), other models are applied such as zero-order, second-order, and third-order models, as well as mixed order models, etc., as summarized by Wang and Guo (2020b)). The same models are also applied for the desorption of adsorbed substances (Jalili et al. 2019).

It is interesting to mention the pseudo-second-order model which is widely applied since 1996 (Ho et al. 1996). The importance of this model is illustrated by the citation number of a paper published by Ho and McKay (1999), which is more than 9000 (data from Science Direct, April 2020). The authors provide no theoretical explanation, but they state, based on "unpublished data", that this model is the best to describe adsorption kinetics. There is a previous publication (Blanchard et al. 1984), the kinetics of ion exchange of bivalent metal ions to ammonium ion on clinoptilolite, that is described by a pseudo-second-order model. The paper suggests that the ion exchange is relatively slow, that is, the concentration of the bivalent ions changes slowly and a bivalent ion occupies two exchange sites, the exchange of monovalent ions to bivalent ions occurs. Using these assumptions, the kinetic equation is (converted to our notations):

$$-\frac{da_t}{dt} = k(a_\infty - a_t)^2 \tag{1.139}$$

The solution of this equation is:

$$\frac{1}{a_\infty - a_t} - const = kt \tag{1.140}$$

The constant can be determined assuming that at $t = 0$ $a_t = 0$. Thus $const = 1/a_\infty$. From here, the pseudo-second-order kinetic equation is:

$$\frac{1}{\left(a_\infty - a_t\right)} = \frac{1}{a_\infty} + kt \tag{1.141}$$

or

$$\frac{t}{a_t} = \frac{1}{ka_\infty^2} + \frac{1}{a_\infty}t \tag{1.142}$$

This means, that using Equation 1.141 and 1.142, the rate constant and the equilibrium sorbed quantity can be obtained.

Azazian (2004) mathematically evaluated the first-order- and pseudo-second-order kinetics and provided the conditions under which these models can be used. In addition, Azazian indentified the meaning of the observed rate coefficients. Accordingly, when the concentration of the adsorptive in the solution is high, that is, it does not change significantly during the sorption, the first-order kinetics is valid. The rate coefficient is the combination of the rate constants of adsorption and desorption processes. When the concentration of the solution is relatively low and significantly changes during the adsorption process, the pseudo-second-order model is valid. The observed rate constant is a complex function of the initial concentration of the solution. When evaluating experimental kinetic curves, "with increasing initial concentration of the solute, the correlation of the experimental data to the pseudo-second-order kinetic model decreases while that to the pseudo-first-order model increases".

As mentioned previously, the pseudo-second-order kinetic model is widely applied in the litereture to describe sorption kinetics. The authors, however, often neglect to consider if the theoretical or experimental conditions under which this approach can be used are satisfied (such as bivalent-monovalent exchange (Blanchard or et al.

1984) or the concentration of adsorptive (Azazian 2004). Consequently, the physical meaning of the rate constants is questionable.

Some authors, such as Seidel et al. (2005) divide the adsorption process into time periods and use different models for each.

1.3.6.2   Kinetics of Heterogeneous Isotope Exchange

The kinetics of heterogeneous isotope exchange reactions, that is, the process when the ions or molecules participating in the exchange reaction are chemically the same but of different isotopic composition, are treated by the same models as usual for interfacial processes (empirical, first-order, second-order, parabolic, etc.) (Gerischer and Vielstich 1952; Kosmulski et al. 1985). In addition, some authors treat the kinetics using the McKay equation originally derived for homogeneous isotope exchange (McKay 1971) which is also a first-order kinetic model.

In addition to information provided by the equilibrium of the heterogeneous isotope exchange (Section 1.3.4.4), the transport rate of any substances between the phases can be determined without disturbing the chemical equilibrium or steady state. This is the most important aspect and advantage of the radiotracer methods: systems in thermodynamic equilibrium or steady state can be quantitatively studied without disturbing the process. No other method is known to be able to provide such information (Kónya and Nagy 2012, 2018).

The application of heterogeneous isotope exchange for the study of phosphate ion uptake by soils will be discussed in Section 3.5.

1.3.6.3   Migration of Water-Soluble Substances in Rocks and Soils

Transport in the pores of rocks and soil occurs by migration of water-soluble materials. The migration in porous solid matrixes such as rocks and soils are determined by hydrological processes as well as the interaction of the soluble substances and geological formations. When describing the migration of a substance in a porous solid matrix, we have to examine the flowing medium (groundwater), the chemical species, and their sorption properties. The factors affecting the migration in groundwater are:

1. Advection: the migration of dissolved components with flowing water.
2. Mixing of solutions in macropores, which is a consequence of different flowing rate of solutions in pores with different size.
3. Diffusion of dissolved components in water.

In addition, the interactions of solid matrixes and dissolved substances are as follows:

4. Adsorption and ion exchange.
5. Precipitation.
6. Structural modification and destruction of materials.

When the dissolved components are not sorbed by solid matrixes, the transport is determined by processes 1–3, that is, the dissolved components migrate together with water. The flux of flow can be described by different migration equations, for example, by Equation 1.143:

$$J_0 = -\left[\Theta D_h + \Theta D_{eff} \right] \frac{\delta c}{\delta x} + v\Theta c = -\Theta D \frac{\delta c}{\delta x} + qc \tag{1.143}$$

where J_o is the flowing rate of water (flux), Θ is the humidity of medium, D_h is the hydrodynamic dispersion coefficient, D_{eff} is the effective diffusion coefficient, c is the concentration of flowing material at place, x, $q = v\Theta$ is the volume flowing in a unit time, and v is the linear flowing rate.

When solid matrixes adsorb the dissolved components, its flowing rate decreases:

$$\frac{\delta(\Theta c)}{\delta t} = \frac{\delta}{\delta x}\left[D \frac{\delta c}{\delta x} \right] - \frac{\delta qc}{\delta x} - \frac{\delta \rho a}{\delta t} \tag{1.144}$$

ρ is the density of the matrix, a is the adsorbed amount.

In the study of transport processes, the hydrological properties of the flowing medium (water) as well as the chemical species and adsorption properties of the dissolved components have to be studied. In this book, the hydrological properties are not discussed in detail; it is just mentioned that the activation energy of the diffusion of water in clay minerals is 10–20 kJ/mol. Swelling minerals had slightly larger activation energy values than bulk water (17 kJ/mol); non-swelling minerals (illite, kaolinite) have lower values than that of bulk water. The low activation energies may be related to weak H-bonds between water and the clay surfaces compared to those in bulk water (Sánchez et al. 2008).

When the dissolved components are not adsorbed on the solid matrix and do not precipitate, they migrate together with water (e.g., chloride ion). Of course, every part of the solution must be electrically neutral so that the migration of negative chloride ions is followed by the migration of cations, or the migration of cations decreases the migration of anions. In spite of this fact, most literature discusses the diffusion of ions on the basis of chemical potentials, not electrochemical potential, and does not take into account the effect of electroneutrality on the diffusion rate. Theoretically, however, the migration rate of water is the maximum migration rate of water-soluble substances.

When the dissolved components are adsorbed on the solid matrix, take place in ion exchange reactions, or precipitate, their migration rates can significantly decrease. The degree of decrease is determined by the chemical species of the given substance under chemical conditions characteristic of the solutions in geological formations (groundwater).

The migration equations (Equations 1.143 and 1.144) cannot be solved generally; only partial solutions can be obtained at certain boundary and initial conditions. Neglecting the advection, from Equation 1.144 we obtain

$$\frac{\delta(\Theta c)}{\delta t} = \frac{\delta}{\delta x}\left[D \frac{\delta c}{\delta x} \right] - \frac{\delta \rho a}{\delta t} \tag{1.145}$$

Adsorption and ion exchange, for example, can be treated by an adsorption or ion-exchange isotherm (Sections 1.3.4.1 and 1.3.4.2). The ion-exchanged quantity, for example, can be expressed from Equation 1.95, and substituted into Equation 1.145. In real geological conditions, the volume of the solution (V) is substituted by the free pore volume (w). Taking into account that the humidity (Θ) and the density of the solid matrix (ρ) remain constant, the differential equation can be further simplified. The result is as follows:

$$\frac{\delta c}{\delta t} + \frac{w}{m} k \frac{\delta c}{\delta t} = D \frac{\delta^2 c}{\delta x^2} \tag{1.146}$$

where m is the mass of the rock (g), w is the free pore volume saturated with groundwater (dm^3), y and x are the relative quantity of the dissolved and sorbed substance, and k is the distribution coefficient ($k = x/y$). After equivalent mathematical transformations, an equation similar to Fick's second law is obtained:

$$\frac{\delta c}{\delta t} = \frac{D}{1 + \frac{w}{m} k} \frac{\delta^2 c}{\delta x^2} = D_m \frac{\delta^2 c}{\delta x^2} \tag{1.147}$$

where D_m is the migration coefficient. This equation is very similar to those obtained in the non-equilibrium models, namely, the diffusion coefficient is corrected by an equilibrium parameter (distribution coefficient) and an apparent diffusion coefficient, called the migration coefficient here.

There are models which take into consideration the diffusion of water-soluble substances in macropores and micropores, for example, in the interlayer space of smectites (two-compartment model; Bourg et al. 2008).

1.3.6.4 Disposal of Nuclear Waste in Geological Formations

A special case of migration in geological formation is the migration of radioactive isotopes with long half-life which are mainly produced during nuclear energy production (Kónya and Nagy, 2012, 2018). Even if the importance of nuclear energy in the electric energy production sector has been decreasing, the dismantling of these power plants produces nuclear waste as a result of previous operations. The spent fuel elements are nowadays stored in temporary disposals; their reprocessing and/or final deposition in geological disposals is required. The list of radioactive elements and isotopes produced in nuclear energy production is very long; the radioisotopes of about 50 elements are present in the nuclear waste, including the fission products and activation products of the fuel, structural material, and environmental components. The chemical species of these isotopes is highly variable, influenced by chemical conditions, inactive inorganic and organic substances.

The other source of radioactive waste is from nuclear medicine, an expanding field of diagnostics and therapy. Some radioisotopes with long half-life and with radioactive daughter product (such as Tc-99 produced from Tc-99m used in nuclear medicine) are also produced here and emitted into the environment.

In addition to the above, the research and industrial applications of radionuclide also produce radioactive waste to a small degree.

Independent of the method of production and the treatment of the radioactive and nuclear waste, geological disposal is always necessary (www.world-nuclear.org/information-library/nuclear-fuel-cycle/nuclear-waste/storage-and-disposal-of-radioactive-waste.aspx). This means that all radioactive nuclei will eventually enter the geological environment.

The geological repositories of nuclear and radioactive waste repositories must isolate the nuclear waste from the biosphere for as long as ten thousand to one million years (Alexander et al. 2015). For the storage of radioactive waste, the geological formations are usually planned on sites where water-soluble compounds have been accumulated for millions of years, such as salt mines, clay rocks, and granite. In these geological formations, further engineering barriers are constructed. The isolation ability is characterized by the migration rate of the isotopes through the barriers and depends on many factors, including the sorption of radioactive species. The geological formations (natural barrier) can sorb the cationic radionuclides because of their negative surface and layer charge, inhibiting their migration. However, the sorption of anionic species can be provided by engineering barrier.

Bentonite clay rock is considered a barrier material for the deep geological disposal of high activity nuclear waste (Sellin and Leupin, 2013). The main mineral component in bentonite rocks is montmorillonite. The swelling of montmorillonite, as well as the isolation properties, are influenced by the entities of the exchangeable cations. The radionuclide species, depending on their charges, migrate in the different water spaces of bentonite, namely in the free pore water, water of the electrostatic double layer, and in the montmorillonite interlayer water (Bradbury and Baeyens, 2003; Tournassat and Appelo, 2011; van Loon et al., 2007). The migration pathways of the cations and anions are different: the cations can be present in all water types; the anions, however, are excluded from the interlayer water and their quantity is less in the water of the electrostatic double layer than that of cations. This is caused by the electrostatic repulsion between the anions and the negative layer and surface charges of bentonite. The net migration coefficient is composed from the migration coefficient in the water spaces (Kozaki et al, 2008):

$$D_{tot} = x_p g_p D_p + x_s g_s D_s + x_i g_i D_i \qquad (1.148)$$

where x_p, x_s, and x_i are the relative quantities of the migrating ion in the given pathway, g_p, g_s, and g_i are the geometrical factors of the pathways, D_p, D_s, and D_i are the apparent migration coefficients on the pathways, and p, s, and i mean the pore water in the pores, on the outer surfaces and in the interlayer. For anions $x_i = 0$.

Moreover, the cations can be sorbed in the interlayer of montmorillonite, inhibiting the migration. The entity of the interlayer cation greatly affects the swelling of montmorillonite, and as a result the ratio of the water spaces. Many publications deal with the factors influencing the types of water spaces, the differences in the migration of cations and anions in sodium-bentonite e.g., Kozaki et al. 1998, 2008). The migration models include the porosity (ε) and density (ρ) of the dry bentonite, and the distribution coefficient of the migrating cation (K_d). The following is an example for the apparent migration coefficients (D_a) for cations and anions:

$$D_{tot} = \varepsilon \Psi D_0 / \left(\varepsilon + \rho K_d \right) \tag{1.149}$$

$$D_{tot} = \varepsilon D_p = \varepsilon D_0 / \Phi^2 \tag{1.150}$$

where D_0 is the diffusion coefficient of the cation in pure water, Ψ is the formation factor, Φ is the turtuosity factor, and D_p is the diffusion coefficient of the anion in pure water (Kozaki et al., 1998, Molera et al., 2003). In clayey media, however, the models overestimate the migration coefficient of anions and underestimate the migration coefficients of cations.

In the real systems, however, the bentonites are wet and the interlayer sodium cations can exchange with cations in groundwater and the nuclear waste. As is well known from the natural transformation of the sodium-bentonites with marine origin into calcium-bentonite on contact with groundwater with calcium content, the bivalent and trivalent cations bind stronger to the layer charge of montmorillonite and modify the swelling properties as well as the ratio of water space types.

In addition to the natural cation exchange processes, the interlayer cations of montmorillonite can exchange with other cations under laboratory or industrial conditions. Independent of the method of cation exchange, the products of these processes can effectively inhibit the migration of anionic species so that they decrease the concentration of these anions in groundwater or surface water. This is true for all anionic species, independent of the radioactivity. An example is the application of lanthanum-bentonite for the sorption of phosphate ions and preventing eutrophication (Haghseresht et al. 2009). Recently, lanthanum-modified bentonite was tested to remove As(III) and As(V) (Cui et al. 2020). There are studies to prepare other cation-exchanged bentonites, including organic and inorganic cationic species, for environmental protection tools, such as for the sorption of arsenic or some other anionic species (Aghaii et al. 2013; Haciyakupoglu and Orucoglu, 2013; Hua 2018; Kamble et al. 2009; Li et al. 2020; Riebe et al. 2001; Yang et al. 2020a; Yang et al. 2020b; Zhu et al. 2016).

For the successful sorption of anions on modified bentonite, cations precipitating with the desired anion are encapsulated into the interlayer space of montmorillonite. The effect of chemical modification of bentonite on sorption properties and the suggested mechanism of the anion uptake by modified bentonite is shown is Chapter 2, Section 2.11.

REFERENCES

Adams, J.M. 1987. Synthetic organic chemistry using pillared, cation-exchanged and acid-treated montmorillonite catalysts – A review. *Applied Clay Science* 2: 309–342.

Adams, J.M., and R.W. McCabe. 2008. Clay minerals as catalysts. In *Handbook of clay science*, 3rd ed., eds. F. Bergaya, B.K.G. Theng, and G. Lagaly, 541–581. Amsterdam: Elsevier.

Aghaii, M.D., M. Pakizeh, and A. Ahmadpour. 2013. Synthesis and characterization of modified UZM-5 as adsorbent for nitrate removal from aqueous solution. *Separation and Purification Technology* 113: 32–24

Alexander, W.R., H.M. Reijonen, and I.G. McKinley. 2015. Natural analogues: Studies of geological processes relevant to radioactive waste disposal in deep geological repositories. *Swiss Journal of Geosciences* 108: 75–100. doi: 10.1007/s00015-015-0187-y.

Anderson, G. 1960. Factors affecting the estimation of phosphate esters in soil. *Journal of the Science of Food and Agriculture* 11: 497–503. doi: 10.1002/jsfa.2740110902

Argersinger, W.J., A.W. Davidson, and O.D. Bonner. 1950. Thermodynamics and ion exchange phenomena. *Transactions of the Kansas Academy of Science* 53: 404–410.

Atterberg, A. 1905. Die rationalle Klassifikation der Sande und Kiese. *Chemiker Zeitung* 29: 195–198.

Azizian, S. 2004. Kinetic models of sorption: A theoretical analysis. *Journal of Colloid and Interface Science* 276: 47–52.

Baeyens, B., and M.H. Bradbury. 1997. A mechanistic description of Ni and Zn sorption on Na-montmorillonite. Part I: Titration and sorption measurements. *Journal of Contaminant Hydrology* 27: 199–222.

Baeyens, B., and M.H. Bradbury. 2009. Sorption modelling on illite Part I: Titration measurements and the sorption of Ni, Co, Eu and Sn. *Geochimica et Cosmochimica Acta* 73: 990–1003.

Barrer, R.M., and J. Klinowski. 1972. Ion exchange involving several groups of homogeneous sites. *Journal of the Chemical Society, Faraday Transactions 1: Physical Chemistry in Condensed Phases* 68: 73–87.

Barrow, N.J., J.W. Bowden, A.M. Posner, and J.P. Quirk. 1980. An objective method for fitting models of ion adsorption on variable charge surfaces. *Australian Journal of Soil Research* 18: 37–47.

Barrow, N.J., J.W. Bowden, A.M. Posner, and J.P. Quirk. 1981. Describing the adsorption of copper, zinc and lead on a variable charge mineral surface. *Australian Journal of Soil Research* 19: 309–321.

Berend, I., J.M. Cases, M. Francois, J.P. Uriot, L.J. Michot, H. Maison, and F. Thomas. 1995. Mechanism of adsorption of water vapor by homoionic montmorillonite.2. The Li+, Na+, K+, Rb+ and Cs+ exchanged forms. *Clays and Clay Minerals* 43: 324–336.

Bergaya, F., A. Aouad, and T. Mandalia. 2008a. Pillared clays and clay minerals. In *Handbook of clay science*, 3rd ed., eds. F. Bergaya, B.K.G. Theng, and G. Lagaly, 393–421. Amsterdam: Elsevier.

Bergaya, F., G. Lagaly, and M. Vayer. 2008b. Cation and anion exchange. In *Handbook of clay science*, 3rd ed., eds. F. Bergaya, B.K.G. Theng, and G. Lagaly, 979–1001. Amsterdam: Elsevier.

Blanchard, G., M. Maunaye, and G. Martin. 1984. Removal of heavy metals from waters by means of natural zeolites. *Water Research* 18(12): 1501–1507.

Blesa, M.A., A.J.G. Maroto, and A.E. Regazzoni. 1984. Boric acid adsorption on magnetite and zirconium dioxide. *Journal of Colloid and Interface Science* 99: 32–40.

Bockris, O'M., M.A.V. Devanathan, and K. Müller. 1963. On the structure of charged interfaces. *Proceedings of the Royal Society (London) A* 274: 55–79.

Bonleter, O., and G. Bonn. 1991. Ion exchange chromatography. In *Ion exchangers*, ed. K. Dorfner, 1187–1241. Berlin: Walter de Gruyter.

Borkovec, M., and G.J.M. Koper. 1994. Affinity distribution of polyampholytes with interacting acid-base groups. *Langmuir* 10: 2863–2865.

Bouabid, R., M. Badraoui, and P. Bloom. 1991. Potassium fixation and charge characteristics of soil clays. *Soil Science Society of America Journal*, 55: 1493–1498.

Bourg, I.C., G. Sposito, and A.C.M. Bourg. 2008. Modeling the diffusion of Na$^+$ in compacted water-saturated Na-bentonite as a function of pore water ionic strength. *Applied Geochemistry* 23: 3635–3641.

Bowden, J.W., S. Nagarajah, N.J. Barrow, A.M. Posner, and J.P. Quirk. 1980. Describing the adsorption of phosphate, citrate and selenite on a variable-charge mineral surface. *Australian Journal of Soil Research* 18: 49–60.

Bowden, J.W., A.M. Posner, and J.P. Quirk. 1977. Ionic adsorption on variable charge mineral surfaces. Theoretical charge development and titration curves. *Australian Journal of Soil Research* 15: 121–136.

Boyd, G.E., A.W. Adamson, and L.S. Myers, Jr. 1947. The exchange adsorption of ions from aqueous solutions by organic zeolites. II. Kinetics. *Journal of the American Chemical Society* 69: 2836–2848.

Boyd, G.E., J. Schubert, and A.W. Adamson. 1947. The exchange adsorption of ions from aqueous solutions by organic zeolites. I. Ion-exchange equilibria. *Journal of the American Chemical Society* 69: 2818–2829.

Brack, A. 2008. Clay minerals and the origin of life. In *Handbook of clay science*, 3rd ed., eds. F. Bergaya, B.K.G. Theng, and G. Lagaly, 379–391. Amsterdam: Elsevier.

Bradbury, M.H., and B. Baeyens. 2003. Porewater chemistry in compacted re-saturated MX-80 bentonite. *Journal of Contaminant Hydrology* 61: 329–338.

Brunauer, S., P.H. Emmett, and E. Teller. 1938. Adsorption of gas multi-molecular layers. *Journal of the American Chemical Society* 60: 309–319.

Buffle, J., and W. Stumm. 1994. General chemistry in aquatic systems. In *Chemical and biological regulation of aquatic systems*, eds. J. Buffle and R.R. DeVitre, 1–42. Boca Raton: CRC Press.

Bujdak, J., K. Faybikova, A. Eder, Y. Yongyai, and B.M. Rode. 1995. Peptide chain elongation: A possible role of montmorillonite in prebiotic synthesis of protein precursors. *Origins of Life and Evolution of the Biosphere* 25: 431–441.

Bujdak, J., and B.M. Rode. 1997. Silica, alumina, and clay catalyzed alanine peptide bond formation. *Journal of Molecular Evolution* 45: 457–466.

Bujdak, J., H. Slosiarikova, N. Texler, M. Schwendinger, and B.M. Rode. 1994. On the possible role of montmorillonite in prebiotic peptide formation. *Monatshefte für Chemie* 125: 1033–1039.

Cases, J.M., I. Berend, M. Francois, J.P. Uriot, L.J. Michot, and F. Thomas. 1997. Mechanism of adsorption of water vapor by homoionic Montmorillonite. 3. The Mg2+, Ca2+, Sr2+ and Ba2+ exchanged forms. *Clays and Clay Minerals* 45: 8–22.

Cassie, A.B.D. 1945. Multimolecular absorption. *Journal of the Chemical Society, Faraday Transactions 1: Physical Chemistry in Condensed Phases* 41: 450–458.

Cernik, M., M. Borkovec, and J.C. Westall. 1995. Regularized least-squares methods for the calculation of discrete and continuous affinity distributions for heterogeneous sorbents. *Environmental Science and Technology* 29: 413–425.

Cernik, M., M. Borkovec, and J.C. Westall. 1996. Affinity distribution of competitive ion binding to heterogeneous materials. *Langmuir* 12: 6127–6137.

Charberek, S., and A.E. Martell. 1959. *Organic sequestering agents.* New York: Wiley.

Charlet, L., P.W. Schindler, L. Spadini, G. Furrer, M. Zysset, and S.M. McLennan. 1994. Rare earth element geochemistry and the "Tetrad" effect. *Geochimica et Cosmochimica Acta* 58: 2025–2033.

Chien, V.A., and W.R.Clayton. 1980. Application of Elovich equation to the kinetics of phosphate release and sorption in soil. *Soil Science Society of America Journal* 44: 265–268.

Chiou, C.T., and D.W. Rutherford. 1993. Sorption of N$_2$ and EGME on some soils, clays and minerals oxides and determination of sample surface area by use of sorption data. *Environmental Science & Technology* 27: 1587–1594.

Chiou, C.T., and D.W. Rutherford. 1997. Effect of exchanged cation and layer charge on the sorption of water and EGME vapors on montmorillonite clays. *Clays and Clay Minerals* 45: 867–880.

Christidis, G.E. 2008. Validity of the structural formula method for layer charge determination of smectites: A re-evaluation of published data. *Applied Clay Science* 42: 1–7.

Churchman, G.J., C.M. Burke, and R.L. Parfitt. 1991. Comparison of various methods for determination of specific surfaces of subsoils. *Journal of Soil Science* 42: 449–461.

Chute, J.H., and J.P. Quirk. 1967. Diffusion of potassium from mica-like materials. *Nature* 213: 1156–1157.

Conway, B.E., J.O'M. Bockris, and I.A. Ammar. 1951. The dielectric constant of the solution in the diffuse and Helmholtz double layers at a charged interface in aqueous solution. *Journal of the Chemical Society, Faraday Transactions 1: Physical Chemistry in Condensed Phases* 47: 756–766.

Crank, J. 1975. *The mathematics of diffusion.* 2nd ed. Oxford: Clarendon Press.

Cui, J., D. Wang, J. Lin, Y. Wang, M. Ren, Y. Yang, and P. Shi. 2020. New application of lanthanum-modified bentonite (Phoslock®) for immobilization of arsenic in sediments. *Environmental Science and Pollution Research.* 28(2):2052–2062. doi: 10.1007/s11356-020-10565-x.

Davis, J.E., R.O. James, and J.O.Leckie. 1978. Surface ionization and complexation at the oxide-water interface. I. Computation of electrical double layer properties in simple electrolytes. *Journal of Colloid and Interface Science* 63: 480–499.

de Paiva, L.B., A.R. Morales, and F.R.V. Díaz. 2008. Organoclays: Properties, preparation and applications. *Applied Clay Science* 42: 8–24.

De Wit, J.C.M., W.H. Van Riemsdijk, M.M. Nederlof, D.G. Kinniburgh, and L.K. Koopal. 1990. Analysis of ion binding on humic substances and the determination of intrinsic affinity distributions. *Analytica Chimica Acta* 232: 189–207.

Derjaguin, B.V., and L.D. Landau. 1941. Theory of the stability of strongly charged lyophobic sols and of the adhesion of strongly charged particles in solutions of electrolytes. *Acta Physico-Chimica. U.S.S.R.* 14: 663–662.

Dodge, C.J., A.J. Francis, J.B. Gillow, G.P. Halada, C. En, and C.R. Clayton. 2002. Association of uranium with iron oxides typically formed on corroding steel surfaces. *Environmental Science and Technology* 36: 3504–3511.

Ekedahl, E., E. Högfeldt, and L.G. Sillén. 1950. Activities of components in ion exchangers. *Acta Chemica Scandinavica* 4: 556–558.

El-Sofany, E.A., A.A. Zaki, and H.S. Mekhamer. 2009. Kinetics and thermodynamics studies for the removal of Co^{2+} and $Cs+$ from aqueous solution by sand and clay soils. *Radiochimica Acta* 97: 23–32.

Everett, D.H. 1972. Manual of symbols and terminology for physicochemical quantities and units, appendix ii: Definitions, terminology and symbols in colloid and surface chemistry. *Pure and Applied Chemistry* 31: 577–638.

Ferris, J.P. 2002. Montmorillonite catalysis of 30–50mer olegonucleotides: Laboratory demonstration of potential steps int he origin of the RNA world. *Origins of Life and Evolution of the Biosphere* 32: 311–322.

Ferris, J.P., A.R. Hill Jr., R. Liu, and L.E. Orgel. 1996. Synthesis of long prebiotic oligomenrs on mineral surfaces. *Nature* 381: 59–61.

Fiedler, G., and R. Wagner. 1967. Untersuchungen zur quantitativen Bestimmung von Montmorillonit mit dem Derivatographen. *Zeitschrift für Angewandte Geologie* 13: 262–266.

Fowler, R.M. 1935. A statistical derivation of the Langmuir adsorption isotherm. *Mathematical Proceedings of the Cambridge Philosophical Society* 31: 260.

Franz, W., P. Gunter, and C.E. Hofstadt. 1959. Catalysis based on acid activated bentonites. *Erdöl und Kohle* 12: 335.

Freitas, A.F., M.F. Mendes, and G.L.V. Coelho. 2009. Thermodynamic study on the adsorption of some organic acids from aqueous solutions by unmodified and modified montmorillonite clays. *Chemical Engingineering Communications* 196: 824–840.

Fuller, C.C., J.A. Davis, and G.A. Waychunas. 1993. Surface chemistry of ferrihydrite: Part 2. Kinetics of arsenate adsorption and coprecipitation. *Geochimica et Cosmochimica Acta* 57: 2271–2282.

Gaines, G.L., and H.C. Thomas. 1953. Adsorption studies on clay minerals II. A formation of the thermodynamics of exchange adsorption. *The Journal of Physical Chemistry A* 21: 714–718.

Gaines, R.V., H.C.W. Skinner, E.E. Foord, and A. Rosenzweig. 1997. *Dana's new mineralogy.* 8th ed. New York: John Wiley & Sons.

Gapon, E.N. 1933. On the theory of exchange adsorption is soils. *Journal of General Chemistry. USSR (Eng. Trans.)* 3: 144–160. (K teorim obmennoj adszorpcii pocsbah. *Zs. Obs. Him.* 144–152.)

Gerischer, H., and W. Vielstich. 1952. Untersuchungen mit radioaktiven Indikatoren über Austausch- und Diffusionsvorgänge an Silberelektroden. *Zeitschrift für Elektrochemie* 56: 380–386.

Giammar, D.E., and J.G. Hering. 2001. Time scales for sorption-desorption and surface precipitation of uranyl on goethite. *Environmental Science and Technology* 35: 3332–3337.

Goldberg, S. 1992. Use of surface complexation models in soil chemical systems. *Advances in Agronomy* 47: 233–329.

Grahame, D.C. 1947. The electrical double layer and the theory of electrocapillarity. *Chemical Reviews* 41: 441–501.

Grahame, D.C. 1950. Effects of dielectric saturation upon the diffuse double layer and the free energy of hydration of ion. *Journal of Chemical Physics* 18: 903–909.

Grahame, D.C. 1957. Capacity of the electrical double layer between mercury and aqueous sodium fluoride. ii. effect of temperature and concentration. *Journal of the American Chemical Society* 79: 2093–2098.

Gülen, J., and S. Aslan. 2020. Adsorption of 2,4-dichlorophenoxyacetic acid from aqueous solution using carbonized chest nut as low cost adsorbent: Kinetic and thermodynamic. *Zeitschrift für Physikalische Chemie* 234: 461–484. doi: zpch-2019-0004.

Haciyakupoglu, S., and E. Orucoglu. 2013. Se-75 radioisotope adsorption using Turkey's Resadiye modified bentonites. *Applied Clay Science* 86: 190–198.

Haghseresht, F., S. Wang, and D.D. Do. 2009. A novel lanthanum-modified bentonite, Phoslockm for phosphate removal from wastewaters. *Applied Clay Science* 46: 369–375.

Hahn, O. 1926a. Gesetzmäßigkeiten bei der Fällung und Adsorption kleiner Substanzmengen und ihre Beziehung zur radioaktiven Fällungsregel. (Nach gemeinsam mit Hrn. O. Erbacher und Frl. N. Feichtinger ausgeführten Versuchen) *Chemische Berichte* 59: 2014–2025. Published online: 23 January 2006. doi: 10.1002/cber.19260590855.

Hahn, O. 1926b. Über die neuen Fällungs- und Adsorptionssätze und einige ihrer Ergebnisse. *Naturwissenschaften* 14: 1196–1199.

Hahn, O., and L. Imre. 1929. Über die Fällung und Adsorption kleiner Substanzmengen. III. Der Adsorptionssatz, Anwendungen, Ergebnisse und Folgerungen. *Zeitschrift für Physikalische Chemie* 144: 161–186.

Haissinsky, M. 1964. *Nuclear chemistry and its applications.* Reading: Addison-Wesley Publishing Company.

Havlin, J.L., and D.G. Wesfall. 1985. Potasium release kinetics and plant response in calcerous soils. *Soil Science Society of America Journal* 49: 366–370.

Hayes, K.F., G. Redden, W. Ela, and J.O. Leckie. 1991. Surface complexation models: An evaluation of model parameter estimation using FITEQL and oxide mineral titration data. *Journal of Colloid and Interface Science* 142: 448–469.

Heller-Kallai, L. 2002. Clay catalysis in reactions of organic matter. In *Organo-clay complexes and interactions*, eds. S. Yariv, H. Cross, 567–613. New York: Marcel Dekker.

Herbelin, A.L., and J.C. Westall. 1996. *FITEQL3.2, A program for the determination of chemical equilibrium constants from experimental data.* Corvallis: Oregon State University.

Hiemstra, T., J.C.M. de Wit, and W.H. van Riemsdijk. 1989. Multisite proton adsorption modeling at the solid/solution interface of (hydr)oxides: A new approach. *Journal of Colloid and Interface Science* 133: 105–117.

Hill, T.L. 1946a. Theory of multimolecular adsorption from a mixture of gases. *The Journal of Chemical Physics* 14: 268–275.

Hill, T.L. 1946b. Statistical mechanics of multimolecular adsorption II. Localized and mobile adsorption and absorption. *The Journal of Chemical Physics* 14: 441–453.

Hill, T.L. 1949. Statistical mechanics of adsorption. VII. Thermodynamic functions for the B.E.T. theory. *The Journal of Chemical Physics* 17: 772–774.

Ho, Y.S, and G. McKay. 1999. Pseudo-second order model for sorption processes. *Process Biochemistry* 34: 451–465.

Ho, Y.S., D.A.J. Wase, and C.F. Forster. 1996. Kinetic studies of competitive heavy metal adsorption by Spagnum Moss Peat. *Environmental Technology* 17(1), 71–77.

Högfeldt, E. 1984. A useful method for summarizing data in ion exchange. I. Some illustrative examples. *Reactive & Functional Polymers* 2, 19–30.

Högfeldt, E. 1988. Application of a simple thermodynamic model to various ion exchange data. *Reactive & Functional Polymers* 7, 81–87.

Högfeldt, E. 1990. A comparison of methods for estimating equilibrium constants in ion exchange. *Journal of the Chemical Society, Dalton Transactions* 1627–1628.

Högfeldt, E., E. Ekedahl, and L.G. Sillén. 1950. Activities of the components in ion exchangers with multivalent ions. *Acta Chemica Scandinavica* 4: 828–829.

Horst, J., W.H. Höll, and S.H. Eberle. 1990. Application of the surface complex formation model to exchange equilibria on ion exchange resins. Part I. Weak-acid resins. *Reactive Polymers* 13: 209–231.

Howery, D.G., and H.C. Thomas. 1965. Ion exchange on mineral clinoptilolite. *The Journal of Physical Chemistry A* 69: 531–537.

Hua, J. 2018. Adsorption of low-concentration arsenic from water by co-modified bentonite with manganese oxides and poly(dimethyldiallylammonium chloride). *Journal of Environmental Chemical Engineering* 6: 156–168.

Imre, L. 1933a. Grenzflachengleichgewichte und innere Gleichgewichte in heterogenen Systemen. I. *Zeitschrift für physikalische Chemie A* 164: 343–363.

Imre, L. 1933b. Zur Kinetik der Oberflächenvorgänge an Kristallgittern. II. Die Elementarprozesse bei der Ausbildung einer aus mehreren Komponenten bestehenden Grenzschicht. *Zeitschrift für physikalische Chemie A* 164: 327–342.

Jalili, M., M. Majeri, and S. Najafi. 2019. Kinetic release and fractionation of cobalt in some calcaroues soils. *Journal of Geochemical Exploration* 204: 131–141. doi: 10.1016/j.gexplo.2019.02.006.

James, R.O., and G.A. Parks. 1982. Characterization of aqueous colloids by their electrical double layer and intrinsic surface chemical properties. In *Surface and colloid science*, Vol. 12, ed. E. Matijevic, 119–216. New York: Plenum Press.

Jang, E.J., K.H. Lee, J.S. Lee, and Y.G. Kim. 1999. Hydrocarboxylation of 1-(4-isobutyl-phenyl) ethanol catalyzed by heterogeneous palladium catalysts. *Journal of Molecular Catalysis A* 144: 431–440.

Jardine, P.M., and D.L. Sparks. 1984. Potassium-calcium exchange in a multireactive soilsystem. I. Kinetics. *Soil Science Society of America Journal* 48: 39–45.

Kabata-Pendias, A., and H. Pendias. 1985. *Trace elements in soils and plants*. Boca Raton: CRC Press, Inc.

Kahr, G., and F.T. Madsen. 1995. Determination of cation exchange capacity and surface area of bentonite, illite, and kaolinite by methylene blue adsorption. *Applied Clay Science* 9: 327–336.

Kamble, S.P., P. Dixit, S.S. Rayalu, and Labhsetwar, N.K. 2009. Defluoridation of drinking water using chemically modified bentonite clay. *Desalination* 249: 687–693. doi: 10.1016/j.desal.2009.01.031.

Kellomaki, A., P. Kuula-Vaisanen, and P. Nieminen. 1989. Sorption and retention of ethylene glycol monoethyl ether (EGME) on silicas. *Journal of Colloid and Interface Science* 129: 373–378.

Kerr, H.W. 1928. The nature base exchange on soil acidity. *Journal of American Society of Agronomy* 20: 309–355.

Kinniburgh, D.G., J.A. Barker and M. Whitfield. 1983. A comparison of some simple adsorption isotherms for describing divalent cation adsorption by ferrihydrite. *Journal of Colloid and Interface Science* 95: 70–384.

Komadel, P.J., J. Madejova, J. Hrobarikova, M. Janek, and J. Bujdak. 2003. Fixation of Li^+cations in montmorillonite upon heating. *Solid State Chemistry Solid State Phenomena* 90–9: 497–502.

Kónya, J., and N.M. Nagy. 1998. The effect of complex-forming agent (EDTA) on the exchange of manganese ions on calcium-montmorillonite, I. Reaction scheme and Ca-montmorillonite/Na_2EDTA system. *Colloids and Surfaces* 136: 297–308.

Kónya, J., and N.M. Nagy. 2009. Isotherm equation of sorption of electrolyte solutions on solids: How to do heterogeneous surface from homogeneous one? Periodica Polytechnica Chemical Engineering 53(2): 55–60.

Kónya, J., and N.M.. Nagy. 2012. *Nuclear and radiochemistry*. Oxford: Elsevier, Hardcover ISBN: 9780123914309; Paperback ISBN: 9780323282451; eBook ISBN: 9780123914873.

Kónya, J., and N.M. Nagy. 2013. Misleading information on homogeneity and heterogeneity obtained from sorption isotherms. *Adsorption* 19: 701–707.

Kónya, J., and N.M. Nagy. 2018. *Nuclear and radiochemistry*, 2nd ed. Oxford: Elsevier. Paperback ISBN: 9780128136430; eBook ISBN: 9780128136447.

Kónya, J., N.M. Nagy, and E. Högfeldt. 1989. Zinc-calcium exchange on bentonite. Application of a simple three-parameter model. *Acta Chemica Scandinavica* 43: 612–614.

Kónya, J., N.M. Nagy, and Z. Nemes. 2005. The effect of mineral composition on the adsorption of caesium ion on geological formations. *Journal of Colloid and Interface Science* 190: 350–356.

Kosmulski, M., M. Jaroniec, and J. Szczypa. 1985. Liquid/solid interfaces: Studies of kinetics of isotope exchange. *Adsorption Science and Technology* 2: 97–119. doi: 10.1177/026361748500200204.

Kozaki, T., J. Liu, and S. Sato. 2008. Diffusion mechanism of sodium ion sin compacted montmorillonite under different NaCl concentration. *Physics and Chemistry of the Earth* 33: 957–961.

Kozaki, T., N. Saito, A. Fujishima, S. Sato, and H. Ohashi. 1998. Activation energy for diffusion of chloride ion sin compacted sodium montmorillonite. *Journal of Contaminant Hydrology* 35: 67–75.

Krishnamoorthy, C., and R. Overstreet. 1949. Theory of ion exchange relationships. *Soil Science* 68: 307–315.

Ksiezopolska, A. 2000. The surface properties of humus acids fractions extracted from the selected soils. *Acta Agrophysica* 38: 139–148.

Kuo, S., and E.G. Lotse. 1974. Kinetics of phosphate adsorption and desorption by lake sediments. *Soil Science Society of America, Proceedings* 38: 50–54.

Lagaly, G., and I. Dékány. 2005. Adsorption on hydrophobized surfaces: Clusters and self-organization. *Advances in Colloid and Interface Science* 114–115: 189–204.

Lagergren, S. 1898. Zur theorie der sogenannten adsorption geloster stoffe. *Kungliga Svenska Vetenskapsakademiens. Handlingar* 24: 1–39.

Laird, D.A. 2004. Reactions of organic molecules with smectite surfaces in aqueous systems. *Acta Mineralogica-Petrographics, Abstract Series* 4, p.64. University of Szeged, Hungary.

Langmuir, I. 1918. The adsorption of gases on plane surface of glass, mica, and platinum. *Journal of the American Chemical Society* 40: 1361–1382.

Laszlo, P. 1986a. Chemical reactions on clays. *Science* 235: 1473–1477.

Laszlo, P. 1986b. Catalysis of organic reactions by inorganic solids. *Accounts of Chemical Research* 19: 121–127.

Latimer, W. 1952. *Oxidation potentials.* New York: Prentice-Hall.

Leroy, P., and A.J. Revil. 2003. A triple-layer model of the surface electrochemical properties of clay minerals. *Journal of Colloid and Interface Science* 270: 371–380.

Li, G., J. Zhang, J. Liu, S. Chen, and H. Li. 2020. Investigation of the adsorption characteristics of Cr(VI) onto fly ash, pine nut shells, and modified bentonite. *Desalination and Water Treatment* 195: 389–402.

Lindsay, W.L. 1979. *Chemical equilibria in soils.* New York: Wiley.

Liu, D.C., C.N. Hsu, and C.L. Chuang. 1995. Ion-exchange and sorption kinetics of cesium and strontium in soils. *Applied Radiation and Isotopes* 46: 839–846.

Liu, R., and H. Tang. 2002. Surface acid–base characteristics of natural manganese mineral particles. *Colloids and Surfaces A* 197: 47–54.

Lomenech, C., R. Drot, and E. Simoni. 2003. Speciation of uranium(VI) at the solid/solution interface: Sorption modeling on zirconium silicate and zirconium oxide. *Radiochimica Acta* 91: 453–461.

Mackenzie, R.C. 1975. Classification of soil silicates and oxides. In *Soil component. Vol. 2.* ed. J. Gieseking. Berlin: Springer-Verlag. pp. 1–25.

Mackenzie, R.C., and B.D. Mitchell. 1966. Clay mineralogy. *Earth-Science Reviews* 2: 47–91.

Marcus, Y., and A.S. Kertes. 1969. *Ion exchange and solvent extraction of metal complexes.* London, New York Wiley-Interscience, pp. 321–340.

Markham, E.C., and A.F. Benton. 1931. The adsorption of gas mixtures by silica. *Journal of the American Chemical Society* 53: 497–507.

McCabe, R.W. 1996. Clay chemistry. In *Inorganic materials*, 2nd ed., eds. D.W. Bruce, and D. O'Hare, 313–376. Chichester: Wiley.

McDonough, W.F., and S.-s. Sun. 1995. The composition of the Earth. *Chemical Geology* 120: 223–253.

McKay, H.A.C. 1971. *Principles of radiochemistry.* London: Butterworths. ISBN: 0 408 70161 7.

Michot, L.J., and F. Villiéras. 2008. Surface area and porosity. In *Handbook of clay science*, 3rd ed., eds. F. Bergaya, B.K.G. Theng, and G. Lagaly, 965–978. Amsterdam: Elsevier.

Miller, C.T., and J.A. Pedit. 1992. Use of reactive surface-diffusion model to describe apparent sorption-desorption hysteresis and abiotic degradation of lindane in subsurface material. *Environmental Science & Technology* 26: 1417–1427.

Missana, T., M. Garcia-Gutierrez, and V. Fernandez. 2003. Uranium(VI) sorption on colloidal magnetite under anoxic environment: Experimental study and surface complexation modelling. *Geochimica et Cosmochimica Acta* 67: 2543–2550.

Molera, M, T. Eriksen, and M. Jansson. 2003. Anion diffusion pathways in bentonite clay compacted to different dry densities. *Applied Clay Science* 23: 69–76.

Nagy, N.M., and J. Kónya. 1988. The interfacial processes between calcium-bentonite and zinc ion. *Colloids and Surfaces* 32: 223–235.

Nagy, N.M., and J. Kónya. 1992. The exchange of ^{54}Mn and ^{45}Ca ions on montmorillonite. *Reactive Polymers* 17: 9–13.

Nagy, N.M., and J. Kónya. 2007. Study of pH-dependent charges of soils by surface acid-base properties. *Journal of Colloid and Interface Science* 305: 94–100.

Nagy, N.M., J. Kónya, M. Beszeda, I. Beszeda, E. Kálmán, Z. Keresztes, K. Papp, and I. Cserny. 2003. Physical and chemical formations of lead contaminants in clay and sediment. *Journal of Colloid and Interface Science* 263: 13–22.

Nagy, N.M., J. Kónya, and T. Budai. 1998. Mn^{2+}/^{54}Mn^{2+} heterogeneous isotope exchange reaction on montmorillonite in the presence of complex-forming agents. *Colloids and Surfaces* 138: 81–89.

Nagy, N.M., E.M. Kovács, and J. Kónya. 2016. Ion exchange isotherms in solid: Electrolyte solution systems. *Journal of Radioanalytical and Nuclear Chemistry* 308: 1017–1026.

Nemecz, E. 1973. *Agyagásványok (Clay minerals, in Hungarian language)*. Budapest: Akadémiai Kiadó.

Nemes, Z., N.M. Nagy, and J. Kónya. 2005. Kinetics of strontium ion adsorption on Carpathian Basin bentonite samples. *Journal of Radioanalytical and Nuclear Chemistry* 266: 289–293.

Neuman, W.F., and M.W. Neuman. 1958. *The chemical dynamics of bone minerals*. Chicago: University of Chicago Press.

O'Donoghue, M. 1990. *American nature guides – rocks and minerals*. New York: Gallery Books.

Paneth, F. 1922. Über eine Methode zur Bestimmung der Oberflache adsorbierender Pulver. *Z. Electrochem.* 28: 113–160.

Pavlidou, S., and C.D. Papaspyrides. 2008. A review on polymer–layered silicate nanocomposites. *Progress in Polymer Science* 33: 1119–1198.

Peak, D., G.W. Luther III., and D.L. Sparks. 2003. ATR-FTIR spectroscopic studies of boric acid adsorption on hydrous ferric oxide. *Geochimica et Cosmochimica Acta* 67: 2551–2560.

Pearson, R.G. 1963. Hard and soft acids and bases. *Journal of the American Chemical Society* 85: 3533–3539.

Pearson, R.G. 1968. Hard and soft acids and bases, HSAB.Part 1. Fundamental principles. *Journal of Chemical Education* 45: 581–587.

Pedit, J.A., and C.T. Miller. 1995. Heterogeneous sorption process in subsurface systems. 2. Diffusion modeling approaches. *Environmental Science & Technology* 29: 1766–1772.

Porter, T.L., M.P. Eastman, R. Whitehorse, E. Bain, and K. Manygoats. 2000. The interaction of biological molecules with clay minerals: A scanning force microscopy study. *Scanning* 22: 1–5.

Pourbaix, M. 1966. *Atlas of electrochemical equilibria in aqueous solutions*. Oxford: Pergamon Press.

Richards, L.A. 1957. *Diagnosis and improvement of saline and alkaline soils*. U.S. Dept. Agr. HandbookN. 60., United State Department of Agriculture, Washington DC.

Riebe, B., J. Bors, and S. Dultz. 2001. Retardation capacity of organophilic bentonite for anionic fission products. *Journal of Contaminant Hydrology* 47: 255–264.

Ross, C.S., and S.B. Hendricks. 1945. *Minerals of the montmorillonite group, their origin and relation to soils and clays*. U.S. Geol. Surv. Tech. Pap. 205B. United States Department of the Interior, Wahington DC.

Rouquerol, J., D. Avnir, C.W. Fairbridge, D.H. Everett, J.M. Haynes, N. Pernicone, J.D.F. Ramsay, K.S.W. Singm and K.K. Unger. 1994. Recommendations for the characterization of porous solids (Technical Report). *Pure and Applied Chemistry* 66: 1739–1758.

Sánchez, F.G., L.R. Van Loon, T. Gimmi, A. Jacob, M.A. Glaus, and L.W. Diamond. 2008. Self-diffusion of water and its dependence on temperature and ionic strength in highly compacted montmorillonite, illite and kaolinite. *Applied Geochemistry* 23: 3840–3851.

Schnitzer, M., and E.H. Hansen. 1970. Organo-metallic interactions in soils: 8. An evaluation of methods for the determination of stability constants of metal-fulvic acid complexes. *Soil Science* 109: 333–340.

Seidel, H., K. Görsch, K. Amstatter, and J. Mattusch. 2005. Immobilization of arsenic in a tailing material by ferrous iron treatment. *Water Research* 39: 4073–4082. doi: 10.10106/j.watres.2005.08.001.

Sellin, P., and O.X. Leupin. 2013. The use of clay as an engineered barrier in radioactive-waste management a review. *Clays and Clay Minerals* 61: 477–498.

Shah, B.A., A.V. Shah, and R.R. Singh. 2009. Sorption isotherms and kinetics of chromium uptake from wastewater using natural sorbent material. *International Journal of Environmental Science and Technology* 6: 77–90.

Solomon, D.H., and D.G. Hawthorne. 1983. *Chemistry of pigments and filter.* New York: Wiley.

Somorjai, G.A. 1994. *Introduction to surface chemistry and catalysis.* New York: John Wiley and Sons, Inc., p. 48.

Sparks, D.L. 1985. Kinetics of ionic reactions in clay minerals and soils. In *Advances in agronomy*, 38., ed. N.C. Brady, 231–266. Orlando: Academic Press.

Sparks, D.L. 1989. *Kinetics of soil chemical processes.* San Diego: Academic Press.

Sparks, D.L. 1999. *Soil physical chemistry.* Boca Raton: CRC Press, Inc.

Sparks, D.L. 2003. *Environmental soil chemistry.* Amsterdam: Academic Press.

Sperry, J.M., and J.J. Peirce. 1999. Ion exchange and surface charge on montmorillonite clay. *Water Environment Research* 71: 316–322.

Sposito, G. 1981. *Thermodynamics of soil solutions.* London: Oxford Clarendon Press.

Stadler, M., and P.W. Schindler. 1993a. Modeling of H^+ and Cu^{2+} adsorption on calcium-montmorillonite. *Clays and Clay Minerals* 41: 288–296.

Stadler, M., and P.W. Schindler. 1993b. The effect of dissolved ligands upon the sorption of Cu(II) by Ca-montmorillonite. *Clays and Clay Minerals* 41: 680–692.

Stevenson, F.J. 1982. *Humus chemistry.* New York: Wiley and Sons.

Strawn, D.G., and D.L. Sparks. 1999. The use of XAFS to distinguish between inner- and outer-sphere lead adsorption complexes on montmorillonite. *Journal of Colloid and Interface Science* 216: 257–269.

Strunz, H. 1941. Mineralogishe tabellen. http://webmineral.com/strunz.shtml.

Stucki, J.W., and K. Lee. 1999. Improving soil tests for potassium: Fundamental considerations for partitioning between fixed and exchangeable forms and redox effects. *Illinois Fertilizer Conference Proceedings.* http://frec.cropsci.uiuc.edu/1999/report2/.

Stumm, W. 1992. *Aquatic surface chemistry.* New York: Wiley.

Stumm, W., and R.Wollast. 1990. Coordination chemistry of weathering. Kinetics of the surface-controlled dissolution of oxide minerals. *Reviews of Geophysics* 28: 53–69.

Theng, B.K.G. 1974. *Thechemistry of clay-organic reactions.* London: Adam Hilger.

Theng, B.K.G. 1982. Clay-activated organic reactions. In *International clay conference 1981. Developments in sedimentology*, eds. van Olphen, H., Veniale, F. 35: 197–238. Amsterdam: Elsevier.

Tiller, K.G., and L.H. Smith. 1990. Limitation of EGME retention to estimate the surface area of soils. *Australian Journal of Soil Research* 28: 1–26.

Tonkin, J.W., L.S. Balistrieri, and J.W. Murray. 2004. Modeling sorption of divalent metal cations on hydrous manganese oxide using the diffuse double layer model. *Applied Geochemistry* 19: 29–53.

Tournassat, C., and C.A.J. Appelo. 2011. Modelling approaches for anion-exclusion in compacted Na-bentonite. *Geochimica et Cosmochimica Acta* 75: 3698–3710.

Tournassat, C., E. Ferrage, C. Poinsignon, and L. Charlet. 2004b. The titration of clay minerals II. Structure-based model and implications for clay reactivity. *Journal of Colloid and Interface Science* 273: 234–246.

Tournassat, C., J-M. Greneche, D. Tisserand, and L. Charlet. 2004a. The titration of clay minerals: I. Discontinuous backtitration technique combined with CEC measurements. *Journal of Colloid and Interface Science* 273: 224–233.

Van Loon, L.R., M.A. Glaus, and W. Müller. 2007. Anion exclusion effect in compacted bentonites: Towards a better undarstanding of anion diffusion. *Applied Geochemistry* 22: 2536–2552.

van Olphen, H. 1977. *An introduction to clay colloid chemistry*, 2nd ed. New York: Wiley and Sons.

van Riemsdijk, W.H., G.H. Bolt, L.K. Koopal, and J. Blaakmeer. 1986. Electrolyte adsorption on heterogeneous surfaces: Adsorption models. *Journal of Colloid and Interface Science* 109: 219–228.

Vanselow, A.P. 1932. Equilibria of the base exchange reactions of bentonites, permutites, soil colloids and zeolites. *Soil Science* 33: 95–113.

Verweij, E.J.W., and J.T.G. Overbeek. 1948. *Theory on the stability of lyophobic colloids*. Amsterdam: Elsevier.

Wahba, M.M., and A.M. Zaghloul. 2007. Adsorption characteristics of some heavy metals by some soil minerals. *Journal of Applied Sciences Research* 3: 421–426.

Wahl, C.A., and N.A. Bonner. 1951. *Radioactivity applied to chemistry*. New York: John Wiley and Sons, Inc.

Wang, J., and X. Guo. 2020a. Adsorption isotherm models: Classification, physical meaning, applications and solving method. *Chemosphere* 258: 127279. doi: 10.1016/j.chemosphere.2020.127279.

Wang, J., and X. Guo. 2020b. Adsorption kinetic models: Physical meanings, applications, and solving methods. *Journal of Hazardous Materials* 390: 122156. doi: 10.1016/j.jhazmat.2020.122156.

Wanner, H., Y. Albinsson, O. Karnland, E. Wieland, P. Wersin, and L. Charlet. 1994. The acid/base chemistry of montmorillonite. *Radiochimica Acta* 66/67: 157–162.

Wazne, M., G.P. Korfiatis, and X.G. Meng. 2003. Carbonate effects on hexavalent uranium adsorption by iron oxyhydroxide. *Environmental Science & Technology* 37: 3619–3624.

West, J.M. 1965. *Electrodeposition and corrosion processes*. London: D. van Nostrand Company Ltd.

Wollast, R. 1967. Kinetics of the alteration of K-feldspar in buffered solution at low temperature. *Geochimica et Cosmochimica Acta* 31: 635–648.

Wu, S., and P.M. Gschwendt. 1986. Sorption kinetics of hydrophobic organic compounds to natural natural sediments and soils. *Environmental Science & Technology* 20: 717–725.

www.world-nuclear.org/information-library/nuclear-fuel-cycle/nuclear-waste/storage-and-disposal-of-radioactive waste.aspx (accessed 18.09.2020).

Yang, J., K. Shi, X. Gao, X. Hou, W. Wu, and W. Shi. 2020b. Hexadecylpyridinium (HDPy) modified bentonite for efficient and selective removal of ^{99}Tc from wastewater. *Chemical Engineering Journal* 382: 122894.

Yang, J., K. Shi, X. Sun, X. Gao, P. Zhang, Z. Niu, and W. Wu. 2020a. An approach for the efficient immobilization of Se-79 using Fe-OOH modified GMZ bentonite. *Radiochimica Acta* 108: 113–126.

Yildrim, H.E. 2006. *Surface chemistry of solid and liquid interfaces*. Oxford: Blackwell Publishing Ltd.

Zdravkov, B.D., J. Cermak, M. Sefara, and J. Janku. 2007. Pore classification in the charac-
    terization of porous materials: A perspective. *CEJC* 5: 385–395.
Zhu, R., Q. Chen, Q. Zhou, Y. Xi, J. Zhu, and H. He. 2016. Adsorbents based on montmo-
    rillonite for contaminant removal from water: A review. *Applied Clay Science* 123:
    239–258.
Zimens, K.-E. 1945. Zur Kinetik heterogener Austauschreaktionen. *Arkiv für Kemi,
    Minerologi och Geologi* 20A: 1–26.

2 Interfacial Processes in Geological Systems

Studies on Montmorillonite Model Substance

Natural rocks and soils are usually complex, multi-component systems. Besides this inherent complexity, the geological and environmental conditions that they are subjected to may also be very diverse. As a result, the interfacial reactions are complicated, and strongly dependent on the substances present and environmental conditions. Therefore, studies of interfacial processes of rocks and soils may have different scopes and objectives. First, there are studies that aim to understand the underlying principles behind the reactions. For these, the natural samples in their original forms are mostly not suitable because, due to the great number of components and other factors, the results obtained are difficult to interpret. For example, the components and factors may mask the effects of one another. So, these studies require the use of model substances and conditions that are representative but somewhat simplified over those in natural samples and enable us to concentrate on the characteristic properties and to avoid disturbing effects. It is important to note that, depending on the objective of the study, the model substance must mimic the original sample as much as possible. Some authors use very pure model substances and strict experimental conditions that vary the parameters in very narrow ranges. As a result, the data obtained in these experiments cannot be adequately used for natural rock and soil systems.

Another way to study the interfacial processes of rocks and soils is the investigation of a sample from the original rock and soil. In the case of real geological or environmental problems, it is essential to do these studies even if a theoretical interpretation cannot be given for the results, and several factors must be neglected. Of course, the results are valid only for the studied system. However, when the samples are well defined from chemical, mineralogical, or soil science points of view, general conclusions may be drawn.

Under natural conditions, the geological and environmental parameters usually vary in a relatively narrow range. Therefore, the changes are too small to provide information on interfacial processes and underlying factors. However, using wider concentration and pH ranges than those characteristic under geological conditions, we can obtain significant results.

This book contains two chapters describing interfacial reactions in geological formations (rocks and soils). Chapter 2 shows the results obtained on a model clay

DOI: 10.1201/9781003020080-2

(montmorillonite). Chapter 3 contains studies on real geological systems, that is, rocks and soils, going from a simple to gradually more complex systems (clay mineral – clay rocks – soils). In this way, the results obtained in simple systems may be used for the interpretation of the results in complex systems. The results shown in this book may assist in a fair understanding of the interfacial processes of rocks and soils in natural environment and may provide a sound approach to interpreting the anthropogenic effects of these processes.

2.1 USE OF MONTMORILLONITE AS A MODEL SUBSTANCE: THE IMPORTANT INTERFACIAL PROCESSES ON MONTMORILLONITE

The montmorillonite (Chapter 1, Table 1.2) clay mineral can be used as a model substance in the study of the interfacial processes of rocks and soils. It is a layer silicate, a member of the smectite group. Its structure is appropriate for modeling the most important interfacial processes in geological formations. Besides, it is a fairly widespread mineral in rocks and soils and plays an important role in the nutrient cycle of soils. In addition, it has many agricultural, industrial, and environmental applications.

2.1.1 CRYSTAL STRUCTURE AND LAYER CHARGE OF MONTMORILLONITE

Montmorillonite has some important characteristics that justify its use as a model substance for the study of the interfacial processes of rocks and soils. It is a dioctahedral three-layer clay (2:1 clays, TOT): an $AlO(OH)$ octahedral sheet is between two tetrahedral SiO_4 layers (Chapter 1, Table 1.2). The distance between the layers is not fixed (–O–O-bonds): the layers can be expanded. Because of the layered structure, it has two surface types: external and internal surfaces. The external surface is the surface of the particles (edge surface), and its size depends on the particle size distribution. Its area can be measured using the Brunauer–Emmett–Teller (BET) method, usually by the adsorption of nitrogen gas at the temperature of liquid nitrogen (Chapter 1, Section 1.1.3). The internal surface is the surface between the layers (interlayer surface), and its size can be determined by introducing substances into the interlayer space (e.g., water) (Chapter 1, Section 1.1.3). The internal surface area is independent of particle size distribution.

As usual for clay minerals, the internal surface area of montmorillonite is about 80–95% that of the total surface area. For different montmorillonite samples, the external surface area ranges from 20–100 m^2/g (Stadler and Schindler 1993b; Wanner et al. 1994; Barbier et al. 2000; Nagy and Kónya 2004, 2006). The internal specific surface area (Chapter 1, Section 1.1.3) is about 700–850 m^2/g (Eltantawy and Arnold 1993; Michot and Villiéras 2008; Table 2.1). Interfacial processes take place on particular surfaces depending on the reactants.

The crystal lattice of montmorillonite, similarly to other 2:1 phyllosilicates, may have isomorphic substitutions both in the tetrahedral and octahedral positions. In

TABLE 2.1

Basal (d_{001}) Spacing, Internal and Total Specific Surface Area, and Cation-Exchange Capacity of Natural Bentonite-Montmorillonite Samples

Type	Interlayer cation	Basal spacing (nm)	Internal specific surface area (m²/g)	Total specific surface area (m²/g)	Cation-exchange capacity CEC (meq/g)
Natural bentonites[a]	Na	1.19–1.25			
Natural bentonites (average of 16 samples)[b]	Na	1.24			
SWy-1[c]	Na		21.4	661.5 (ethylene glycol monomethyl ether)	93
SWy-2[d]	Na		29.4		26
MX-80[e]	Na	12.45	31.53		108
BSAB[d]			87		48
FEBEX[f]	Ca	1.52	62	725	100
Istenmezeje, Hungary[g]	Ca	1.533	93.5		104
Serrata de Nijar, Spain[h]	Mixed Ca, Mg, Na, K		57	649	106
GMZ, China[i]	Mixed Ca, Na, Mg, K			570	77.3

[a] Aceman et al. 2000, [b] Földvári 2008, unpublished data, [c] Stadler and Schindler 1993b, [d] Barbier et al. 2000, [e] Wanner et al. 1994, Bradbury and Baeyens 2004, [f] Fernandez et al. 2004, [g] Nagy and Kónya 2004, [h] Huertas et al. 2001, [i]Chen et al. 2018

the tetrahedral positions, the central tetravalent silicon can be substituted by trivalent aluminum ions, in the octahedral positions the trivalent aluminum ions can be substituted by bivalent (usually magnesium and iron(II)) cations of similar sizes. As a result, layer charge is negative. The number of isomorphic substitutions can vary, depending on the origin of the sample, so the chemical composition and layer charge also vary. The ideal structure of montmorillonite is shown in Figure 2.1.

The negative layer charge is mostly neutralized by the hydrated cations in the interlayer space. These cations are bonded to the internal surfaces by electrostatic forces, and they are exchangeable with other cations. The interaction strength between the hydrated cation and the layers (the internal surface) increases when the charge of the cation increases and the hydrated ionic radius decreases. Cations with hydrate shell can be considered as outer-sphere complexes. Cation exchange is the determining interfacial process of the internal surfaces of montmorillonite.

In montmorillonite, similar to other minerals, when the size of the exchanged cation is similar to the pore sizes in the crystal lattice, cations can build into the crystal lattice and, consequently, they reduce the negative layer charge (Chapter 1, Section 1.3.3.2). Other neutral molecules or cationic substances (Chapter 1, Sections 1.3.3.1 and 1.3.3.2) can also be sorbed in the interlayer space and on the external surfaces as well. They play an important role in defining the internal and total surface area and catalytic properties, and they may have an effect on the hydrophobicity of the mineral, as well as playing an important role in the production of pillared materials, etc.

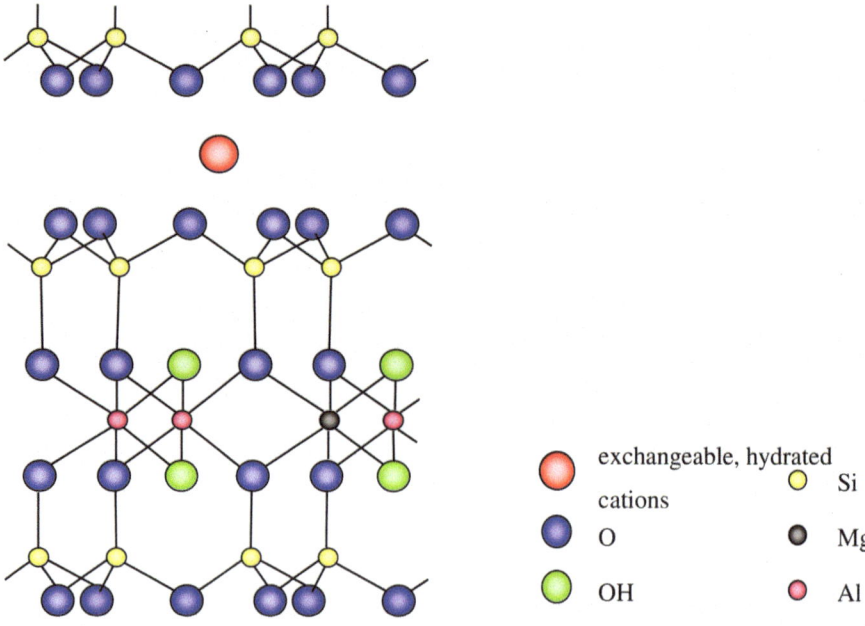

FIGURE 2.1   Ideal structure of montmorillonite.

2.1.2   CLAY–WATER INTERACTIONS

Since the size of the crystal lattice (TOT layer) is well defined by chemical bond lengths, the net basal (d_{001}) space of montmorillonite only depends on the size of the interlayer space. As mentioned previously, the interlayer space contains hydrated cations and adsorbed water molecules (interlayer water). Many of the important properties in clay minerals (including montmorillonite), such as swelling, plasticity, catalytic processes, etc., can be explained by the interactions between the clay and water.

The clay–water interactions and the quantity of water present in the interlayer space strongly depend on the nature of exchangeable cations. The water in the clay may be acting as hydrate water or additional absorbed water.

1. Similar to the interaction strength between the hydrated cation and the layers, the hydration energy of exchangeable cations is determined by electrostatic forces. In other words, the greater the charge and the smaller the radius are, the greater the hydration energy is. Thus, the hydration energy is greater for bivalent cations than for monovalent ones.
2. In addition, the quantity of adsorbed water also depends on the exchangeable cations. X-ray diffraction (XRD) studies show that, for air-dried samples, the basal (d_{001}) spacing is about 1.5 nm for bivalent montmorillonite (i.e., montmorillonite in which the exchangeable cation is bivalent), and about 1.2 nm for monovalent montmorillonite (in which the exchangeable cation is monovalent; Table 2.1). This approximately 0.3 nm corresponds to the diameter of water molecules. It means that bivalent montmorillonite contains one more layer of water than monovalent montmorillonite, that is, monovalent montmorillonites contain one monolayer of water, and bivalent montmorillonites contain two monolayers of water. However, water can more easily be introduced into the interlayer space in sodium-montmorillonite because the electrostatic attraction of the bivalent ion and the negatively charged layers is greater than in the case of monovalent sodium-montmorillonite (Helmy 1998). The number of water monolayers in sodium-montmorillonite can increase into four, and the basal spacing increases to 2 nm. As a result, sodium-montmorillonite can swell to several times its original volume. The swelling of calcium-montmorillonite is limited. Water permeability is inversely proportional to swelling. So, sodium-montmorillonite can be applied for barriers, for example, in waste disposals, but its low water permeability is disadvantageous in the soil.

The d(001) basal spacing of trivalent montmorillonite is slightly larger that that of bivalent montmorillonite as shown in Table 2.3. The swelling is similar to bivalent montmorillonite.

As a result of swelling, the basal spacing strongly depends on the water content and conditions of production and storage (Oueslati et al. 2009). Therefore, X-ray diffraction studies have to be complemented by other analytical methods (thermal analysis, infrared spectroscopy).

Some basic properties, such as basal (d_{001}) spacing, internal and total specific surface area, and cation-exchange capacity (CEC) of some natural montmorillonite or bentonite with high montmorillonite content are listed in Table 2.1. Similar characteristics of different cation- exchanged montmorillonites are given in Section 2.3.

The clay–water interactions can be studied by thermal analysis and infrared spectroscopy. Thermal analytical studies enable us to measure the quantity and the binding energy of water in the interlayer space and the crystal lattice. The elimination of water coordinated to cations and other forms of water in the interlayer space (adsorbed water) can be distinguished on differential thermal analytical (DTA) and differential thermogravimetric (DTG) curves only when the hydration energy of the cation is sufficiently higher than the binding energy of the water to clay. In the case of monovalent (e.g., sodium) montmorillonite, the binding energy of the adsorbed water and the hydration energy of cations is approximately equal, so the total water content of the interlayer is eliminated together in the range of 100–120°C. There has been some attempt to separate the two processes using the second derivative of the low-temperature water elimination reaction (Földvári et al. 1998; Földvári and Kovács-Pálffy 2005). For bivalent (e.g., calcium) and trivalent montmorillonites, the hydration energy of cations is greater than the binding energy of adsorbed water, so the water elimination peak is divided into two parts. In this way, the hydration energy of cations can be measured. Besides the interlayer water, the elimination of the structural water can also be seen at high temperatures (above 500°C).

The infrared spectra of clays are characterized by absorption bands corresponding to the vibration of water molecules and OH groups of the crystal lattice. Namely, bands near 3620 cm^{-1} and 3550 cm^{-1}, as well as 1630 cm^{-1}, have been assigned to hydrate water of cations. These absorption bands overlap with those that characterize the OH groups of the silicate framework of the mineral. Absorption bands near 3450 cm^{-1} and 3350 cm^{-1} have been assigned to additional adsorbed water molecules (Theng 1974; Bishop et al. 1994; Peker et al. 1995). The assignment of the bands differs ±10–20 cm^{-1} depending on the report. The vibrational bands of the crystal lattice are also present in infrared (IR) spectra; bands in the 1200–800 cm^{-1} range belong to the tetrahedral layers; bands in the 680–500 cm^{-1} wave number range are assigned to the octahedral layer.

2.1.3 EDGE CHARGES OF MONTMORILLONITE

While on the internal surface cation exchange is the most significant interfacial process, on the external surfaces it is the formation of the edge charges (depending on pH) and the different sorption processes on the protonated and deprotonated surface sites. The formation of pH-dependent charges on aluminosilicates was discussed in Chapter 1, Section 1.3.2.1. Montmorillonite, as aluminosilicates, has similar protonation and deprotonation processes on aluminol sites and deprotonation on silanol sites (Chapter 1, Equations 1.53–1.55). The protonation of silanol groups takes place at pH<2. However, this process can be neglected since the crystal lattice of montmorillonite is destroyed under very acidic conditions.

When studying the surface acid–base properties of montmorillonite, it is essential to understand that hydrogen ions and cations of the support electrolyte can also participate in cation exchange processes. The processes on the internal and external surfaces have to be taken into consideration simultaneously, and they both have to be included into the equilibrium thermodynamical models.

The edge charges can also bond ions with opposite charges. This process, however, is not directed clearly by electrostatic forces; chemical properties play an important role. The ions are sorbed with no hydrate shell, that is, inner-sphere complexation occurs. These reactions and the surface complexation models for their quantitative treatment are shown in general in Chapter 1, Table 1.7.

An interesting consequence of the ion sorption on the edge sites is that different transformation processes can be initiated by the sorbed ion, for example, heteronucleation and surface precipitation (Section 2.10.4).

2.1.4 MONTMORILLONITE AS A MODEL SUBSTANCE

As seen earlier, montmorillonite has all surface properties characteristic of mineral components of rocks and soil, namely, ion adsorption and exchange processes in the internal and external surfaces, adsorption of neutral molecules, and structural changes of the crystal lattice. Even some interfacial reactions of humic substances may be modeled using montmorillonite (cation exchange or inner-sphere complexation of the functional groups of humic substances). Additional information on the behavior of functional groups of humic substances can be obtained by adding complex forming agents to the solution phase of the rock/soil/solution systems. Thus, montmorillonite is a very suitable model substance to study the interfacial process of rocks and soils in aqueous media characteristic of geological systems.

Before discussing the interfacial reactions of montmorillonite, we have to define what we call "montmorillonite in practice". As is known, most minerals are not found in pure forms; they are present in rocks together with other minerals. Montmorillonite is the main component of bentonite rocks, so bentonites with high montmorillonite content are frequently called montmorillonite. For scientific studies, we use bentonite with high montmorillonite content in natural form after purification with sedimentation. Usually, the fraction <2 μm is the clay fraction containing montmorillonite.

In the literature, different commercial montmorillonites are used, especially for catalytic purposes (KSF or K10 montmorillonites). They are usually acid-treated clays, the montmorillonite content of which is rather low. Our X-ray diffraction studies show 44% Ca-montmorillonite content of K10 montmorillonite, and 53% Na-montmorillonite content of KSF montmorillonite; the CEC of KSF montmorillonite was found to be 30 mequ/100 g by ammonium acetate method (Richards 1957). A similar value has been given by Abollino et al. (2003). So, in a strict clay science sense, they cannot be considered as montmorillonite. Naturally, this causes no problems in organic chemistry when the main objective is the catalysis of a given reaction.

In our studies, the model substance (montmorillonite) was a calcium-bentonite (Istenmezeje, Hungary), the characteristic features are given here. X-ray diffraction (intensity of the basal reflection) and thermoanalytical (weight loss upon heating) data show 91% montmorillonite content. The other constituents are 5% calcite, 3% kaolinite, 1% X-ray amorphous silicates, and a trace of quartz. The amorphous phase is silicate particles, which are not crystalline for X-ray diffraction. The basal spacing (d_{001}) of the air-dried sample was 1.533 nm. The average elemental composition calculated from X-ray spectrometric data was $(Si_{7.7}Al_{0.3})(Al_{2.6}Fe(III)_{0.7}Fe(II)_{0.1}Mg_{0.5})$ $O_{20}(OH)_4Ca_{0.45}$. CEC is 104 meq/100g, the internal specific surface area is 93.5 m^2/g, and the total specific surface area was 660 m^2/g (Nagy et al. 2004).

2.2   CATION EXCHANGE: OUTER-SPHERE COMPLEXATION

As mentioned earlier, one of the most important interfacial processes of rocks and soils is cation exchange. Although all cations participate in cation exchange, its importance in geological media is highly variable and depends on the concentration of cations. On the basis of concentration, sodium, potassium, magnesium, and calcium are considered as the main cations of groundwater. The trace cations of groundwater are manganese, iron, cobalt, nickel, copper, zinc, and strontium ions. By anthropogenic activity, different cations, for example, chromium, cadmium, mercury, lead, etc., can contaminate the rocks and soils. The term *"polluting ion"*, however, is concentration-dependent: a cation can be an essential ion in low concentration, but it is toxic in high concentration. As an example, while the level of copper ion in soils varies from 6 to 60 ppm, in contaminated surface soils it can be 3700 ppm (Kabata-Pendias and Pendias 1985, pp. 77 and 78). Other cations, or cations of certain radioactive isotopes (Cs-137, transuranium elements, etc.), can be considered as contaminating ions even in a very low concentration due to a dangerous characteristic (e.g., radioactivity).

As usual, hydrogen ion plays a special role in the interfacial reactions of geological systems. The characteristic pH range of groundwater ranges from 6 to 8, that is, the solution is close to neutral. Water molecules, however, provide an unlimited source of hydrogen ions, and, of course, hydroxide ions, which have an effect on the solid and solution phases, as well as on the interface.

Hydrogen ions participate in the cation-exchange processes of the interlayer space. As will be seen later (Section 2.7.1), they have a very large affinity to the layer charge. Hydrogen and hydroxide ions are potential-determining ions of the external surfaces via the protonation and deprotonation processes of aluminol and silanol sites. In acidic media, the degradation of aluminosilicates can be observed.

In addition, water molecules may cause the hydrolysis of some cations, and the formation of oxide, hydroxide colloids, and precipitates. Consequently, the cations in the solution precipitate, forming an independent solid phase. In this case, the solubility of the precipitate will determine the metal ion concentration in the solution. Thus, caution should be exercised in the so-called adsorption edges at pH values when the hydrolysis of the dissolved cations takes place (Schindler et al. 1976; McKenzie 1980) since the process is not related to the adsorption but to hydrolysis.

A special phenomenon is observed in the case of mercury ion (Kónya and Nagy 2011). Due to the chemical species of mercury(II), cation exchange of mercury ions and the cations in the interlayer space of montmorillonite can be expected at pH<3 since the dominant species is Hg^{2+} ion in this pH range. At pH~3, the hydrolysis of mercury(II) ion starts, and at the same time, Hg^{2+} and OH– ions may form $Hg(OH)_2$ molecules. The solubility product of $Hg(OH)_2$ is $3.6*10^{-26}$ (www.princetonschools .net/site/handlers/filedownload.ashx?moduleinstanceid=714&dataid=1472&FileN ame=Solubility_Product_Constants.pdf). This means very low Hg^{2+} concentration. However, $Hg(OH)_2$ molecules are fairly soluble in water, the concentration of the saturated $Hg(OH)_2$ solution is 3.19e-4 mol/dm^3. The solubility is constant within a wide pH range (from 3 to 14) (Pourbaix 1966; VisualMinteq Database). When the concentration of mercury(II) is higher than 3.19e-4 mol/dm^3, or $Hg(OH)_2$ eliminates water, solid HgO precipitates. The solid $Hg(OH)_2$ compound is not known (Wang and Andrews 2005). The chemical species of mercury ions in aqueos solution significantly influences the sorption on clay minerals. Generally, during the interaction of an electrolyte solution and an ion exchanging clay mineral, the cations of the phases exchange when the thermodynamic properties of the solution allow this. The exchange process is derived by electrostatic forces between the cations and the layer charge. When increasing pH, cations can be hydrolyzed and precipitated as insoluble hydroxides. When, however, a soluble hydroxide is formed, as in the case of mercury(II) forms, a special behavior is observed. Namely, the sorption is nearly independent of pH in the range of 3.5 < pH <6.5 and does not reach the total sorption usual for other cations. It is caused by the relatively high solubility of mercuric hydroxide. When the mercury(II) concentration of the system is below the solubility of $Hg(OH)_2$ molecules, no precipitation of mercury hydroxide occurs. Sorption equilibrium between the dissolved mercuric hydroxide and the clay is formed, independently of pH in this pH range.

Since the cation-exchange processes of the main cations (sodium, potassium, magnesium, and calcium) have a significant role in the nutrient cycle of soils, the classical literature has discussed their cation-exchange processes in detail (e.g., Boyd et al. 1947; Gaines and Thomas 1953; Howery and Thomas 1965; Sposito 1981; Filep 1999). As described earlier, cation exchange is characterized by the selectivity coefficients and the equilibrium constants obtained by the integration of the selectivity coefficient versus surface concentration of cations (Chapter 1, Section 1.3.4.2.1).

Beside the cations that are abundant in the natural environment, others, natural or polluting, and cationic organic substances can participate in ion-exchange reactions in the interlayer space. When the concentration of these cations is low, as generally it is under geological conditions, the monocationic form obviously does not form since the cation is not present in equivalent quantity to exchange all exchangeable cations present in the clay. For this reason, the sorption of cation in low concentration is treated as "adsorption" by some authors. The mechanism, however, is cation exchange because, if these cations are applied in high concentration or after several ion exchanges, cation-exchanged monoionic montmorillonites can be produced. The aim of making monoionic montmorillonites is to study of the structural changes that occur as the result of ion exchange and also the production of clay catalysts (e.g., Kukovskij 1966; Barshad 1969; Patel 1982; Lacher et al.1993; Lahav et al.1993; Stadler and Schindler 1993b; Adams 1987; Adams et al. 1983; 1994; Tateiwa et al.1994a;1994b, 1994c; Yariv et al. 1994; Heller-Kallai and Mosser 1995; Földvári et al. 1998; Nagy and Kónya 2004; Nagy et al. 1998a, 2002, 2003a, 2003b, 2004, Komlósi et al. 2006, 2007). Recently, monoionic montmorillonite/bentonite are used for new applications such as the sorption of anionic species (Chapter 1, Section 1.3.6.4, Chapter 2, Section 2.11). Monoionic montmorillonites are also produced fom commertial montmorillonites with low montmorillonite content (Kannan et al. 1997; Ramchandani et al. 1997; Waterlot et al. 2000).

Another characteristic of some of these cations compared to those of the main element is that they can be sorbed on the deprotonated edge sites by inner-sphere complexation (Gaskova and Bukaty, 2008). Some examples will be discussed in detail in Section 2.5.

Cation exchange can be conveniently studied by radioisotopic tracer methods because of the fairly broad concentration range that can be used, from tracer (carrier-free isotopes) to high (applying an inactive carrier) concentrations. By this method, the ion-exchange isotherms and selectivity coefficients can be precisely determined in a wide surface concentration range, which allows the construction of ion-exchange isotherm and selectivity function, and the integration of the selectivity function (Chapter 1, Section 1.3.4.2.1, Equation 1.80). An example of cation- exchange isotherm and the isotherm parameters are shown in Figure 2.2 for the cation exchange of cobalt ions and calcium-montmorillonite.

The cation exchange of cobalt and calcium ions on montmorillonite was studied by dual radioisotopic tracer method using ^{60}Co and ^{45}Ca isotopes. The isotopes have different radiation properties: ^{60}Co has gamma and beta radiation with high energy,

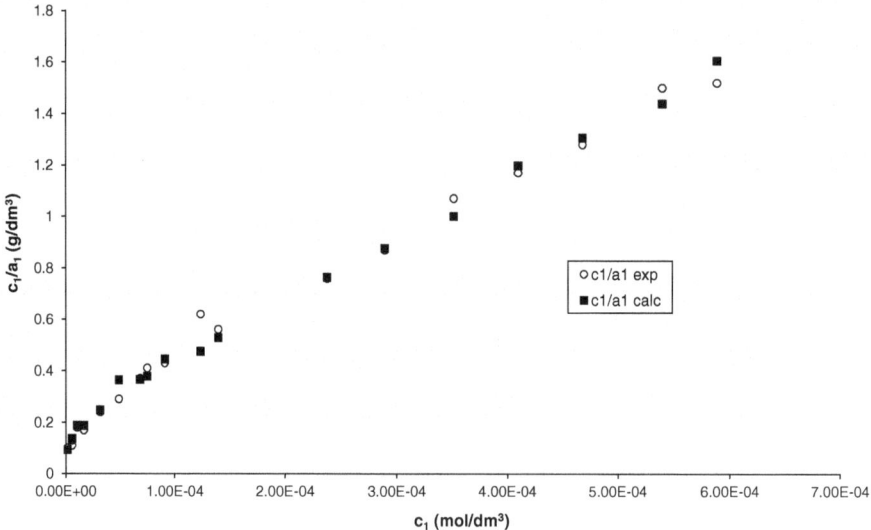

FIGURE 2.2 The equilibrium distribution of cobalt ions as a function of equilibrium cobalt ion concentration for cobalt-calcium ion exchange at a constant pH value. $T = 33°C$; pH 6.5 (Reprinted from Nagy et al. 1997, with permission from Elsevier.)

and ^{45}Ca has a beta radiation with low energy; so, they can be measured separately. The different radiation properties of the isotopes allow us to determine the quantity of both ions during the cation exchange. The result shows that cobalt and calcium ions exchanges happen in equivalent quantities when the concentration of cobalt ions is not too high (<6e-4 mol/dm^3). It means that there is no adsorption of cobalt ions, so the isotherm can be treated by Equation 1.94 (Chapter 1) in this concentration range (Figure 2.2). In Chapter 1, Equation 1.94, the concentration of cobalt and calcium ions (c_1 and c_2), as well as the sorbed quantity of cobalt ions (a_1) are experimentally determined by the radioisotopic tracer method. From the isotherm equation, ζ = 4.65e-4 mol/g and $K_1/K_2 = 0.65$ were obtained. The evaluation can be made by a parameter-estimating program (e.g., Scientist). As seen in Equation 1.103, Chapter 1, Section 1.3.4.2.3 the ratio of K_1/K_2 is equal to the reciprocal of the equilibrium constant of the exchange process. By the integration of the selectivity function $K_{Co,Ca}$ = 1.5 (Table 2.2) was obtained, so the two values are in a fairly good agreement (1/0.64 = 1.54). The number of exchange sites can also be compared to the CEC (5.2e-4 mol/g for divalent cations). The difference is due to the presence of hydrogen ions that can also neutralize the layer charge. The number of exchange sites neutralized by hydrogen ions at the pH of the experiments (pH 6.5) is in good agreement with the number of exchange sites (Section 2.7.1). This fact, similarly to the experimental data proving the equivalent exchange of cobalt and calcium ions, shows that cobalt ions do not exchange with hydrogen ions in the applied concentration range.

The results of the cobalt and calcium ion exchange on montmorillonite show that the ion exchange isotherm equation (Chapter 1, Equation 1.94) can be applied well; in this case, the estimated isotherm parameters (ζ and K_1/K_2) have real physical meaning and their values can be confirmed by independent experimental procedures.

TABLE 2.2

Equilibrium Constants of Cation-Exchange Processes of Calcium-Montmorillonite (Istenmezeje, Hungary)

	Equilibrium constant
$K_{Na,Ca}$	0.11
$K_{Cs,Ca}$	2
$K_{Sr,Ca}$	2
$K_{Mn,Ca}$	1
$K_{Co,Ca}$	1.5
$K_{Cu,Ca}$	0.16
$K_{Zn,Ca}$	0.8
$K_{Cd,Ca}$	0.3
$K_{Hg,Ca}$	0.5
$K_{Ag,Ca}$	0.5
$K_{Pb,Ca}$	About1.2, but it is uncertain because of the sorption on the edge sites (Sections 2.5.1.4 and 2.10.4)
$K_{H,Ca}$	$1.6*10^4$

The equilibrium constants of cation exchange processes of calcium-montmoril-lonite (Istenmezeje, HU) determined by the integration of the selectivity function are shown in Table 2.2.

On the basis of the equilibrium constants determined in this way, the relative affinity of the cations to the layer charge can be given as follows:

$$H^+ > Sr^{2+} \cong Cs^+ > Co^{2+} > Pb^{2+} > Ca^{2+} \cong Mn^{2+} > Zn^{2+} > Ag^+ \cong Hg^{2+} > Cd^{2+} > Cu^{2+} > > Na^+$$

In the literature, the affinity order of smectites is usually given for alkali and alkali earth cations, in that order, as follows (Laudelout et al. 1968; Sawhney 1972; Maes and Cremers 1977; Bergaya et al. 2008):

$$Cs^+ > Rb^+ > K^+ > Na^+ > Li$$

$$Ba^{2+} > Sr^{2+} > Ca^{2+} > Mg^{2+}.$$

The two affinity orders are in good agreement. The affinity order of the cations is in agreement with the aforementioned charge/size ratio. The following discrepancies are seen in the preceding affinity order:

1. It is important to emphasize the high affinity of hydrogen ion (Section 2.7.1).
2. Cesium ion has a small charge but a very thin hydrate shell, so the affinity to the negative layer charge is high.
3. The equilibrium constant of lead is uncertain because of the confounding effects of the sorption of lead ion on edge charges (Sections 2.5.1.4 and 2.10.4).
4. On the basis of ionic radius, cobalt ion should be between copper and cad-mium ions, and thus it has a higher affinity than expected. Perhaps it sub-stitutes the iron(II) of crystal lattice since the two-ionic radius is about the same (0.84 nm).
5. As mentioned in Chapter 1, Section 1.3.3.2, potassium and lithium ions can fit into the cavities of the silicate crystal lattice and decrease the layer charge. It is also the case for montmorillonite (Gast 1972; Maes and Cremers 1977; Eberl 1980; Goulding and Talibudeen 1980; Bouabid et al. 1991; Komadel et al. 2003), so the affinity order is modified.

The lists above contain no trivalent cations. The experience, however, shows that the pH does not have an effect on the exchange of calcium ions and trivalent lanthanum and ittrium ions (Kovács et al. 2019) revealing the great affinity of these trivalent cations to layer charge.

The cations in the interlayer space can spontaneously transform into other chemi-cal species; they can undergo oxidation or reduction, and thus they can exert an effect on the redox conditions of soils and rocks. The new species can react fur-ther, forming precipitates and nano layers in the interlayer space. In some cases, the basal spacing (d_{001}) remains the same, and the thickness of the interlayer does not change. The transformation processes of the interlayer cations will be shown in Section 2.10.1.

The other type of transformation processes of the interlayer cation is the pillaring (Chapter 1, Section 1.3.5). In this process, metal oxide chains are formed in the interlayer space upon thermal treatment of different cation-exchanged montmorillonites. As a result, the basal spacing, that is, the size of the interlayer space, increases. The pillared montmorillonites have been widely applied in catalytic reactions, as evidenced by over 1,400 scientific publications in the last 20 years (e.g., Fetter et al. 2000; Johnson and Brody 1988; Perez Zurita et al. 1996). Pillared montmorillonites are used for the preparation of porous systems and as templates of catalysts also (Barakan et al. 2019; Chmielarz et al. 2017; Han et al. 2018; Kang et al. 2018; Rangel-Rivera et al. 2018; Rubiyanto et al. 2020; Silva et al. 2019; Yang et al. 2020; Zhao et al. 2020).

As mentioned in Chapter 1, Section 1.3.5, organic cations (tensides, amino acids, polymers, etc.) can be used to increase the size of the interlayer space (Fudala et al. 1999). In the increased interlayer space, metallic and metal oxides nano layers can be produced by cation exchange and heat treatment (Mogyorósi et al. 2003; Németh et al. 2003; Papp and Dékány 2002). The process is frequently carried out in organic media.

The pillared and nano particle-containing montmorillonites open up new venues for the application of clay minerals in organic syntheses and production of composite materials.

2.3 SYNTHESIS AND CHARACTERIZATION OF CATION-EXCHANGED MONTMORILLONITES

There are some factors that must be taken into consideration when synthesizing cation-exchanged montmorillonites. First of all, the relative affinities of the cations to the layer charge are important. As seen in Table 2.2 and the affinity order of cations, sodium ion is the cation that is the easiest to exchange. So, sodium-montmorillonite (bentonite) should be used as starting material if possible. However, precisely because of this characteristic of readily undergoing cation exchange, the natural sodium-montmorillonite is rare. It is present in zeolitic riolit tuffs; however, in contact with the calcium ions present in groundwater, sodium-montmorillonite transforms into calcium-montmorillonite.

The synthesis of cation-exchanged montmorillonite is performed through the following reactions.

From sodium-montmorillonite:

$$zNa - mm + Me^{z+} \rightleftarrows Me - mm + zNa^{+} \qquad (2.1)$$

or calcium-montmorillonite:

$$\frac{z}{2}Ca - mm + Me^{z+} \rightleftarrows Me - mm + \frac{z}{2}Ca^{2+} \qquad (2.2)$$

During the preparation of cation-exchanged montmorillonite, especially when calcium-montmorillonite is used as a starting material, the cation (Me^{z+}) should be present in as high a concentration as possible, and the cation-exchange process has to be repeated several times. Depending on the relative affinities of cations (Table 2.2), three to seven cation-exchange processes may be necessary. It is also important that the co-anion of the solution does not form complexes with the cation; that is, the exchanging cation (Me^{z+}) should be present as positively charged aqua complexes. In practice, it means that perchlorate and nitrate salts of the cations have to be used for cation exchange. The application of chloride salts is frequently not desirable because the chloride complexes have smaller positive charge; they can even be neutral or negative, decreasing the efficiency of cation exchange. The complexation of the interlayer cation (especially calcium ion) with the coanions, however, can be desirable. The application of a complex-forming species is a complex problem; the relative stability of the possible complexes should always be taken into account for the given cation pair (Fujikawa and Fukui 1997; Criscenti and Sverjensky 1999).

The other important factor is the pH of the solution or the montmorillonite-solution suspension. The pH range is limited due to

1. The hydrolytic processes of cations, which give the upper value of pH.
2. The acidic destruction of montmorillonite, determining the lower value of the pH range.

For easily hydrolyzing cations (iron, noble metal, tin(II), chromium(III), etc.), the suitable pH range is very narrow, or it does not exist. In this case, the direct cation exchange from solution can hardly be done; short exchange times (Buzetzky et al. 2019b) or some by-pass methods have to be applied. One of these methods is the cation exchange between solid phases. The clay mineral and an insoluble compound of the exchanging cation (e.g., a metal oxide, hydroxide) are in suspensions separated by a semipermeable layer, for example, by a membrane filter. Both cations (sodium/calcium and Me^{z+}) can transfer the membrane filter, and the cation exchange can take place. To minimize the hydrolysis of cations, organic solvents may be employed. For example, iron(III)-montmorillonite can be prepared from acetoneous $FeCl_3$ solution (Li et al. 1998). In this case, caution must be exercised because the organic substances can cause the dehydration of montmorillonite (Komlósi et al. 2007). Furthermore, once the cation is in the interlayer space, it can undergo hydrolyzation by the water molecules present in the clay (Section 2.10.2).

Another way to get around the problem is complex formation; that is, a positively charged complex of the exchanging cation (Me^{z+}) is formed at a pH where the degradation of the clay mineral is avoided. The disadvantage of this method is that the size of the complex is greater than that of the hydrated cations, and they cannot, or can only very slowly, be introduced into the interlayer space. As a result, the exchanging cations (Me^{z+}) may get sorbed on the external surfaces. An example of cation exchange through complex formation is the synthesis of palladium-montmorillonite from its amin and ethylene diamin complexes (Nagy and Kónya 2007, 2008; Section 2.5.1.3).

The most important industrial example of cation exchange is the preparation of sodium-montmorillonite/bentonite from calcium-bentonite. As seen in Table 2.2, calcium ions have greater affinity to the layer charge than sodium ions, so the calcium-sodium cation exchange must be performed in the presence of carbonate ions. It means that calcium-montmorillonite/bentonite is suspended in sodium carbonate solution. Calcium ions precipitate with carbonate ions, so sodium ions can occupy the interlayer space. This process is known as *soda activation of bentonite*. The disadvantage of soda activation is that sodium-montmorillonite is contaminated with calcium carbonate.

To avoid calcium carbonate contamination of montmorillonite, the sodium salt of compounds forming very stable complexes with calcium ions, such as ethylene diamin tetraacetic acid (EDTA), can be used for the exchange of calcium ions to sodium ions. The product, calcium-EDTA complex, is soluble in water, thus it can be washed out of montmorillonite (Section 2.9.1).

A special case of cation exchange is the isotope exchange when both ions are the isotopes of the same elements, for example, a stable and a radioactive isotope. Calcium-montmorillonite can be labeled with ^{45}Ca isotope when calcium-montmorillonite is suspended in the solution containing a radioactive ^{45}Ca isotope with known specific activity (activity/mass). The radioactive calcium ions exchange with the stable calcium ions in the interlayer space. Naturally, any other labeled montmorillonite can be produced if there is a suitable isotope (e.g., ^{54}Mn, ^{65}Zn, or ^{137}Cs, etc.) (Nagy et al. 1998a, 2003b). The applications of labeled montmorillonites allow us to follow the cation exchange or other interfacial processes from the solid to the solution phase. By the dual tracer method (radioactive tracers both in solid and solution), both directions can simultaneously be studied. So, the equivalency of the ion exchange can be verified or, when there is a lack of equivalency, other interfacial processes can be observed (Section 2.6). Another possibility of the application of radioactive calcium-montmorillonite is the control of calcium uptake by animal tissues or animals (Szántó et al. 1999).

The structural parameters of cation-exchanged montmorillonites prepared from calcium-montmorillonite (Istenmezeje) are listed in Table 2.3.

As seen in Table 2.3, the basal spacing of monovalent montmorillonite is approximately 1.25 nm, and the water content is approximately 7%. It means that there is one layer of water in the interlayer space. For bivalent montmorillonite, both the basal spacing (>1.5 nm) and the water content (>10%) is higher, showing two layers of water molecules in the interlayer space. The basal spacing of Pb-montmorillonite is 1.254 nm, which is similar to the value characteristic of monovalent montmorillonite (1.241 nm). However, it does not mean that lead is sorbed on the surface of montmorillonite as monovalent cation since the other parameters that are determined by the distance between the layers (hydration entropy, charge/ion radius value, water content in the interlayer space) lie between the values for bivalent and monovalent cations (Földvári et al. 1998).

The basal spacing of lanthanoid-montmorillonites is similar to other trivalent montmorillonites and varies between 1.5 and 1.6 nm. For example, the basal spacing of Fe(III)-bentonite is about 1.6 nm (Komlósi et al. 2007; Izumi et al., 2005; Kong et

TABLE 2.3

The Structural Parameters of Cation-Exchanged Montmorillonites Prepared from Calcium-Montmorillonite (Istenmezeje, Hungary)

Interlayer cation	Valency of interlayer cation	Basal spacing (nm)	Water content, (%)	H_2O/OH
Na	1	1.256	6.9	1.63
Cs	1	1.243	6.91	
Ag	1	1.248		
Ca	2	1.533	10.21	2.80
Mn	2	1.512		
Co	2	1.542		
Cu	2	1.522	15.98	
Zn	2	1.533		
Cd	2	1.544	16.88	
Hg	2	1.554	17.78	
Pb	2	1.254		
Cr(III)	3	1.412		
Sn(II)/IV)	2 and 4	1.351		
Al	3	1.576	19.47	5.81
Fe(III)	3	1.567	9.3	2.69
La	3	1.556	13.37	3.23
Ce	3	1.561	13.59	4.21
Pr	3	1.568	14.41	5.29
Nd	3	1.582	11.21	2.47
Eu	3	1.541	12.83	2.84
Gd	3	1.515	12.61	2.67
Tb	3	1.543	13.36	4.02
Dy	3	1.343	11.06	4.13
Ho	3	1.525	12.36	4.83
Er	3	1.516	11.97	4.53
Tm	3	1.515	12.03	3.17
Yb	3	1.522	14.35	4.69
Lu	3	1.516	13.05	5.08
Ce	4	1.523	9.3	4.2

al. 2005), and that of Al-bentonite is 1.576 nm; the ionic radius of Fe(III) and Al is, however, much smaller. For the lighter lanthanoids this value is above 1.55 nm, and for the heavier lanthanoids it is between 1.51–1.53 nm which follows the well-known size contraction of lanthanoids; that is, the greater the atomic number of lanthanoid, the smaller the size of the ion (Figure 2.30, Kovács et al. 2017). In addition, after the ion exchange with light lanthanoid ions and ittrium, the iron content of montmorillonite decreases as will be discussed in Section 2.10.5.

A little bit smaller d(001) values are given for Cr(III) bentonite as expected from the valency. However, in order to avoid the hydrolysis of Cr(III) ion, chromium-bentonite was prepared under acidic conditions (pH $\approx$ 1) at very short exchange times (3*5 min), which can influence the basal spacing (Buzetzky et al. 2019).

Sometimes, to achieve a particular objective, the conditions of the correct preparation of cation-exchanged montmorillonites can be neglected. For example, Fe(III)-montmorillonite have been prepared for application in the deep geological repository of high-level nuclear waste (Manjanna et al. 2009). Acidic pH has been applied, destroying the crystal lattice; however, a part of iron(III) ion has been precipitated as oxide and hydroxide, desirable for the sorption of radioactive matter.

2.4 SURFACE ACID–BASE PROPERTIES OF MONTMORILLONITE

2.4.1 FORMATION OF EDGE SITES ON MONTMORILLONITE

Besides the cation exchange in the interlayer space, cations and anions can also undergo sorption on the edge charges of montmorillonite. The edge charges are formed by the protonation and deprotonation of silanol and aluminol sites, and thus they depend on the pH.

The formation of edge charges of minerals has been discussed in Chapter 1, Section 1.3.2.1. It has been shown that aluminosilicates (including montmorillonite) have two types of surface (aluminol and silanol) sites, and their protolytic processes have been expressed by Chapter 1, Equations 1.54–1.56. For simplicity, the reaction equations are repeated here. For aluminol sites,

$$-AlOH + H^+ \rightleftarrows -AlOH_2^+ \qquad K^{int}_{AlOH_2} = \frac{\left[AlOH_2^+\right]}{\left[AlOH\right]\left[H^+\right]exp\left(-\dfrac{F\Psi}{RT}\right)} \qquad (2.3)$$

$$-AlOH \rightleftarrows -AlO^- + H^+ \qquad K^{int}_{AlO} = \frac{\left[AlO^-\right]\left[H^+\right]exp\left(-\dfrac{F\Psi}{RT}\right)}{\left[AlOH\right]} \qquad (2.4)$$

For silanol sites,

$$-SiOH \rightleftarrows -SiO^- + H^+ \qquad K^{int}_{SiO} = \frac{\left[SiO^-\right]\left[H^+\right]exp\left(-\dfrac{F\Psi}{RT}\right)}{\left[SiOH\right]} \qquad (2.5)$$

The symbols were defined earlier (Chapter 1, Section 1.3.2.1.1).

As seen earlier, depending on pH, positively or negatively charged edge sites are formed. The study of surface acid–base properties, that is, the determination of the intrinsic stability constants and the number of edge sites, means a titration process by an acid and/or a base. The titration is usually made under a nitrogen atmosphere to avoid the formation of carbonates. The acid or base solution has to be added in small volumes to the suspension of montmorillonite (or other solid), and pH is usually measured by a suitable electrode under continuous mixing. At equilibrium, a given reading should not differ more than 0.01 pH unit from the average of the six foregoing readings. In the absence of layer charge, the difference between the total added quantity of hydrogen/hydroxide ions and the hydrogen ion concentration (pH) of the solution equals to the quantity of sorbed hydrogen ions on edge sites. The characteristic acid–base properties of edge sites can be calculated from these values by an equation system containing the mass balance, thermodynamic equilibrium constants, and the surface electric properties. The equation systems can be solved by different commercial computer programs (Chapter 1, Section 1.2.3).

2.4.2 Effect of Permanent Charge on the Study of Edge Charges

In the case of montmorillonite, however, besides edge sites, there is a permanent negative layer charge that can sorb cations in the interlayer space. During the titration process, hydrogen ions are added to montmorillonite, and also the proton (as a cation) consumption of the negative layer charge has to be taken into account. So, the quantitative treatment of protonation or deprotonation processes is more complex. In the following text, how to take into account the effect of layer charge when studying the acid–base properties of edge charges is discussed.

The layer charges (X^-) of montmorillonite are neutralized by cations, namely, sodium or calcium ions in natural minerals or any other cations in cation-exchanged montmorillonites. As our model substance is calcium-montmorillonite, the neutralization of the layer charge can be described as follows:

$$Ca^{2+} + 2X^- \rightleftarrows CaX_2 \quad K_{CaX_2} = \frac{\left[CaX_2\right]}{\left[Ca^{2+}\right]\left[X^-\right]^2} \quad (2.6)$$

The hydrogen ions added by the titration can also neutralize the layer charge:

$$H^+ + X^- \rightleftarrows HX \quad K_{HX} = \frac{\left[HX\right]}{\left[H^+\right]\left[X^-\right]} \quad (2.7)$$

Since charged particles involve all of these processes, including the formation of edge charges (Equations 2.3–2.5), first, the electric properties of the interfaces have to be determined. A simple way to do so is the application of a support electrolyte in high concentration. The electric double layer, in this case, behaves as a plane, and, as a first approach, the Helmholtz model, that is, the constant capacitance model, can be used (Chapter 1, Section 1.3.2.1, Table 1.7). It is important to note, that the support electrolyte has to be inert. A suitable support electrolyte (such as sodium perchlorate) does not form complexes (e.g., with chloride ions; Section 2.3) and does not cause the degradation of the montmorillonite (e.g., potassium fixation in the crystal cavities).

In this case, however, cations of the support electrolyte, usually sodium ions, can also neutralize the layer charges:

$$Na^+ + X^- \rightleftarrows NaX \quad K_{NaX} = \frac{[NaX]}{[Na^+][X^-]} \tag{2.8}$$

In Equations 2.6–2.8 K's are the stability constants characteristic of the neutralization reactions of layer charges by cations, [ ]'s mean the concentration of the species, [X$^-$] means the number of the layer sites.

Since the values of K's in Equations 2.6–2.8 cannot be determined directly for a given cation, they are calculated from the equilibrium constants (Chapter 1, Equation 1.77) of cation exchange reactions (Equations. 2.9 and 2.10):

$$CaX_2 + 2H^+ \rightleftarrows 2HX + Ca^{2+}, \quad K_{Ca,H} = \frac{\overline{a_{HX}}^2 * a_{Ca^{2+}}}{\overline{a_{CaX_2}} * a_{H^+}^2} \approx \frac{K_{HX}^2}{K_{CaX_2}} \tag{2.9}$$

$$CaX_2 + 2Na^+ \rightleftarrows 2NaX + Ca^{2+}, \quad K_{Ca,Na} = \frac{\overline{a_{NaX}}^2 * a_{Ca^{2+}}}{\overline{a_{CaX_2}} * a_{Na^+}^2} \approx \frac{K_{NaX}^2}{K_{CaX_2}} \tag{2.10}$$

where $\overline{a}$ and a mean activities. As seen from Equations 2.9 and 2.10, the equilibrium constants give only the ratio of the stability constants of the neutralization reactions of the layer charges. So, the calculation of K's of Equations 2.6–2.8 from the equilibrium constants of ion-exchange reactions (Equation 2.9 and 2.10) requires a known value of any K of Equations 2.6–2.8. This value can be selected arbitrarily; the only consideration is that the known value must be so high that it expresses that every layer charge is neutralized. In the literature, K_{CaX2} is usually chosen to be 10^{20} (Stadler and Schindler 1993b), and the values of the stability constants of the neutralization reactions of the layer charges are calculated using this value. However, there are some fundamental problems with this approach: in Equations 2.6–2.8 concentration is used, while in Equations 2.9 and 2.10, activities are used. This problem has not been discussed in the literature so far, and the equations are used in such a form. The corresponding error is not significant if the potentiometric titration of montmorillonite is done in a relatively concentrated supporting electrolyte solution.

To precisely determine the zero point of charge of oxide surfaces (no permanent charge), the support electrolyte is usually applied in several concentrations. This adds another uncertainty to the studies on montmorillonites because cations of the support electrolyte participate in exchange reactions with the cations in the interlayer space, thus the application of several concentration effects the cation composition of the interlayer space. So, when studying the acid–base properties of the edge sites at one or several concentrations, the reactions of both types of charges (layer charge and edge) have to be taken into account.

2.4.3 ACID–BASE PROPERTIES OF CATION-EXCHANGED MONTMORILLONITES

The acid–base properties of calcium-, copper-, zinc-, and manganese-montmorillonite produced from calcium-montmorillonite were studied by potentiometric titration

(Nagy and Kónya 2004; Nagy et al. 2004). The neutralization reactions of Cu, Zn, and Mn ions can be described by Equations 2.6–2.8:

$$Cu^{2+} + 2X^- \rightleftarrows CuX_2 \quad K_{CuX_2} = \frac{[CuX_2]}{[Cu^{2+}][X^-]^2} \tag{2.13}$$

$$Zn^{2+} + 2X^- \rightleftarrows ZnX_2 \quad K_{ZnX_2} = \frac{[ZnX_2]}{[Zn^{2+}][X^-]^2} \tag{2.14}$$

$$Mn^{2+} + 2X^- \rightleftarrows MnX_2 \quad K_{MnX_2} = \frac{[MnX_2]}{[Mn^{2+}][X^-]^2} \tag{2.15}$$

The stability constants of the neutralization reactions can be calculated from the equilibrium constants of the cation-exchange reactions in the same way as in the case of sodium and hydrogen ions (Equations 2.9 and 2.10):

$$CaX_2 + Cu^{2+} \rightleftarrows CuX_2 + Ca^{2+} \quad K_{Ca,Cu} = \frac{a_{CuX_2} * a_{Ca^{2+}}}{a_{CaX_2} * a_{Cu^{2+}}} \approx \frac{K_{CuX_2}}{K_{CaX_2}} \tag{2.16}$$

$$CaX_2 + Zn^{2+} \rightleftarrows ZnX_2 + Ca^{2+} \quad K_{Ca,Zn} = \frac{a_{ZnX_2} * a_{Ca^{2+}}}{a_{CaX_2} * a_{Zn^{2+}}} \approx \frac{K_{ZnX_2}}{K_{CaX_2}} \tag{2.17}$$

$$CaX_2 + Mn^{2+} \rightleftarrows MnX_2 + Ca^{2+} \quad K_{Ca,Mn} = \frac{a_{MnX_2} * a_{Ca^{2+}}}{a_{CaX_2} * a_{Mn^{2+}}} \approx \frac{K_{MnX_2}}{K_{CaX_2}} \tag{2.18}$$

The equilibrium constants of the cation-exchange reactions were determined earlier by the radioisotopic tracer method (Table 2.3; Nagy and Kónya 1988; Kónya et al. 1988; Nagy et al. 1997) by the Gaines and Thomas method (Chapter 1, Section 1.3.4.2.1; Gaines and Thomas 1953; Howery and Thomas 1965), and the stability constants of the neutralization reactions was calculated from these values. The K values obtained from the equilibrium constants of the cation- exchange reactions can be used as constant starting parameters in the interpretation of titration curves during the study of the acid–base properties of the edge sites.

In the potentiometric titrations, a 0.1 mol/dm³ NaClO₄ solution was used as a supporting electrolyte. Because of the relatively high concentration of the support electrolyte, the constant capacitance model (Chapter 1, Section 1.3.2.1, Table 1.7) can be used in equations describing the formation of edge charges. It means that the surface potential (ψ) in Equations 2.3–2.5 was approximated according to Schindler et al. (Stadler and Schindler 1993b; Schindler and Gamsjager 1972):

$$\Psi = \frac{\sigma_0 F}{SmC} \tag{2.19}$$

where σ_0 is the surface charge on the edges, S is the specific external surface area (m^2/g), m is the concentration of montmorillonite (g/dm^3), and C denotes the double-layer capacitance. The surface charge σ_0 can be calculated from the potentiometric titration points (Davis et al. 1978) as follows:

$$\sigma_0 = \frac{F\left(c_{H^+} - \left[H^+\right]\right)}{S} \tag{2.20}$$

where F is the Faraday constant, c_{H^+} is the total added concentration of hydrogen ions during the potentiometric titration, and $[H^+]$ is the hydrogen ion concentration of the solution. $S = 93.5$ m^2/g (Table 2.1) and $C = 1.2$ C/Vm^2 were used as the starting values. In addition, the acid–base properties of a commercial montmorillonite, KSF montmorillonite, was also studied in the same way. The specific surface area of KSF montmorillonite was 10 m^2/g, and CEC was $3.04*10^{-4}$ mol/g for monovalent ions.

As a conclusion, the parameters of the protonation/deprotonation of edge sites, namely, the number of edge sites and intrinsic stability constants, can be estimated if the parameters of the cation-layer charge interactions (number of layer charge [cation-exchange capacity] and the specific surface area) are known from independent experimental data and they can be included into the equation describing the system as constant values.

A typical titration curve is shown in Figure 2.3 for copper-montmorillonite.

As seen in the figure, the layer and edge charges can be separated using the surface complexation model. The left-lower figure shows that, at pH>4, the layer charge is compensated by cations of the support electrolyte (sodium ions), so protons react only with the edge charges. At pH = 3, the interlayer charge is compensated by sodium and hydrogen ions with a ratio 1:1 at 0.1 mol/dm^3 sodium ion concentration. Silanol sites are deprotonated and neutral, and their ratio depends on pH. The number of deprotonated and neutral silanol sites is equal at about pH 4.5. Aluminol sites are protonated up to pH ~ 5.3; at higher pH values, the ratio of the neutral aluminol sites increases.

The results of potentiometric titration, the number of the edge sites, and intrinsic stability constants of the protonation and deprotonation reactions of calcium-, copper-, zinc-, manganese(II)-montmorillonites, and KSF montmorillonite are shown in Table 2.4. As a comparison, some similar data of other montmorillonites are also listed.

The data in Table 2.4 show that calcium-montmorillonite (Istenmezeje, HU) and the cation-exchanged montmorillonites produced from it usually show the same number of aluminol and silanol sites. The intrinsic stability constants of the edge charge reactions are usually the same, except for manganese-montmorillonite, where the absolute values of the intrinsic stability constants are slightly lower. The protonation of aluminol sites shows the greatest difference (about 1.5 in ln K $_{AlOH2+}$). It can be related to the redox reactions of manganese(II) ions in the interlayer space, which will be discussed in Section 2.10.1.

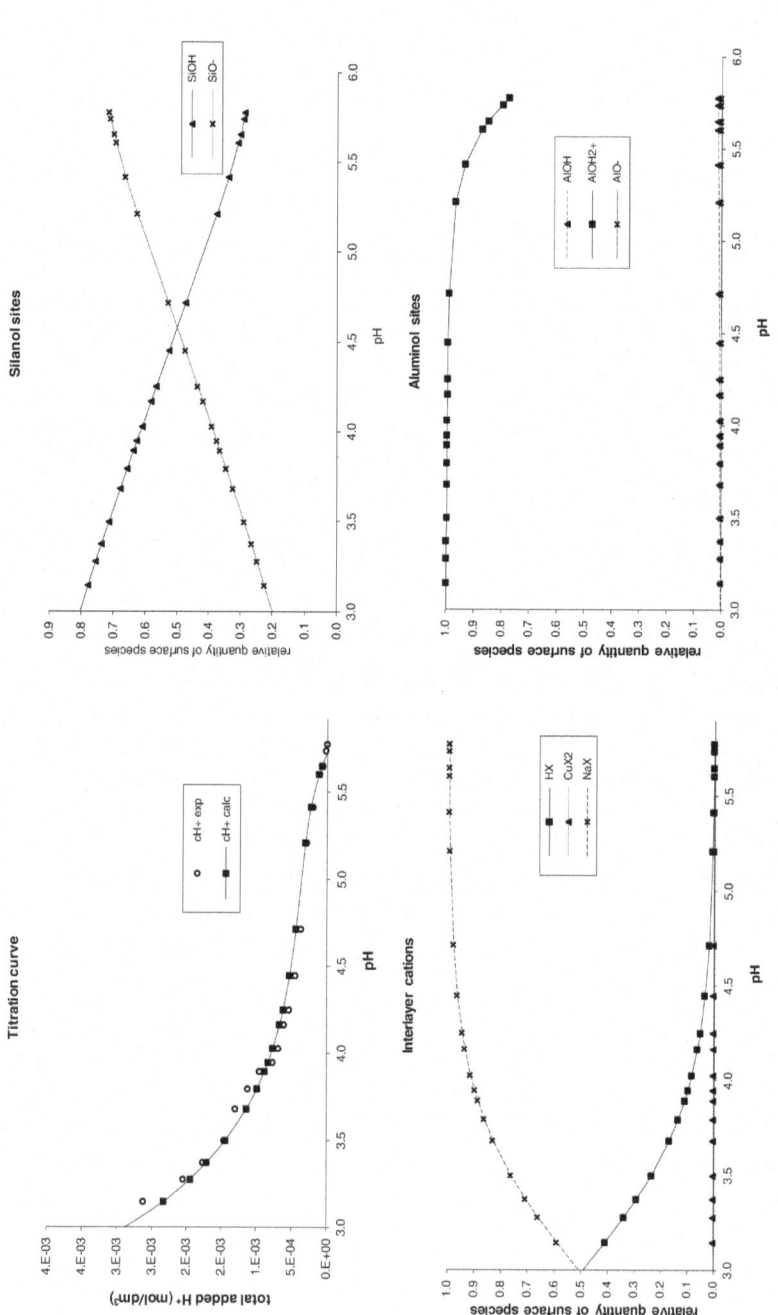

FIGURE 2.3   Potentiometric titration curve of copper-montmorillonite in 0.1 mol dm^{-3} NaClO$_4$ solution. $m = 50$ mg, $V = 20$ cm^3 (upper left). ◆s are the experimental points; line is the fitted curve by the surface complexation model. The concentration of surface sites-lower left: interlayer cations; upper right: silanol sites; lower right: aluminol sites (Nagy and Kónya 2004).

TABLE 2.4

The Concentration of Edge Sites and Intrinsic Stability Constants of Protonation and Deprotonation of Silanol and Aluminol Sites of Montmorillonite Samples Calculated by the Surface Complexation Model

Sample	AlOH mol/g	SiOH mol/g	lgK $AlOH_2^+$	lgK AlO^-	lgK SiO^-
Ca-mont.[a]	2.72E-04	2.32E-04	8.82	−7.97	−6.68
Cu-mont.[a]	2.72E-04	2.32E-04	8.82	−7.97	−6.68
Zn-mont.[a]	2.72E-04	2.32E-04	8.82	−7.97	−6.68
Mn-mont.[b]	2.72E-04	2.32E-04	7.37	−7.18	−5.86
Mont. KSF	$2.72*10^{-5}$	$2.34*10^{-5}$	8.82	−7.97	−6.68
SWy-1[c]	5.98E-05	3.5E-05	8.16	−8.71	−5.77
SWy-2[d]	3.68E-05		4.38	−5.26	
BSAB[d]	2.47E-04		3.70	−4.34	
MX-80[e]	2.8E-05		5.4	−6.7	
SWy-3[f*]	8.48E-5		5.79	−7.32	

[a] Nagy and Kónya 2004, [b] Nagy et al. 2004, [c] Stadler and Schindler 1993b, [d] Barbier et al. 2000, [e] Wanner et al. 1994, [f] Jeon and Nam 2019. Because of the error in a reaction equation in the paper, the data can be reversed.

For KSF montmorillonite, the number of silanol and aluminol sites was found to be less by an order of magnitude. It is in accordance with the ratios of the specific surface areas (10 m²/g for KSF montmorillonite, and 93.5 m²/g for montmorillonite [Istenmezeje]). This is an interesting observation because KSF montmorillonite is an acid-treated substance. Thus, it seems that acidic treatment causes the decrease of the layer charges (the CEC decreases; montmorillonite content of Ca-, Cu-, and Zn-montmorillonite is 91%, and that of KSF montmorillonite is 53%). The acidic treatment, however, does not change the nature of silanol and aluminol sites, the stability constants of the edge charge reactions remain the same, and the number of edge sites is proportional to the specific surface area.

The parameters obtained by others for SWy-2, BSAB, and MX-80 cannot be compared to the previously discussed data because the silanol and aluminol sites as well as the deprotonation processes (Equations 2.4–2.5) were treated together. Calcium-bentonite (Istenmezeje, HU) shows a similar intrinsic stability constant to SWy-1 bentonite, but the number of the edge site is different. Note, however, that the specific external surface areas are also very different: 21.4 m²/g for SWy-1, and 93.5 m²/g for Istenmezeje montmorillonite (Table 2.1). The ratio of the specific surface area (Istenmezeje/ SWy-1) is 4.4, the ratio of the total number of edge sites (silanol+aluminol) is 5.3, which are in a fairly good agreement, if the surface charge density is the same.

2.5   ION ADSORPTION ON THE EXTERNAL SURFACES

Similarly to other minerals, montmorillonite can also sorb ions on the external surfaces. This process, however, is masked by the cation exchange in the interlayer space because the number of layer charges is usually, by an order of magnitude, higher than the number of edge charges. The surface complexation models (Chapter 1, Table 1.7) take into account ion sorption on edge sites, called inner- and outer-sphere complexation, depending on whether the ions are sorbed without or with the hydrate shell. Inner-sphere complexation is usually described as cation-hydrogen or anion-hydroxide ion exchange on the surface sites, supposing that the ions are in the inner Helmholtz layer. Some models allow the sorption of ions of the supporting electrolyte as outer-sphere complexes, entering into the outer Helmholtz layer (Chapter 1, Table 1.7).

Cation sorption, therefore, can be considered in most cases as dominant process. Montmorillonite is an exception because cation exchange in the interlayer space is the dominant process. Ion sorption on edge sites can be observed only when the quantity of the sorbed cations is higher than the CEC. The excess quantity is frequently called *specific adsorption*, neglecting the real mechanism of sorption. Another possibility of observing the sorption on edge sites is when both directions of the ion-exchange process are studied, that is, the quantity of both cations entering and leaving the solid are simultaneously measured and compared. It gives information on the different processes even when the quantity of the entering ion is below the CEC. When a portion of the sorbed cations cannot be desorbed, that is, sorption is not reversible, the sorption on the edge sites should be assumed because it can indicate the formation of a chemical bond between the cations and oxygen, or the anion and silicon/aluminum atoms of the edge sites, respectively.

2.5.1 SOME EXAMPLES OF ION SORPTION PROCESSES ON MONTMORILLONITE

2.5.1.1 Sorption of Zinc Ion on Montmorillonite

These parallel ion sorption processes of montmorillonite will now be demonstrated through some examples. First, let us see the sorption of zinc ion on calcium-montmorillonite. The process can easily be examined by a dual radioisotopic tracer method, in which zinc ions and calcium ions are labeled by ^{65}Zn and ^{45}Ca isotopes, respectively (Kónya et al. 1988; Nagy and Kónya 1988). The radiation properties of these isotopes are different, so they can be measured separately. At the beginning of the interfacial processes, zinc ions are in solution, so they can be labeled by a solution containing ^{65}Zn ions. ^{45}Ca ions, however, must be added to the solid phase to form labeled calcium-montmorillonite (produced by heterogeneous isotope exchange; Section 2.3). The experimental results show that the quantity of zinc ions entering into the montmorillonite phase is greater than the quantity of calcium ions leaving the solid phase. Some data are shown in Table 2.5. The initial and equilibrium pH of the solution as well as the quantity of dissolved hydrogen ions are also listed here.

The data in Table 2.5 show that, besides Zn^{2+}/Ca-montmorillonite cation exchange, zinc ions are sorbed on montmorillonite through other than Zn^{2+}/Ca^{2+} cation-exchange processes. The ion exchange of zinc and hydrogen ions is of a small degree, the quantity of the dissolved hydrogen ions count for less than 10% of the difference between the quantity of zinc ions entering into the montmorillonite

TABLE 2.5

The Quantity of Zinc and Calcium Ions in the Process Zn^{2+}/Ca–Montmorillonite, and the Equilibrium pH of the Solution, the Difference between the Quantities of Zinc Entering and Calcium Ions Leaving the Montmorillonite Phase, the Quantity of Dissolved Hydrogen Ions

10^{5}*initial zinc ion concentration (mol/dm³)	10^{7}*quantity of entering zinc ions (mol)	10^{7}*quantity of leaving calcium ions (mol)	Equilibrium pH	10^{7}*$(Zn_{in}-Ca_{out})$	10^{7}*quantity of dissolved hydrogen ions
1.44	2.1	1.9	6.15	0.2	0.02
2.44	2.6	2.8	5.85		0.16
6.44	7.7	5.2	5.76	2.5	0.22
11.4	12.8	7.4	5.63	5.4	0.34
51.4	29	16	5.63	13	0.34
303	35	25	5.66	10	0.31
506	38	27	5.62	11	0.35
1000	58	29	5.65	29	0.32
3000	84	32	5.60	52	0.38
5000	140	28	5.58	112	0.40

Note: $m = 10$ mg Ca-montmorillonite, $V = 20$ cm³, $T = 20°C$, initial pH 6.2. The standard deviation of zinc cr calcium quantity is ±3% or 5%, respectively.

phase and the quantity of calcium ions leaving the solid phase. The remaining (90%) sorbed zinc ions get into the solid as a result of another process. There are the following possibilities:

The exchange of zinc and hydrogen ions: the data in Table 2.5 show a low contribution.

The exchange of monovalent complexes of zinc ions: it can be excluded by thermodynamic calculations. At pH<6 an insignificant portion (<0.1) of zinc ion is present as $ZnOH^+$ complexes (Pourbaix 1966). Neither can other anionic complexes be present in the applied concentration ranges (Horne et al. 1957).

The formation of zinc polycations: no polycations are present in the solution (Matijevic et al. 1962).

Surface precipitation of zinc ions (Shukla and Mittal 1980; Siskens et al. 1975).

Adsorption process: both surface precipitation and adsorption can be described by the adsorption equations (Delgado et al. 1986).

As seen earlier, zinc ions can be a part of different interfacial processes on calcium-montmorillonite. The ion exchange in the interlayer space and the other sorption on the external surfaces can be separated by measuring the ions entering and leaving the montmorillonite phase. The quantitative separation of sorption isotherms will be discussed in Section 2.6.

2.5.1.2   Sorption of Manganese Ion on Montmorillonite

Similar experiments with dual radioisotopic labeling show that there is a small manganese(II) ion excess in the manganese(II) ion/calcium-montmorillonite (labeled with ^{54}Mn and ^{45}Ca isotopes) interfacial reaction. It means that there is an adsorption reaction besides ion exchange, but it has a very low contribution. For this reason, the presence of adsorption can be observed, but its quantitative treatment is difficult.

2.5.1.3   Sorption of Palladium Ion on Montmorillonite

The other example on the ion sorption on the edge sites is the sorption of palladium ions (Nagy and Kónya 2007) on montmorillonite. As known, palladium ions are hydrolyzed at such a low pH value where the lattice of clay minerals is destroyed. For this reason, during the preparation of palladium-montmorillonite, we have to avoid the hydrolysis of Pd(II) and the acidic destruction of clay at the same time. We can do so by applying Pd(II) as positively charged complexes at neutral pH. For example, $[Pd(NH_3)_2]^{2+}$ complexes can be prepared from metallic palladium as follows:

$$Pd(s) \xrightarrow{\text{aqua regia}} H_2[PdCl_6](aq) \xrightarrow{H_2O}$$

$$H_2[PdCl_4](aq) \xrightarrow{\text{filtration}} \xrightarrow{\text{evaporation}} \xrightarrow{\text{cc.NH}_3 \text{solution}}$$

$$[Pd(NH_3)_4]^{2+}(aq) \xrightarrow{H_2O} \xrightarrow{\text{pH adjustment}}$$

$$[Pd(NH_3)_4]^{2+} \text{solution}, pH = 8$$

(2.21)

Similarly, the ethylene diamin complex of palladium ($[Pd(en)_2]^{2+}$) can be prepared. These positively charged complexes can be used for the study of interfacial reaction with montmorillonite.

The interfacial reaction of palladium ions and montmorillonite can be studied very well by using sodium-montmorillonite, because monovalent (sodium)-bivalent (palladium) cation exchange can easily be followed by X-ray diffraction, which indicates the presence of palladium ions in the interlayer space (Table 2.3). When bivalent cations are in the interlayer space, the basal spacing is about 1.5 nm, while this value for monovalent montmorillonites is about 1.25 nm. X-ray diffraction studies, however, show that the basal spacing of palladium-montmorillonite, obtained with the reaction of $[Pd(NH_3)_2]^{2+}$, is 1.20 nm, and that of the original sodium-montmorillonite is 1.24 nm. It means that the basal spacing slightly decreases during the sorption of palladium(II), so palladium(II) ions hardly ever enter into the interlayer space. Therefore, we can conclude that most Pd(II) ions are sorbed on the edge sites. At pH = 8 of the experiments, the silanol sites are totally deprotonated and so can sorb the positively charged palladium ions. In addition, chemical properties can also lead to the sorption on edge sites, that is, the high affinity of palladium to oxygen atoms (a reason of the hydrolysis of palladium ions are at low pH).

It should be also noted that a high excess of ammonium ions will interfere with the experiments using $[Pd(NH_3)_2]^{2+}$. (Ammonium ions are added to solution during the preparation of the complex.) Since ammonium ions are also positively charged, they can also participate in exchange reactions with the sodium ions in the interlayer space. By comparing the concentrations of the sorbed and dissolved cations (sodium, ammonium, and palladium ions), it reveals that a ternary ion-exchange reaction does indeed take place. So, ammonium ions may inhibit the exchange of palladium-sodium ions in the interlayer space.

Furthermore, the sorption isotherm does not show the usual tendencies seen in other cation exchange reactions: the sorbed quantity versus the equilibrium concentration function has an increasing slope instead of the usual asymptotic curve. In Figure 2.4, the quantity of the dissolved sodium ion is plotted as a function of the sorbed quantity of palladium(II) ion. The slope of the straight line is 4.8. In the case of binary Pd(II)–Na(I) exchange, this value would be 2. Since the slope is much higher than 2, that is, the quantity of the dissolved sodium ion is greater than the sorbed quantity of palladium(II), ammonium-sodium ion exchange has to be supposed.

To eliminate this effect of the ammonium ions, $[Pd(en)_2]^{2+}$ can be used. The results again show a basal spacing is 1.200 nm, characteristic of monovalent ions. Therefore, we can conclude that a significant portion of palladium ions are sorbed on the edge sites.

The results of thermal analytical studies of Pd-montmorillonite produced from $[Pd(NH_3)_2]^{2+}$ and $[Pd(en)_2]^{2+}$ further demonstrate that palladium ions are sorbed on the edge sites. Pd-montmorillonite shows the usual thermal processes of montmorillonites, with two new exothermic reactions at 338 and 548°C. At around 338°C (maximum) exothermic reaction a slight weight loss of the sample is also observed

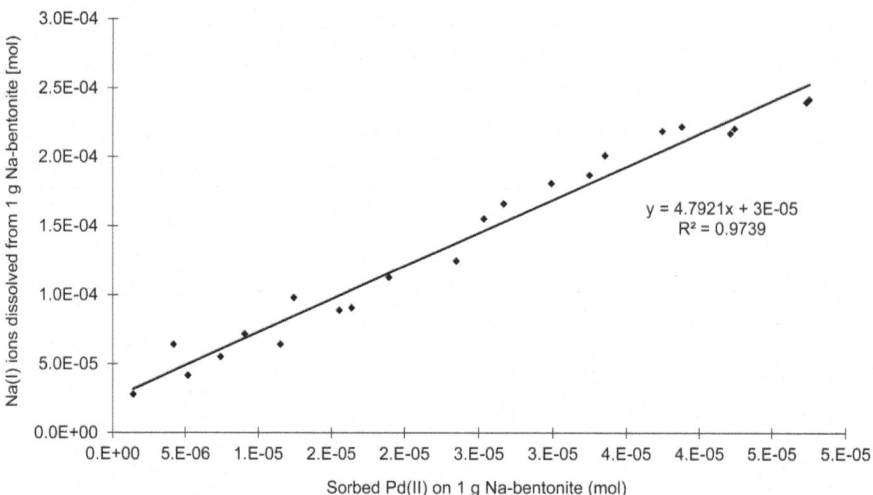

FIGURE 2.4 The dissolved quantity of sodium ion as a function of the sorbed palladium(II) ion. Complex: $[Pd(NH_3)_2]^{2+}$.

in a wide temperature range. It may indicate the partial amorphization of bentonite. At 548°C, the reduction of Pd(II) to Pd(0) may happen.

2.5.1.4 Sorption of Lead Ion on Montmorillonite

Finally, the sorption of lead ions on calcium-montmorillonite is described (Nagy and Kónya 1998; Nagy et al. 1998b, 2001, 2003a). Lead was found to react with the Ca-montmorillonite in two different ways. The first is by the expected mechanism of cation exchange with the calcium in the interlayer space. The second way is the formation of lead enriched area over the surface of the clay particles. The initial step of their formation is the adsorption of lead ions on the edge sites (formation of lead-oxygen bonds) that can act as the initial of a heterogeneous nucleation on particle surface, followed by crystal growth. This process will be discussed in detail in Section 2.10.4.

The two different species of the sorbed lead ions showed different desorption properties. The desorption behavior in solutions of complex forming agents will be discussed in Section 2.8.4.

2.6 SEPARATION OF INTERFACIAL PROCESSES OF MONTMORILLONITE

When a cation (Me_1) can be sorbed on montmorillonite by two processes, ion exchange and adsorption, the total sorbed amount of the cation (a_1) is the sum of the amounts sorbed by ion exchange ($a_{1\,ie}$) and adsorption (a_{1ads}) (Chapter 1, Equation 1.104). The two processes can be separated through studying the ions entering and leaving the solid phase simultaneously. We see in Table 2.6 that there is a significant difference between the sorbed zinc ions and the dissolved calcium ions in the zinc

TABLE 2.6

The Estimated Isotherm Parameters of Zinc Ion/Calcium-Montmorillonite Interfacial Reactions (pH 5.8)

	0°C	20°C	40°C	55°C
Composite isotherm (Chapter 1, Equation 1.129)				
K_1/K_2 ion exchange		0.58		
ζ (mol/g)		2.1e-4		
K_1 (mol/dm^3) adsorption		1e-4		
K_1/K_2 adsorption		1000		
z (mol/g)		4.6e-3		
Ion exchange (Chapter 1, Equation 1.94)				
K_1/K_2 ion exchange	1.6	1.2	1.5	1.3
ζ (mol/g)	2.2e-4	2.8e-4	3e-4	3.9e-4
Adsorption (Chapter 1, Equation 1.73)				
K_1 (mol/dm^3) adsorption		1e-4		
K_1/K_2 adsorption		590		
z (mol/g)		2.3e-3		

ions/calcium-montmorillonite system. When an equivalent ion exchange of zinc and calcium ions is supposed, the quantity of the dissolved calcium ions is considered to be equal to the quantity of zinc ions sorbed by ions exchange (a_{1ie}), and the difference between the quantities of zinc ions and calcium ions is treated as the quantity of the adsorbed zinc ions (a_{1ads}). In this way, the ion-exchange and adsorption processes are quantitatively separated.

It is important to note that this separation, or even the comparison of quantities, must be expressed in mol or in equivalents. It is usual in the literature that the mass of the adsorbed, exchanged, or dissolved substances are expressed in grams (e.g., Bhattacharyya and Gupta 2008; Stuckey et al. 2008; Green-Ruiz 2009). Since the molar mass of the ions is different, the quantities in mass are not suitable for the comparison of quantities.

The simultaneous ion-exchange and adsorption process are characteristic of the zinc ion/calcium-montmorillonite system; they can be observed at different temperatures and zinc ion concentrations in the excess of calcium ions in solution. When the sorbed quantity of zinc ions is separated experimentally, ion-exchange and adsorption isotherms can be constructed. Ion- exchange isotherms can be analyzed using Equation 1.94 (Chapter 1). The interpretation of the adsorption isotherm generally depends on whether the adsorption process is competitive or not. In the case of zinc ions/calcium-montmorillonite interfacial reaction, assuming that calcium ions do not take part in the adsorption process and so zinc ions occupy the "free" edge sites (e.g., deprotonated silanol or aluminol sites), the simple adsorption isotherm (Chapter 1, Equation 1.65) can be satisfactorily used. Theoretically, the simple adsorption isotherm can only be used in surface precipitation. If there was an exchange of zinc and hydrogen ions of the edge site (which is also an ion-exchange process), Equation 1.94 (ion-exchange isotherm equation) should be used. However, it was proved by pH measurements that this is not the case. In addition, water molecules on the external surfaces also lead to competition. When two ions can be adsorbed, Equation 1.73 (Chapter 1) of competitive adsorption is applied. However, in all cases, the same function (c/a versus c) is plotted; only the interpretation of K_1 and K_1/K_2 is different.

In Figure 2.5, three isotherms of zinc sorption are shown. The upper figure illustrates a so-called sorption isotherm obtained from total sorbed amount of zinc ion (a_1) as a function of the zinc concentration of the solution (c_1); the lower figure shows the ion exchange and sorption isotherms obtained after the separation of zinc ion quantities sorbed by ion exchange (a_{1ie}) and adsorption (a_{1ads}).

The composite isotherm contains the total sorbed (ion-exchanged and adsorbed) quantities of zinc ions (a_1). The isotherm can be interpreted by the Equation 1.119 (Chapter 1), with that uncertainty that it is not known whether the adsorption process is competitive or not, and, in the case of competitive adsorption, what the competing substance is. The estimated isotherm parameters, when assuming competitive adsorption with calcium ions, are shown in Table 2.6. In addition, Table 2.6 lists the isotherm parameters of zinc-calcium ion exchange and zinc ion adsorption obtained from the ion-exchanged (a_{1ie}) and adsorbed (a_{1ads}) quantities of zinc ions after the experimental separation of the ion-exchange and adsorption processes by dual radioisotopic labeling (Table 2.5). The isotherm parameters were estimated by Equation 1.94 for ion exchange and Equation 1.73 for adsorption.

As shown in Table 2.6, the estimated isotherm parameters obtained from the composite isotherm, separated ion exchange, and adsorption isotherms are in fairly good agreement, especially if we take into account the uncertainty of the adsorption process mentioned above. The number of exchange sites (ζ) is lower than the CEC, but it increases as temperature increases. The average value of K_1/K_2 obtained at different temperatures is 1.4; the reciprocal of this value is 0.7, which is in very good agreement with the equilibrium constant determined by the integration of the selectivity coefficient vs. surface molar or equivalent fraction (Equation 1.80; Table 2.2). The adsorption parameters indicate that the number of adsorption sites (z) is very large; it is an order of magnitude higher than the number of exchange sites. This could indicate surface precipitation. The value of K_1/K_2 for adsorption is also very high, showing that $K_1 > K_2$, that is, the competing substance (indexed by 2), is more strongly adsorbed than zinc ions ($K_2 < 10^{-6}$ mol/dm^3). Therefore, the competitive sorption model has to be used. It is worth mentioning here that calculation with the simple Langmuir isotherm for the adsorption (Chapter 1, Equation 1.65) or in the second terms of Equation 1.119 provides no adequate parameters, unreal or negative results are obtained.

2.7 ROLE OF HYDROGEN IONS IN THE INTERFACIAL AND DISSOLUTION PROCESSES OF MONTMORILLONITE

As usual in chemistry, hydrogen (and hydroxide) ions play a significant role in geochemical processes, including the solution, solid, and interfacial processes. As discussed in Chapter 1, Sections 1.2.2 and 1.2.3, hydrogen ion concentration or activity,

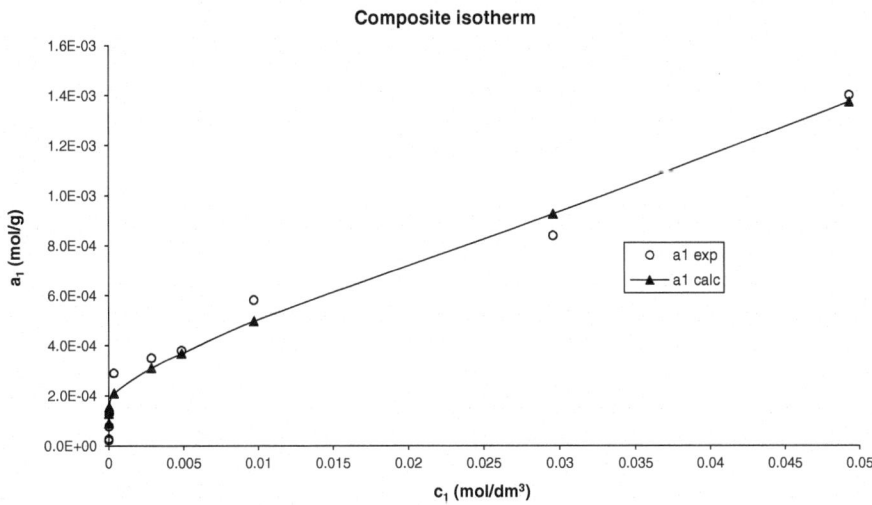

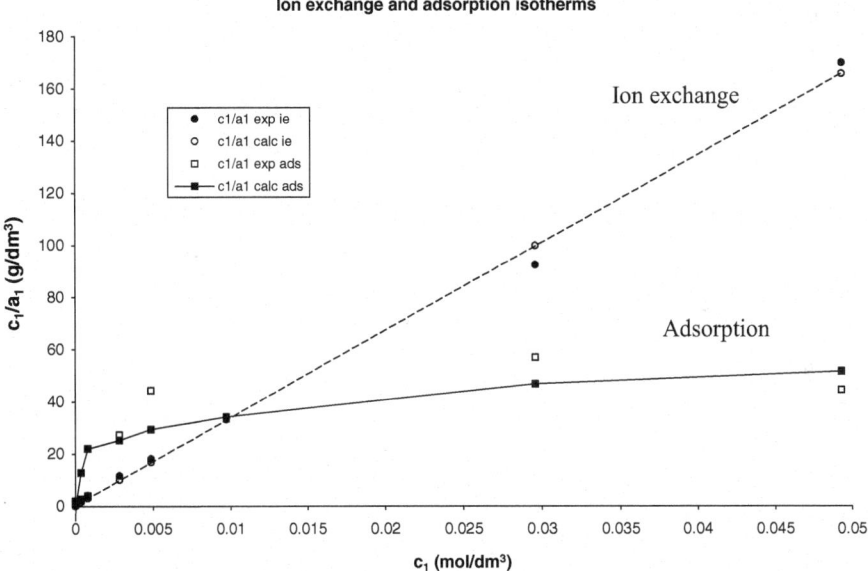

FIGURE 2.5   Composite (upper), ion exchange, and adsorption isotherms (lower) of zinc ion/calcium-montmorillonite interfacial reaction. $T = 20°C$, pH 5.8.

that is pH, of the solution influences the thermodynamically stable chemical species of dissolved substances and directs the dissolution/precipitation processes. The interfacial processes involving hydrogen ions also depend on hydrogen ion concentration. The scale of reactions with hydrogen ions is very broad: at low hydrogen ion concentrations, the surface protolytic reactions are dominant; in a strongly acidic medium, even the destruction of the crystal lattice can occur.

In the case of montmorillonite, the effect of hydrogen ions can be observed in different reactions. The silanol and aluminol sites on the external surfaces can be protonated and deprotonated, depending on pH, which means that the hydrogen ion is the potential-determining ion. This process was discussed in detail in Section 2.4 for different cation-exchanged montmorillonites. It is usual in the literature that the zero point of charge (*pzc*, meaning the pH value where the number of the positive and negative surface sites is the same; Equation 1.57) of silanol and aluminol sites are considered an important parameter of the external surface. For silanol and aluminol sites, *pzc* = 2–4 or 7–9, respectively, are given (e.g., Parks and de Bruyn 1962; Parks 1965; Schwarz et al. 2000; Hlavay and Polyák 2005; Robinson et al. 2006). Montmorillonite, however, has two types of edge sites, so the net zero point of charge of external surfaces is the difference between the total number of negative and positive charges. Since the permanent negative layer charge of montmorillonite due to isomorphic substitutions is always much larger than the external surface charge, the net surface (including external and internal) charge is always negative.

2.7.1 Effect of Hydrogen Ion on the Cation Exchange Processes

The other interfacial process involving hydrogen ion is the cation exchange process in the interlayer space. When montmorillonite is suspended in water or in an electrolyte solution, a part of exchangeable cations can be dissolved. In Table 2.7, the relative quantity of calcium ions dissolved in water or in acidic solutions is shown.

As seen in Table 2.7, the lower the pH is, the greater the quantity of calcium ions dissolved is. The dissolved calcium ions are substituted by hydrogen ions in the interlayer space. In other words, the cations in the interlayer space of montmorillonite react with water itself, and a hydrogen-calcium cation-exchange reaction takes place:

$$\text{Ca-mont} + 2H_2O = 2H\text{-mont} + Ca^{2+} + 2OH^- \tag{2.22}$$

where Ca^{2+} means the hydrated cation, or in other words, calcium aqua complex.

Equation 2.22 is a heterovalent ion exchange process (Chapter 1, Section 1.3.4.2.3); bivalent calcium ions are exchanged for monovalent hydrogen ions. For the distribution of monovalent hydrogen ions, the c_1/a_1 versus c_1 function can be plotted (Figure 2.6).

The function seems to be linear in the visible hydrogen ion concentration range, so the number of exchange sites and the isotherm constant could be determined from the slope and the intercept. However, it was shown in Chapter 1, Section 1.3.4.2.3 (Chapter 1, Equation 1.120 for the monovalent-bivalent cation exchange) that the "intercept" is a complex quantity including many parameters, namely, the concentrations of both ions in the solution and the sorbed quantity of the exchanging ion, and the selectivity coefficient of the ion exchange. Therefore, the treatment of the c_1/a_1 versus c_1 function as an adsorption isotherm is not possible. The function should be evaluated by a parameter-estimating program in its original form of Chapter 1, Equation 1.120. Using such a program raises other problems, however, because of the complexity of the function; also, in the absence of fairly accurate initial parameters

TABLE 2.7
The Quantities of Calcium, Silicon, Aluminum, and Magnesium Dissolved (y) into the Solution from Montmorillonite as Expressed as the Percentages of the Total Calcium, Silicon, Aluminum, and Magnesium Content of Montmorillonite at Different Equilibrium pH values (m = 10 mg Ca-montmorillonite, V = 20 cm³ HClO$_4$ solution with different pH, T = 20°C, t = 1 h)

pH	Dissolved Ca (%)	Dissolved Si (%)	Dissolved Al (%)	Dissolved Mg (%)
7.85	6.3	0.08	0.02	0.04
7.8	7.1	0.08	0.01	0.09
7.6	13.7	0.15	0.03	0.13
7	19			
6	24.5			
5	28.4			
4.35	27.6	0.13	0.007	0.18
4	31.8			
3.35	34.7	0.21	0.03	0.24
3.15	39.8	0.19	0.04	0.3
3.05	78.7	0.59	0.54	0.54
2.8	65.9	0.48	0.25	0.48
2.55	75.0	0.47	0.37	0.48

Source: Reprinted from Nagy et al. 1997, with permission from Elsevier.

for $K_{Me1, Me2}$, and ζ, the plotted parameters will be false. To resolve this, considering that the c_1/a_1 versus c_1 function is linear, a common (all-inclusive) parameter can be deduced as an approximation, and the real parameters can then be estimated by using Equation 2.23 (or Chapter 1, Equation 1.120):

$$\frac{c_{Me_1}}{a_{Me_1}} = \frac{1}{\zeta}\left(c_{Me_1} + C\right) \qquad (2.23)$$

where

$$C = \frac{2a_{Me_1}}{K_{Me_2,Me_1}}\frac{c_{Me_2}}{c_{Me_1}} \qquad (2.24)$$

From the parameter estimation, C can easily be determined, (note, however, that this value has an uncertainty due to the linear curve fitting). Then, substituting the concentrations (a_{Me_1} and c's) into Equation 2.24, the selectivity coefficients can be determined. The selectivity coefficients determined this way can be plotted as a function of surface equivalent fraction of any ions (e.g., hydrogen ion). Similarly, the

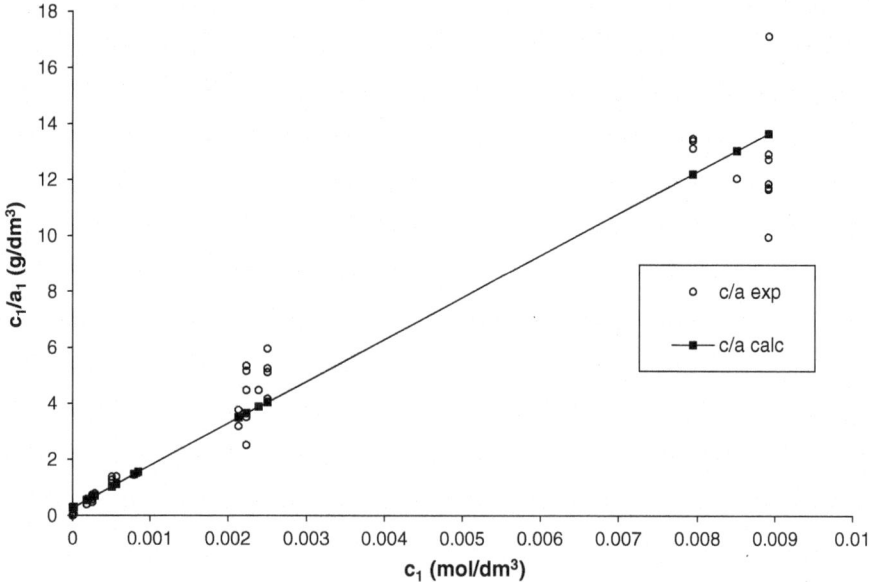

FIGURE 2.6 Hydrogen ion-exchange isotherm in Ca-montmorillonite/hydrogen ion exchange. $T = 20°C$.

selectivity coefficients can directly be calculated from the experimental data, and the values can also be determined as a function of surface equivalent fraction of the ions. Thus, two selectivity functions can be obtained (Figure 2.7).

As seen in Figure 2.7, the selectivity functions obtained by two methods show some differences because of the approximation expressed by Equation 2.23 and 2.24, but the equilibrium constants obtained by the integration of the selectivity function (Equation 1.81) are in a fairly good agreement: 1.6e4 is obtained from the direct experimental data (in the $\ln K_{sel} - X_H$ curve, the solid line in Figure 2.9; Nagy et al. 1997), and 1.8e4 is given from the data calculated from the c_1/a_1 versus c_1 curve (the $\ln K - X_H$ curve, the dotted line in Figure 2.9). Consequently, the two procedures give similar results, proving the equivalency of the two approaches. It is important to note, however, that, in heterovalent ion exchange processes, the correct isotherms are more complicated, the parameters can hardly be estimated; the application of the law of mass action is a more reliable method.

The equilibrium constant of Ca-montmorillonite/hydrogen cation exchange is very high; it is more than 10^4. It is much higher than the typical equilibrium constant of metal-metal cation exchange on montmorillonite (Table 2.2), which is about from 0.1 to 2. It shows that hydrogen ions are very strongly sorbed in the interlayer space of montmorillonite and can hardly be exchanged for metal ions. This has some ramification with regards to the nutrient cycle in soils since in acidic soil the layer charge neutralized by hydrogen ions practically does not take part in cation exchange.

Another consequence of the strong sorption of hydrogen ions in the interlayer space of montmorillonite is that, in diluted solutions (such as in the cobalt ion/

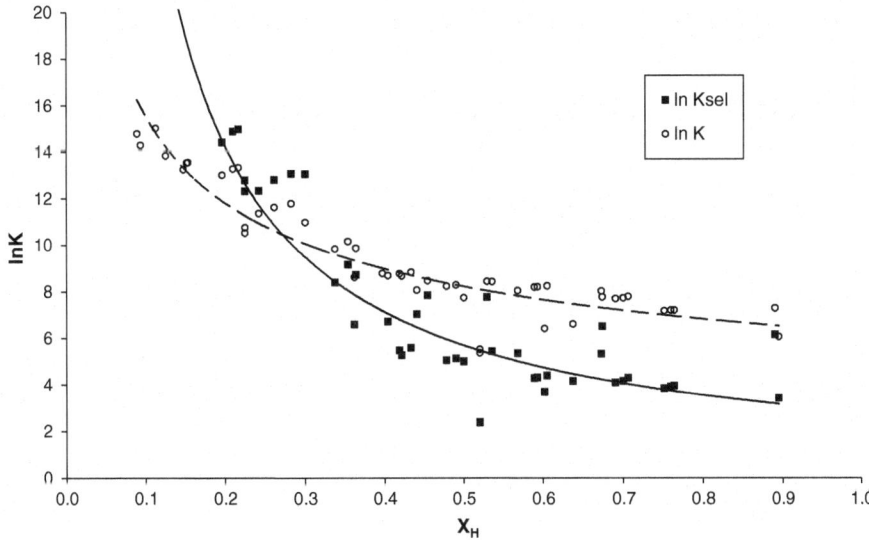

FIGURE 2.7 The selectivity coefficients of hydrogen ion/calcium-montmorillonite cation exchange calculated from the experimental data and on the basis of Equation 2.24.

calcium-montmorillonite cation exchange process [Nagy et al. 1997]), the cation-exchange processes of montmorillonite have to be treated as the exchange of metal ions and hydrogen ions; that is a ternary ion-exchange has to be considered. This process has already been discussed in Section 2.2, where the ion- exchange isotherm has been analyzed (Figure 2.2). It has also been mentioned that pure cobalt-calcium ion exchange takes place when the cobalt ion concentration is not too high (<6e-4 mol/dm³ under the applied experimental conditions). The number of exchange sites has been found to be less than the CEC because the layer charges neutralized by hydrogen ions do not take part in the cation-exchange process.

As seen in Table 2.8, at low initial cobalt concentrations, only calcium-cobalt ion exchange takes place; the equivalent fraction of hydrogen ion remains approximately the same. Cobalt-hydrogen exchange starts only when the equilibrium fraction of calcium is decreased below 0.3 due to the preference of hydrogen ions for the layer charges of montmorillonite. The same conclusion can be drawn from the selectivity function of cobalt-calcium cation exchange (Figure 2.8).

The selectivity coefficient versus surface fraction curve has a minimum at $X_{Ca} =$ 0.3, the same value at which hydrogen-cobalt ion exchange (Table 2.8) starts to take place. The minimum is probably related to the transformation of calcium-cobalt ion exchange to hydrogen-cobalt ion exchange. A similar minimum value on the selectivity function of cobalt/sodium-montmorillonite ion exchange has been found by Singhal and Singh (1973).

Under these experimental conditions, the equilibrium constant of cobalt-calcium ion exchange in montmorillonite cannot precisely be determined for two reasons. Firstly, for the determination of the equilibrium constant, the integration

TABLE 2.8

Equivalent Fractions of Cobalt, Calcium, and Hydrogen Ions on Montmorillonite at Different Initial Cobalt Concentrations

$10^4\ c_0\ (Co^{2+})$	X_{Co}	X_{Ca}	X_H
10^{-5}	0	0.81	0.19
0.13	0.04	0.76	0.20
0.26	0.08	0.66	0.26
0.32	0.10	0.66	0.24
0.65	0.19	0.58	0.23
0.98	0.26	0.51	0.23
1.6	0.36	0.40	0.24
1.7	0.39	0.40	0.21
2.3	0.40	0.43	0.17
2.6	0.46	0.38	0.16
3.2	0.44	0.30	0.26
3.9	0.59	0.30	0.11
4.6	0.64	0.33	0.03
5.2	0.65	0.13	0.22
5.9	0.68	0.18	0.14
6.5	0.70	0.23	0.07
7.8	0.74	0.25	0.01

Note: $m = 10$ mg Ca-mont; $V = 20$ cm³, pH 6.5; $T = 33°C$).

Source: Reprinted from Nagy et al. 1997, with permission from Elsevier.

has to be in the whole range of X_{Ca}. It is seen in Figure 2.8 (and it is in good agreement with the data in Table 2.7) that there is a surface concentration range ($X_{Ca} = 0.8–1$ at pH 6.5) where no experimental points can be found because, at this pH value, about 25% of the montmorillonite internal surface is covered by hydrogen ions. Second, the equivalent fraction of hydrogen ions also changes with increasing cobalt concentration, that is, ionic strength. To avoid the second problem, is to maintain a constant ionic strength, and for this, only the ratio of the two ions studied is varied. In this case, however, the presence of hydrogen ions (or another cation, if any) on the surface is neglected. So, the equilibrium constants obtained have some uncertainties.

In surface chemistry studies, when integrating the selectivity coefficient versus surface concentration function, the integration constant is determined using a so-called standard state (Boyd et al. 1947; Gaines and Thomas 1953). The standard reference state is when the homoionic form of the ion exchanger is in equilibrium with an infinitely diluted solution containing the same ion. The preceding discussion implies that, in the case of montmorillonite as an ion exchanger, a standard state does not exist because, when the solution is dilute, the surface is not homoionic (mixed calcium-hydrogen form); when the surface is converted to homoionic by the addition

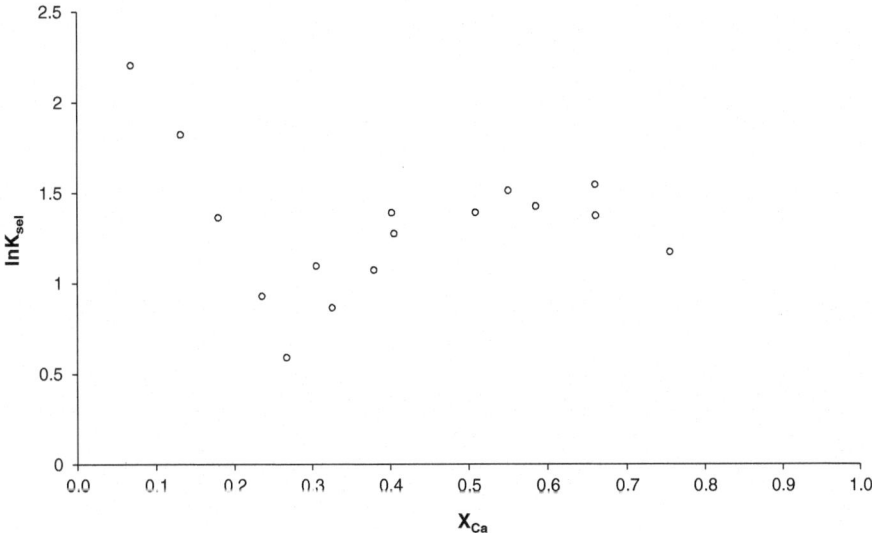

FIGURE 2.8 Logarithm of selectivity coefficient for cobalt-calcium ion exchange on mont-morillonite as a function of equivalent surface fraction of calcium ion (m = 10 mg Ca-mont, V = 20 cm³, T = 33°C; pH 6.5). (Reprinted from Nagy et al. 1997, with permission from Elsevier.)

of calcium ions, the solution is not dilute. Thus, two conditions can never be fulfilled because of the strong preference of montmorillonites for hydrogen ions.

When the calcium ions are exchanged for trivalent lanthanide cations, the pH practically has no effect on the process. This is due to the strong interaction between the trivalent ion and the layer charge of montmorillonite (Section 2.10.5; Kovács et al. 2019).

2.7.2 ACIDIC DESTRUCTION OF MONTMORILLONITE

When the pH of the solution is in the strongly acidic range, that is, the hydrogen ion concentration of the solution increases, the crystal lattice of montmorillonite starts to disintegrate, and the cations dissolved from the crystal lattice (magnesium, iron, and aluminum ions) may enter into the interlayer space and neutralize the negative layer charge. In Table 2.7, the relative quantities of dissolved silicon, aluminum, and magnesium at different pH values are listed. The relative quantity of the dissolved elements is not too high; dissolution time, however, is very short (only 1 h) for the significant decomposition of the crystal lattice. This time is suitable for the equili-bration of the ion exchange in the interlayer space. However, under real geological conditions, for example, in an acidic soil, the time of destruction of the crystal lattice may be much longer, so the quantity of the dissolved species can be much higher.

Obviously, the dissolution of the elements of the crystal lattice leads to change in the crystal lattice and the mineral composition. This can well be seen during the acidic treatment of montmorillonite or bentonite for catalytic purposes (Section 2.1). The treatment is done using concentrated hydrochloric, sulfuric, or phosphoric acid. X-ray diffraction studies show that a commercially available montmorillonite has low montmorillonite content (53%). The other constituents are illite 10%, kaolinite 6%, quartz 10%, plagioclase 5%, gypsum 1%, anhydrite 4%, and amorphous 7%.

The acidic destruction of montmorillonite results in the release of silicon and aluminum. The initial fast exchange of surface cations by hydrogen ions is followed by the release of aluminum and silicon. The dissolution rate of Si is higher than that of Al and is influenced by the relative ratios of basal siloxane and edge surfaces. The shift of pH to more basic values by the ion exchange processes and the hydrolysis of dissolved species induce the formation of secondary amorphous solids, initiating the formation of amorphous aluminosilicates (Sondi et al. 2008). At the same time, the significant portion of the adsorption properties of bentonite remains even after prolonged acidic treatment at elevated temperature (Krupskaya et al. 2017).

2.8 EFFECT OF COMPLEXATION AGENTS

As seen in Chapter 1, Section 1.2.2, complex-forming agents can greatly influence the quantity and ratio of different species of cations in soil solution and groundwater. So, the role of complex forming agents in the interfacial reactions between soils/rocks and the dissolved metal ions is very important. The study of the effect of complex-forming agents on the processes in soils can help us, for example, with practical problems such as washing out polluting metal ions or radioactive fallouts from soil. As a first step, it is important to build a model system and study it. The basic principles for the interactions of cation exchange and complex formation are summarized in Chapter 1, Section 1.2: the anionic and neutral complexes are considered not to take part in cation-exchange reactions.

The effect of a complex-forming agents on the cation-exchange processes of montmorillonite is well demonstrated in calcium-montmorillonite, manganese(II) ion, and sodium salt of a EDTA system (Kónya and Nagy 1998; Kónya et al. 1998). The reactions are illustrated in Figure 2.9.

The solid phase in the figure (Boxes 1 and 3) is relatively simple because, as will be discussed later, EDTA does not enter into the interlayer space of montmorillonite to a significant degree, and so the different EDTA species have to be taken into account only in the solution phases (Boxes 3 and 4).

Iron(III) ions also appears in Figure 2.9. Its source is the montmorillonite (Box 4), the crystal lattice of which is destroyed under the influence of acidic media and complex-forming agent (Section 2.7.2). It will be discussed in detail later (Section 2.8.1).

The system seems at first to be very complicated: all the equilibria possible in the solution can influence the resultant state. These equilibria, however, can be grouped as follows.

Equations 2.25–2.34 represent the reactions between solid and liquid (Boxes 3 and 4). Among them, Equations 2.25–2.28 represent the dynamic heterogeneous

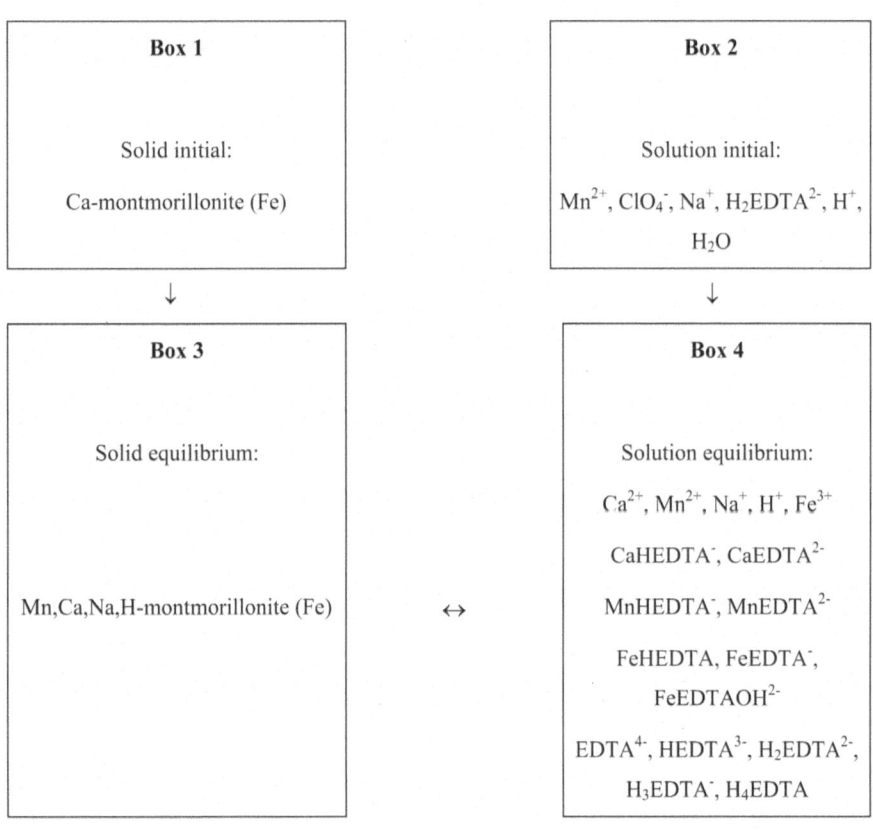

FIGURE 2.9 The scheme of the equilibria in the Ca-montmorillonite-Mn(ClO$_4$)$_2$-Na$_2$EDTA system. (Reprinted from Kónya and Nagy 1998, with permission from Elsevier.)

exchange of ions between montmorillonite and solution; if the ions are labeled by radioactive isotopes, they also include heterogeneous isotope exchange:

$$\text{Ca-mont} + {}^{45}\text{Ca}^{2+} \rightleftarrows {}^{45}\text{Ca-mont} + \text{Ca}^{2+} \tag{2.25}$$

$$\text{Mn-mont} + {}^{54}\text{Mn}^{2+} \rightleftarrows {}^{54}\text{Mn-mont} + \text{Mn}^{2+} \tag{2.26}$$

$$\text{Na-mont} + {}^{22}\text{Na}^{+} \rightleftarrows {}^{22}\text{Na}^{+}\text{-mont} + \text{Na}^{+} \tag{2.27}$$

The exchange of hydrogen ions between the solution and montmorillonite is also possible (Section 2.7.1):

$$\text{H-mont} + \text{H}^{+} \rightleftarrows \text{H-mont} + \text{H}^{+} \tag{2.28}$$

The equilibria 2.29–2.34 are ion-exchange reactions between the surface of montmorillonite and the solution:

$$Ca\text{-mont} + Mn^{2+} \rightleftarrows Mn\text{-mont} + Ca^{2+} \tag{2.29}$$

$$Ca\text{-mont} + 2Na^+ \rightleftarrows 2Na\text{-mont} + Ca^{2+} \tag{2.30}$$

$$Ca\text{-mont} + 2H^+ \rightleftarrows 2H\text{-mont} + Ca^2 \tag{2.31}$$

$$Mn\text{-mont} + 2Na^+ \rightleftarrows 2Na\text{-mont} + Mn^2 \tag{2.32}$$

$$Mn\text{-mont} + 2H^+ \rightleftarrows 2H\text{-mont} + Mn^2 \tag{2.33}$$

$$Na\text{-mont} + H^+ \rightleftarrows H\text{-mont} + Na^+ \tag{2.34}$$

The role of sodium ions is predicted to be small according to Schulze-Hardy's rule, that is, the effectiveness of counter ions in the interfacial phenomena is approximately proportional to the sixth power of the counter ion charge. The formation of complexes between Na^+ ions and anionic sites on the montmorillonite surface is included here for the sake of completeness, although such complexes are subject to reservations because they are unknown in aqueous solutions, and evidence against their physical reality has been presented (Janssen and Stein 1986). Indeed, as will be seen later, they are not prevalent.

The equilibria 2.35–2.45 describe the different complex formation and protonation equilibria with EDTA. These take place only in the solution (Box 4). The complex formation reactions can be characterized by the following equations:

$$Ca^{2+} + H^+ + EDTA^{4-} \rightleftarrows CaHEDTA^- \tag{2.35}$$

$$Mn^{2+} + H^+ + EDTA^{4-} \rightleftarrows MnHEDTA^- \tag{2.36}$$

$$Fe^{3+} + H^+ + EDTA^{4-} \rightleftarrows FeHEDTA \tag{2.37}$$

$$Ca^{2+} + EDTA^{4-} \rightleftarrows CaEDTA^{2-} \tag{2.38}$$

$$Mn^{2+} + EDTA^{4-} \rightleftarrows CaEDTA^{2-} \tag{2.39}$$

$$Fe^{3+} + EDTA^{4-} \rightleftarrows FeEDTA^- \tag{2.40}$$

Especially in the case of iron(III):

$$EDTA^{4-} + OH^- \rightleftarrows FeEDTAOH^{2-} \tag{2.41}$$

The concentration of OH⁻ of course is determined by pH. Finally, we can describe the protonation equilibria of EDTA:

$$EDTA^{4-} + H^+ \rightleftarrows HEDTA^{3-} \tag{2.42}$$

$$EDTA^{4-} + 2H^+ \rightleftarrows H_2EDTA^{2-} \tag{2.43}$$

$$EDTA^{4-} + 3H^+ \rightleftarrows H_3EDTA^- \tag{2.44}$$

$$EDTA^{4-} + 4H^+ \rightleftarrows H_4EDTA \tag{2.45}$$

The stability constants of the complexes are summarized in Table 2.9 (Martel and Smith 1974). The resultant equilibrium state of the system is determined by all the aforementioned equilibria. The total amount of different ions (calcium, manganese, sodium, hydrogen ions) in solid and solution can be measured using radioisotopic labeling, and then the number of different species in solution can be calculated on the basis of total amounts and stability constants (Table 2.9).

TABLE 2.9

Logarithm of Stability Products of the Species Formed in the Solution Phase of the Manganese(II) Ion/EDTA/Calcium-Montmorillonite System

Species	lg K
HEDTA³⁻	10.24
H₂EDTA²⁻	16.40
H₃EDTA⁻	19.06
H₄EDTA	21.06
CaEDTA²⁻	10.69
CaHEDTA⁻	13.18
MnEDTA²⁻	13.82
MnHEDTA⁻	16.19
FeEDTA⁻	25.10
FeHEDTA	26.40
FeEDTAOH²⁻	17.61
CoEDTA²⁻	16.49
CoHEDTA⁻	19.49
PbEDTA²⁻	18
PbHEDTA⁻	20.8

Source: Martell and Smith 1974.

2.8.1   Effect of Complex Formation in the EDTA/ Ca-Montmorillonite System

For the sake of simplicity, we examine at first the EDTA/Ca-montmorillonite system: in the absence of manganese(II) ions in the system, the complex formation of manganese(II) ions can be neglected. EDTA can form complexes with calcium ions and iron ions dissolved from the montmorillonite crystal lattice. The lower the pH, the greater the iron concentration in the solution is. The influence of EDTA is much smaller; it is only significant at pH range where the EDTA species is capable of complex formation with iron(III) ions, and it inhibits the precipitation of iron(III) oxide hydroxide. The highest iron concentration of the solution (at pH = 1) is about $3*10^{-6}$ mol/dm³, which is much lower than the concentration of EDTA and calcium ions. It does not mean that there is an observable change in the crystal structure (less than 0.001 % of the total iron content gets into solution); the stability constant of the different iron-EDTA complexes, however, is fairly high, and the results of computations show that these complexes are present in the solution, the concentration of "free" hydrated iron(III) ion is practically zero. The highest concentrations of FeEDTA⁻ and FeEDTAOH²⁻ are about 10^{-6} mol/dm³, it is cca. 10% of the smallest EDTA concentration, so the different iron(III) complexes slightly influence the concentration of the different protonated EDTA species and calcium complexes. The same stands for aluminum ions, too (Table 2.7).

The experiments with ¹⁴C labeled EDTA show that EDTA is sorbed on montmorillonite in a rather low degree. At pH<4, the sorption of EDTA is less than 10%. As will be seen later, in this pH range, the dominant species of EDTA is H_2EDTA^{2-}, and it sorbs on the protonated edge sites of montmorillonite (Section 2.9.1). Since the complexes of EDTA are anionic, this is in agreement with the general principles discussed in Chapter 1, Section 1.2: the anionic complexes do not take part in cation-exchange reactions. For this reason, the interpretation of the results became relatively easy because the species containing EDTA have to be taken into consideration only in the solution, neglecting the small degree of sorption of EDTA on the edge sites.

The ion-exchange equilibrium of calcium ions is determined by the ratio of calcium ion species in the solution. Only positive calcium ions (calcium aqua complexes) can participate in the ion exchange reaction; the negative CaEDTA²⁻ and CaHEDTA⁻ complexes do not. The ratio of CaEDTA²⁻ and CaHEDTA⁻ depends on the EDTA concentration and pH. When the concentration of EDTA is much smaller than the concentration of calcium ions, the quantity of EDTA complexes is negligible at any pH values. EDTA has an effect when its concentration is more than at least 10% of the concentration of calcium ions. In this case, the ratio of the calcium species depends on pH. In Figure 2.10 the ratio of the calcium species is shown when the ratio of the total concentrations of EDTA and calcium ion is 1:1.

As seen in Figure 2.10, the ratio of the different calcium species is determined by pH. At pH<4 calcium ions are exclusively present as positive aqua complexes involving the ion-exchange reaction, and the protonated species of EDTA (H_3EDTA^- and H_2EDTA^{2-}) are dominant. At about pH = 4, the formation of CaEDTA²⁻ complexes begins, at pH>5 it is the dominant chemical species of

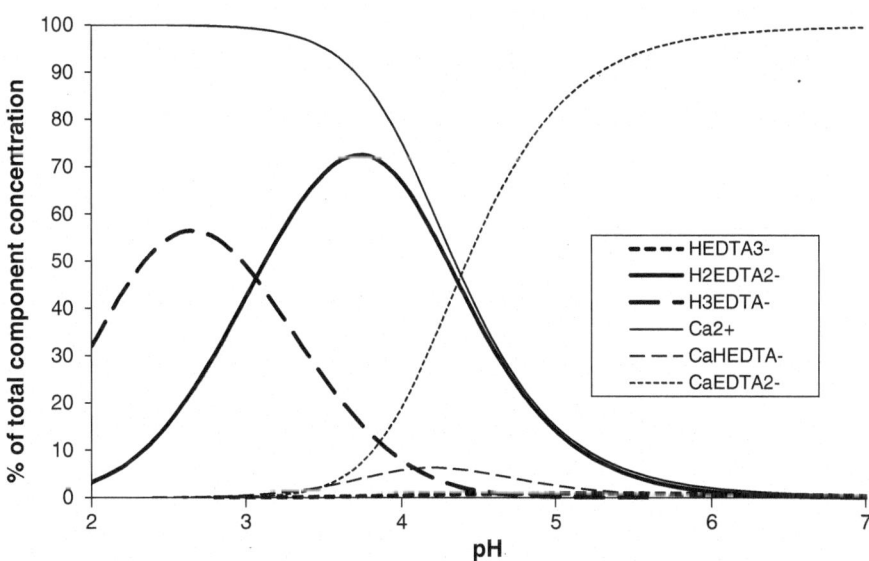

FIGURE 2.10   The ratio of calcium and EDTA species as a function of pH of the solution when the ratio of the total concentrations of EDTA and calcium ion is 1:1.

calcium ion. They remain dissolved and do not participate in ion-exchange reactions on montmorillonite.

To summarize, calcium ions in the solution can be present as hydrated Ca^{2+} ions, and $CaHEDTA^-$ and $CaEDTA^{2-}$ complexes, of which only the positive calcium aqua complexes (Ca^{2+}) participate in the ion-exchange reaction. In the process, calcium ions dissolve from montmorillonite into the solution, and mostly hydrogen ions get into the interlayer space of montmorillonite. In addition, sodium ions are also present in the system (EDTA is added as disodium salt; H_4EDTA is hardly soluble in water), which also affects the ion-exchange process. The ratio of $c_{Ca}{:}a_{Ca}$ can be plotted as a function of the concentration of Ca^{2+} (Figure 2.11).

In Figure 2.11, the isotherm for calcium-hydrogen ion exchange without EDTA (Section 2.7.1) is also plotted. We can see that the points of the two isotherms (with and without EDTA, i.e., below cca. $4*10^{-4}$ mol/dm^3 Ca^{2+}) coincide. This means that the distribution of the Ca^{2+} ions does not depend on the presence of EDTA. EDTA influences only the ratio of Ca^{2+} ions to the total calcium concentration of the solution. In this concentration range, only calcium-hydrogen ion exchange takes place (the sodium concentration is low), characterized with the same $c_{Ca}{:}a_{Ca}$ versus c_{Ca} function, independent of the presence or absence of EDTA. To reach higher Ca^{2+} concentrations in the solution, a higher EDTA concentration is needed since calcium ions dissolve from the interlayer of montmorillonite. The addition of EDTA means an increase in the concentration of sodium ions. Sodium ions tend to exchange with Ca^{2+} in the interlayer space of montmorillonite, but because of the very high difference between the preferences of sodium and hydrogen ions to the interlayer charge of montmorillonite (Table 2.2), it occurs only at high sodium ion concentration. As a

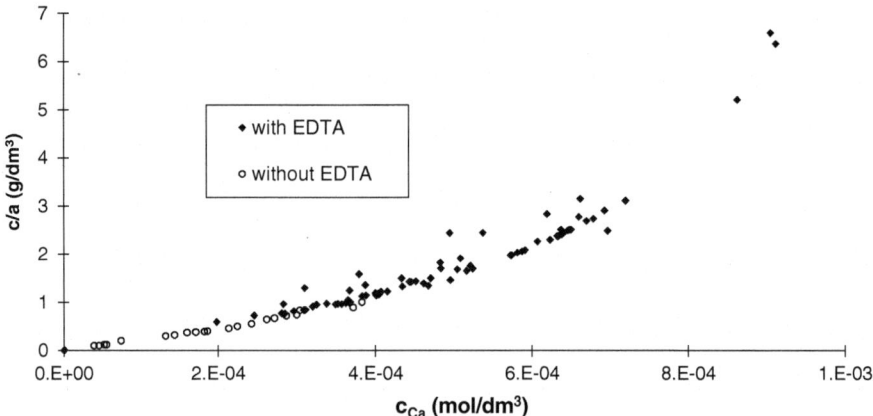

FIGURE 2.11 The ratio of c_{Cd}/a_{Ca} as a function of the concentration of Ca^{2+} on montmorillonite (the exchange isotherm of a calcium-hydrogen-sodium ion exchange) with and without the EDTA complex-forming agent. (Reprinted from Kónya and Nagy 1998, with permission from Elsevier.)

result of sodium-calcium exchange, the ratio of the calcium-hydrogen ion exchange decreases, which causes the increase of the slope of the isotherm.

Therefore, we can conclude that the interfacial equilibria and the complex equilibria are connected by the cations, here by Ca^{2+} and H^+ ions. Both equilibria are characterized by the same Ca^{2+} and H^+ concentrations. In Figure 2.11, we can see that the interfacial equilibrium can be described solely by Ca^{2+} ion concentration. Complex formation influences the interfacial equilibrium only via the decrease of Ca^{2+} ion concentration. The concentration of the Ca^{2+} ion can be calculated from Equations 2.38 and 2.45, or 2.42 and 2.35, respectively. From these equations we obtain:

$$\left[Ca^{2+}\right] = \left(\left[CaEDTA^{2-}\right] * \left[H^+\right]^4 * \beta_{H_4EDTA}\right) \Big/ \left(\left[H_4EDTA\right] * \beta_{CaEDTA}\right) \quad (2.46)$$

$$\left[Ca^{2+}\right] = \left(\left[CaHEDTA^-\right] * \beta_{HEDTA}\right) \Big/ \left(\left[HEDTA^{3-}\right] * K_{CaHEDTA}\right) \quad (2.47)$$

where [ ]'s mean the equilibrium concentrations of the different species in the solution, βs are their stability constants, and:

$$\log K_{CaHEDTA} = \log \beta_{CaHEDTA} - \log \beta_{HEDTA} \quad (2.48)$$

Equations 2.46 and 2.47 can be substituted into Equation 2.48 or other expressions characteristic of ion exchange, and the stability constants or their ratios can be calculated. Schubert's method is also based on similar principles (Marcus and Kertes 1969, Chapter 1, Section 1.2.2). His method to determine stability constants, however, is not widely used, likely because it is strongly affected by experimental conditions.

The determination of the ratio of Ca^{2+} to total calcium concentration seems to be easier. The total calcium concentration in the solution is as follows:

$$Ca_{tot} = \left[Ca^{2+}\right] + \left[CaEDTA^{2-}\right] + \left[CaHEDTA^{-}\right] \tag{2.49}$$

By expressing [CaEDTA^{2-}] and [CaHEDTA^{-}] from Equations 2.46 and 2.47, and substituting into Equation 49, we obtain

$$\frac{\left[Ca^{2+}\right]}{Ca_{tot}} = \frac{1}{1 + \dfrac{\left[H_4EDTA\right]\beta_{CaEDTA}}{\left[H^+\right]\beta_{H_4EDTA}} + \dfrac{\left[HEDTA\right]K_{CaEDTA}}{\beta_{HEDTA}}} \tag{2.50}$$

By substituting the pH-dependent fraction α,

$$\alpha = 1 + \frac{\left[H^+\right]}{\beta_{HEDTA}} + \frac{\left[H^+\right]^2}{\beta_{H_2EDTA}} + \frac{\left[H^+\right]^3}{\beta_{H_3EDTA}} + \frac{\left[H^+\right]^4}{\beta_{H_4EDTA}} \tag{2.51}$$

and

$$\beta_{HEDTA} = \beta'_{HEDTA} * \alpha = \left[HEDTA\right]\frac{\alpha}{\left[H^+\right]} EDTA_{tot} \tag{2.52}$$

where

$$\alpha = \frac{EDTA_{tot}}{\left[EDTA^{4-}\right]} \tag{2.53}$$

into Equation 2.50 and making equivalent mathematical transformations, the following expression is obtained:

$$\frac{\left[Ca^{2+}\right]}{Ca_{tot}} = \frac{1}{1 + \dfrac{\beta_{CaEDTA}}{\alpha} + \dfrac{K_{CaEDTA} * \left[H^+\right]}{\alpha} * EDTA_{tot}} \tag{2.54}$$

where $EDTA_{tot}$ is the total concentration of EDTA in the system.

By means of Equation 2.54, knowing the pH, the total concentration of the complex-forming agent, its pH-dependent α parameter, and the stability constants of its complexes, the ratio of the "free" cation and the total metal ion concentration can be estimated. Using these, the conditions needed for the dissolution of a cation can also be predicted: the ratio in Equation 2.54 must be small. Of course, this equation can also be used for other metal ions, for example, polluting ions, and so it could be important to determine how to decontaminate soils.

Similarly, the conditions needed for the hydrogen-cation exchange can also be predicted (the ratio in Equation 2.54 has to be high). This can prove to be important in the treatment of acidic soils, where the quantity of the exchangeable cations can be increase and so the nutrient cycle in soils can be improved.

Equation 2.54 is valid only when the cations are sorbed only by reversible ion exchange in the interlayer space of montmorillonite. In this case, the sorption strength is determined by electrostatic forces, which are approximately the same for any cations. In this case, the clay itself plays a less important role compared to the complex-forming agents. However, when specific interactions exist between the clay and the cation, these specific interactions may have greater importance than complex formation. For example, cations (lead ion) sorbed by covalent bond on the edge sites cannot be desorbed by complex forming agent (Section 2.8.4).

2.8.2 Effect of Complex Formation in the EDTA/ Ca-Montmorillonite/Manganese(II) Ion System

When the system contains another EDTA-complex-forming cation besides the calcium ion, the ratio of the different complex species in the solution, and consequently in the in interlayer space, depends on the ratio of the stability constants of the complexes. The stability constants of EDTA complexes of transition metal ions, including polluting cations, is usually greater than the stability constant of EDTA complexes of the calcium ion. In the case of manganese(II) ions, the stability constants of the manganese (II) complexes are about three orders of magnitude higher than those of calcium complexes. This means that manganese(II) ions form complexes at lower pH values than calcium ions. In addition, when the concentration of EDTA is equal to the concentration of manganese(II) ions, CaEDTA complexes are not formed (Figure 2.12), and calcium ions are present as positive calcium aqua complexes (Ca^{2+}) in the whole pH range. The formation of manganese(II) EDTA complexes starts at pH = 2; at first the protonated MnHEDTA$^-$ forms, and its concentration is maximum at pH ~ 3.3. At pH>4.5, MnEDTA^{2-} is the dominant species. The ratio of EDTA species is also varied with pH. At pH = 2.5, H_3EDTA^- has maximum concentration. When the pH increases and the formation of the Mn(II)EDTA complexes starts, H_2EDTA^{2-} becomes the dominant protonated EDTA species. Its quantity, however, is smaller than the quantity of H_3EDTA^- at pH = 2.5, because some manganese ions have already used up some EDTA in the complex formation.

The species distribution of the solution determines the cation composition of the interlayer space of montmorillonite. In equilibrium, the cation exchange sites of montmorillonite are covered by calcium, hydrogen, manganese, and sodium ions (Figure 2.9). Figure 2.13 shows the equivalent fractions (X) of these cations as a function of pH at the ratio of Mn:EDTA = 1:1.

As seen in Figure 2.13, the equivalent fraction of manganese(II) ion (X_{Mn}) is determined by the ratio of manganese species in the solution. At pH<2.6, only positive manganese ions (manganese(II) aqua complexes, Mn^{2+}) are present in the solution. They participate in the ion- exchange reaction and enter into the interlayer space. When the negatively charged manganese(II) complexes become dominant,

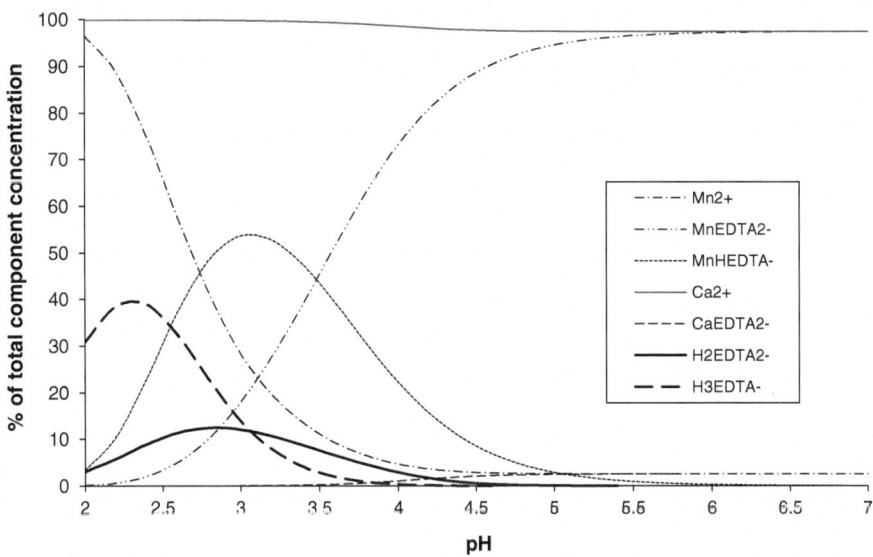

FIGURE 2.12 The ratio of calcium, manganese(II), and EDTA species as a function of the pH of the solution when the ratio of the total concentrations of EDTA, calcium, and manganese ions is 1:1:1.

the quantity of Mn(II) ions decreases in the interlayer space. This is the same process as discussed previously for the Ca ions/EDTA system.

The equivalent fractions of calcium (X_{Ca}) and hydrogen ions (X_H) as a function of pH in the presence of EDTA can be compared to the equivalent fractions of these ions without EDTA (also plotted in Figure 2.13). At pH>4.5, the equivalent fraction of calcium and hydrogen ions are present regardless of the presence of EDTA, which is explained by having a near 100% ratio of Ca^{2+} ion. At pH<4.5, the equivalent fractions of calcium ions are higher than the values obtained without EDTA. The equivalent fraction of hydrogen is smaller than is the case without EDTA, and it shows a slight maximum at pH $\approx$ 3.3.

The increase of the equivalent fraction of calcium at pH<4.5 is unexpected on the basis of the ratio of the calcium species (Figure 2.12) since only hydrated, positively charged calcium cations are dominant in this pH range. In addition, low pH values should favor the sorption of hydrogen ions (Table 2.2, Section 2.7.1) because, among the cations in the solutions (calcium, hydrogen, manganese, and sodium ions) hydrogen ions have the greatest affinity to montmorillonite. The change of the equivalent fraction of calcium ions on the surface of montmorillonite can be interpreted only together with the decrease of the equivalent fraction of hydrogen ions. We can see in Figure 2.13 that, without EDTA, the equivalent fraction of hydrogen ions decreases when pH increases. In the presence of EDTA, at about pH<3.1, it is present as H_3EDTA^-. Since EDTA is added as a disodium salt, a part of the hydrogen ions needed for protonation must come from the interlayer space of montmorillonite.

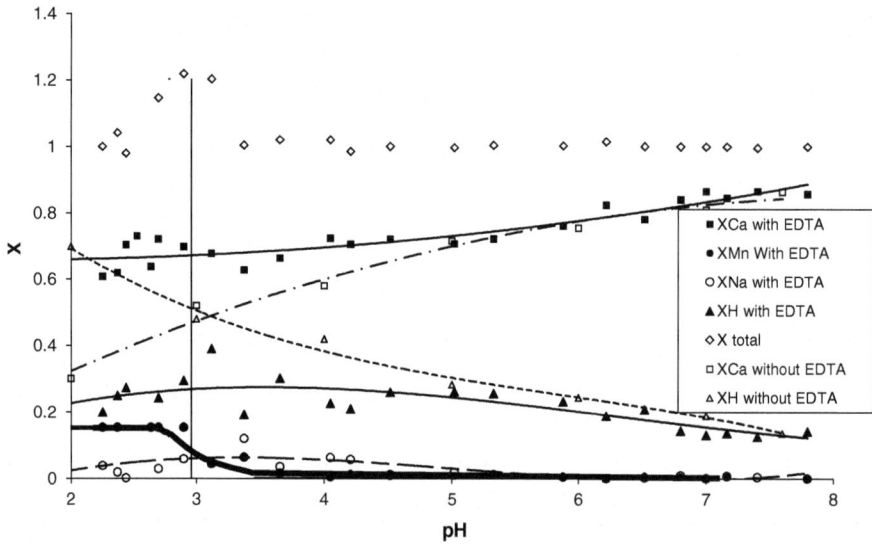

FIGURE 2.13 The equivalent fractions of manganese(II) (X_{Mn}), calcium (X_{Ca}), hydrogen (X_H), sodium ions (X_{Na}), and their sum (X total) in the interlayer space of montmorillonite as a function of pH of the solution. The ratio of the total concentrations of EDTA, calcium, and manganese ions is 1:1:1. The equivalent fractions of calcium and hydrogen ions as a function of pH without complex-forming agent is also shown.

Sodium ions are also present in the interlayer space. Their equivalent fraction (X_{Na}) changes as a function of pH. The sorption of univalent sodium ions is unexpected (Table 2.2), especially at low pH value where the interlayer space is usually occupied by hydrogen ion and bivalent cations. The considerations of the local potential hypothesis seem to be valid (van Diemen and Stein 1978). In an acidic medium, the aluminol edge sites of montmorillonite are positive and can sorb anions (Section 2.4.1). In the system, bivalent protonated EDTA complexes (H_2EDTA^{2-}) are present just at the pH range (Figure 2.12) where the increased quantity of sodium ions is sorbed (Figure 2.13). So, bivalent H_2EDTA^{2-} species can sorb on the positive edge sites of montmorillonite and they stimulate the sorption of sodium ions. Studies with ^{14}C-EDTA show 5%–10% EDTA sorption at pH<4 (Section 2.9.1).

In Figure 2.13, the sum of the equivalent fractions (X total) is also plotted. It has a pH range (2.5–3.5) where X total>1, with a maximum X total $\approx$ 1.2 at pH $\approx$ 2.9. It means that about 20% of the layer charge can sorb EDTA species. In Section 2.4.3 and Table 2.4 we could see that the number of aluminol sites is in a fairly good agreement with this value.

2.8.3 EFFECT OF STABILITY CONSTANTS ON THE CATION COMPOSITION OF THE INTERLAYER SPACE

When the stability of the complexes is greater than the stability constant of manganese(II)-EDTA complexes, the composition of the interlayer space becomes simpler, and the quantity of positive aqua cations in the solution is negligible in the studied pH range. The stability constants of cobalt-EDTA and lead-EDTA complexes are about six or seven orders of magnitude, respectively, higher than the stability constants of calcium-EDTA complexes (Table 2.9). For these ions, the quantity of positive cations (Co^{2+} and Pb^{2+}) is smaller than 10% at pH = 2.3 and 1.8, respectively. So, when the quantity of the EDTA is equal to the quantity of cobalt or lead ions, the system practically behaves as the calcium-montmorillonite-acidic solution system, that is, calcium-hydrogen ion exchange takes place.

Obviously, when the ratio of the cation:EDTA is not equal to 1, the conditions are modified. When the quantity of the complex forming agent is less than the quantity of the exchanging metal ions (manganese(II), cobalt, lead, etc.), only a part of the metal ions can form complexes; the residual part remains as positive cations and participate in cation-exchange reaction as usual. When the quantity of EDTA is higher than the quantity of metal ions, calcium ions can also form negatively charged complexes and dissolve. Again, calcium-hydrogen ion exchange takes place in the interlayer space, depending on pH (Figure 2.10) and the quantity of EDTA. When the EDTA concentration and pH is large enough to complex the total quantity of calcium ions (and the other cations in the system, of course), the latter are dissolved and, practically, hydrogen-montmorillonite (with a small quantity of sodium ions in the interlayer space) is formed.

As a conclusion, the quantity and the stability constants of the complex forming agent determines the quality and quantity of the interlayer cations of montmorillonite. It has very important consequences during the decontamination of soils polluted with metal cations, which should be done as follows. At first, a suitable complex forming agent is chosen, for which the stability constants of the polluting ion are much (at least two to three orders of magnitude) higher than that of the calcium ion. This complex-forming agent is added to the soil suspension in a quantity equivalent to the polluted ions. Calcium ions are also added to the system; its quantity should be equivalent with the quantity of the polluting cations. The pH is adjusted to where the polluting cations form negatively charged complexes. The pH of the suspension should be close to neutral, if possible, but care should be taken to avoid the hydrolysis of the calcium ions. The simplest way of decontamination is to prepare a solution with the adjusted pH and containing $CaEDTA^{2-}$ complexes in the desired concentration and to suspend the soil in it. In this method, calcium (and hydrogen ions, depending on pH) penetrate into the interlayer space, the polluted metal ions are dissolved, and, together with the complex-forming agent, can be separated by filtration, centrifugation, or simply by sedimentation. Thus, the metal-ion-polluted soils can be decontaminated. When the complex-forming agent is in excess, calcium ions may also dissolve, which is undesirable because of the physical structure of clay fails.

A special case is the decontamination of soil polluted with radioactive metal ions (e.g., $^{60}Co^{2+}$). The concentration of the radioactive ion is very low, so a very small quantity of the complex- forming agent should be applied. Experiences show (Nagy and Kónya 1998; Nagy et al. 1999), however, that ^{60}Co ions in carrier-free concentration cannot be desorbed from montmorillonite when the EDTA concentration of the solution is less than the concentration of the iron ion dissolved from montmorillonite.

Since Fe(III)-EDTA complexes have greater stability constants than Co-EDTA complexes (Table 2.9), iron(III) ions will consume EDTA. So, cobalt ions are present as Co^{2+} ions and remain in the interlayer space; the concentration of Co^{2+} determines the equivalent fraction of cobalt in the interlayer space of montmorillonite, exactly in the same way as it would without the complex-forming agent. As a consequence, when cobalt ions have to be removed from montmorillonite (or other clay mineral or soil), the amount of iron also has to be determined and EDTA has to be added in excess, taking into account the concentration of iron. Theoretically, this is true for all of the ions in the soil solution: we have to determine the quantity of ions forming more stable complexes than the decontaminated ion, and the quantity of EDTA should be adjusted consequently.

2.8.4   EFFECT OF COMPLEX FORMATION IN THE EDTA/ CA-MONTMORILLONITE/LEAD(II) ION SYSTEM

As mentioned previously, the considerations about the effect of the complex-forming agents are valid when cations are in such a chemical species in montmorillonite (or soil) where the complex formation can take place freely. For example, they are in the interlayer space of montmorillonite and can participate in reversible cation-exchange reactions. However, cations sorbed by covalent bond on the edge sites cannot be desorbed by a complex-forming agent. Moreover, they can sorb on the edge sites even when they are present as fairly stable dissolved complexes in the solution. The lead ion, which has been studied by some complex-forming agents (EDTA), diethyl triamin pentaacetic acid [DTPA], tartaric acid [Tart], and citric acid [Cit]) are a good example to demonstrate this. The ion-exchange, complex-formation, and protonation reactions can similarly be described as in the case of manganese(II) ions/calcium-montmorillonite/EDTA system (Equations 2.25–2.45; Nagy and Kónya 1998; Nagy et al. 1998b). The stability constants of EDTA complexes are listed in Table 2.9, and those of DTPA, tart and cit complexes are shown in Table 2.10.

As seen in Tables 2.9 and 2.10, lead ion forms very stable complexes with EDTA and DTPA, the difference between the calcium and lead complexes being about seven orders of magnitude for both complex-forming agents. So, lead ions are present as negative EDTA complexes in the whole studied range (pH 2–7) of EDTA when the concentration of EDTA is at least equal to the concentration of lead ions. When lead ions consume all of the complex forming agents (Pb:EDTA/DTPA = 1:1), calcium ions are present as Ca^{2+}. Consequently, according to previous discussions, lead ions should not participate in cation-exchange reactions. In the case of lead ions, however, there is no such strict relationship between the sorbed quantity and the positively charged hydrated ion concentration; the quantity of the sorbed lead never reaches to zero (Figure 2.14). It is likely caused by the formation of Pb–O on the edge sites (Section 2.5.1.4; Hachiya et al. 1979; Strawn and Sparks 1999).

The sorption of calcium ions is as expected: at low pH values, the calcium ions, present as Ca^{2+}, sorb, as in the absence of a complex forming agent the quantity of the sorbed calcium increases when the pH increases. At pH values where negative Ca-EDTA complexes are already formed and the concentration

TABLE 2.10

Logarithm of Stability Constants of Different Complexes and Protonated Ligands in the Presence of DTPA, Tartaric Acid, and Citric Acid (Martell and Smith 1974)

DTPA		Citric acid		Tartaric acid	
Species	lg K	Species	lg K	Species	lg K
$PbDTPA^{3-}$	18	$PbCit^-$	4.48	PbTart	3.12
$HPbDTPA^{2-}$	23.32	HPbCit	8.64	$HPbTart^+$	5.72
Pb_2DTPA^-	22.21	$PbCit_2^{4-}$	5.92		
		$HPbCit_2^{3-}$	10.61		
		H_2PbCit^+	11.70		
$CaDTPA^{3-}$	10.75	$CaCit^-$	3.45	CaTart	1.94
$HCaDTPA^{2-}$	16.86	HCaCit	7.79	$HCaTart^+$	5.06
Ca_2DTPA^-	12.35	H_2CaCit^+	11.00		
$FeDTPA^{2-}$	28.00	FeCit	11.20	$FeTart^+$	6.49
$FeHDTPA^-$	31.56	$FeHCit^+$	17.90		
$FeDTPAOH^{3-}$	18.22				
$HDTPA^{4-}$	10.49	$HCit^{2-}$	5.66	$Htart^-$	3.96
H_2DTPA^{3-}	19.09	H_2Cit^-	10.00	H_2Tart	6.78
H_3DTPA^{2-}	23.37	H_3Cit	12.90		
H_4DTPA^-	26.01				
H_5DTPA	28.01				
$H_6 DTPA^+$	29.61				

Source: Reprinted from Nagy et al. 1998b, with permission from Elsevier.

of hydrated Ca^{2+} ions decreases, the sorption of calcium on the surface of montmorillonite also decreases.

As seen in Table 2.10, the stability constants of tartarate and citrate complexes of lead and calcium ions are much smaller than those of EDTA and DTPA complexes. The calculations show that the dominant species of cations are Pb^{2+} and Ca^{2+} at pH<4, and so the same ion exchange can be expected as would happen without complex-forming agents. At pH>4, the effect of citric acid is significantly higher than expected from stability constants. The structure of citrate complexes, however, is not known precisely, and different structures can be imagined (Nagy et al. 1998b). The sorption of nickel ions in the presence of citric acid was successfully explained by the surface complexation model by Marcussen et al. (2009).

As a conclusion, the ion-exchange processes are greatly influenced by the composition of the solution, pH, and the stability constant of the complex-forming agent. However, it is also influenced if the sorption happens through ion exchange in the interlayer space and covalent bonds on the edge sites. The two types of the sorbed lead behave differently in the presence of complex-forming agents. Lead ions in the interlayer space

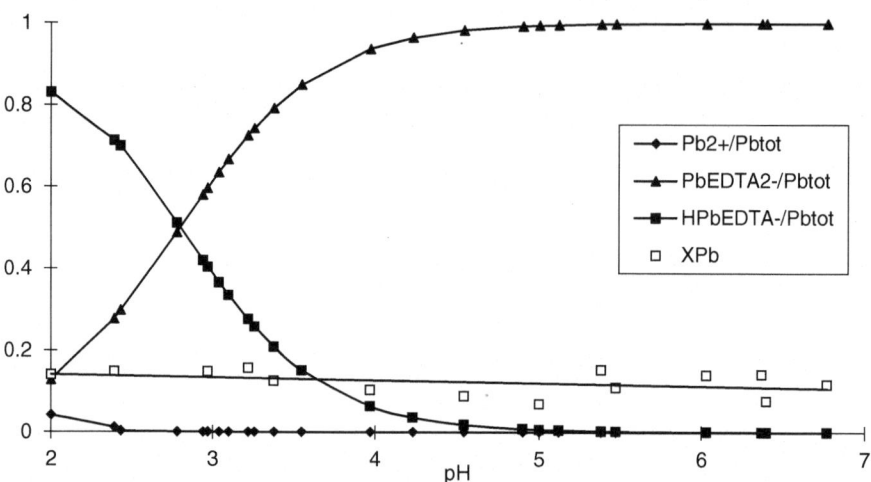

FIGURE 2.14 The equivalent fractions of lead ion on the surface of montmorillonite and the ratios of different lead-EDTA complexes in the solution at Pb:EDTA = 1:1. (Reprinted from Nagy and Kónya 1998, with permission from Elsevier.)

can be desorbed by complex formation. Therefore, the electrostatic forces between the cations and the layer charge do not inhibit the complex-formation reactions. The part of lead ions sorbed on the edge sites, however, always remains on the external surface even when complexes with great stability should form. The Pb-O bond is stronger and cannot be broken by a complex-formation process. As a consequence, lead contamination cannot totally be cleaned up using complex-forming agents.

In this chapter we discussed the effect of artificial (EDTA, DTPA) and natural (tart, cit) complex-forming agents on the cation sorption/desorption processes. These complex-forming agents are applied in diagnostic extraction for micronutrients in soils and in trace element fertilizers (Kabata-Pendias and Pendias 1985). When they form very stable complexes with trace elements, the latter, added with fertilizers can remain in the soil solution for a long time. This can be undesirable because they can penetrate into the deep layer of the soil without being absorbed by plants and can become soil pollution. The natural complex-forming agents responsible for the uptake of trace elements by roots (amino acids, organic acids, hydroxide compounds, etc.) usually form fewer stable complexes with the trace elements cations (similarly to tart and cit) than, for example, EDTA and DTPA. So, adding a complex-forming agent in a fertilizer is an optimization problem. The different sorption possibilities of metal ions as well as complex-forming agents, the stability constants, pH, and concentrations have to be taken into account.

2.9 SORPTION OF ORGANIC MATTER ON MINERALS

As discussed in Chapter 1, Section 1.3.5, clay minerals, including montmorillonite, can sorb different organic substances in the interlayer space and on the external surfaces as well. A large number of papers are published on the various cation-exchanged montmorillonites as catalysts in organic syntheses. On the basis of the wide variety of clay-catalyzed reactions, we can suppose that clay minerals play a role in the decomposition of organic matter and the formation of humus. The first step of catalytic reactions is always the sorption of the organic substance, resulting in the geometrical and energetic modification of the molecular structure. On the other hand, it is important to be aware of the properties of montmorillonite when studying the contamination and decontamination of aquatic and soil systems as well as the interfacial reactions between organic materials and soils (Gratzel and Kalyanasundaram 1991).

In this section, the sorption of two organic substances, EDTA and valine amino acid, are discussed. They were selected for two reasons: as seen in Figure 2.8, EDTA can be sorbed on the edge sites of montmorillonite. The other reason is the presence of characteristic functional groups in both molecules. The carboxylic and amin groups are very important in organic syntheses and in humic substances, too.

2.9.1   Sorption of EDTA on Montmorillonite

As it was seen in Section 2.8.2 and in Figure 2.13, some EDTA species are sorbed on the edge site of montmorillonite, stimulating the sorption of cations also. The sorption of EDTA can be studied by ^{14}C-labeled EDTA, and the sorbed quantity can be compared to the different EDTA species present in the solution (Figure 2.15).

As seen in Figure 2.15, EDTA is sorbed on montmorillonite to a small degree, and there is a flat maximum at pH = 3–3.5. As also seen, EDTA has only negatively

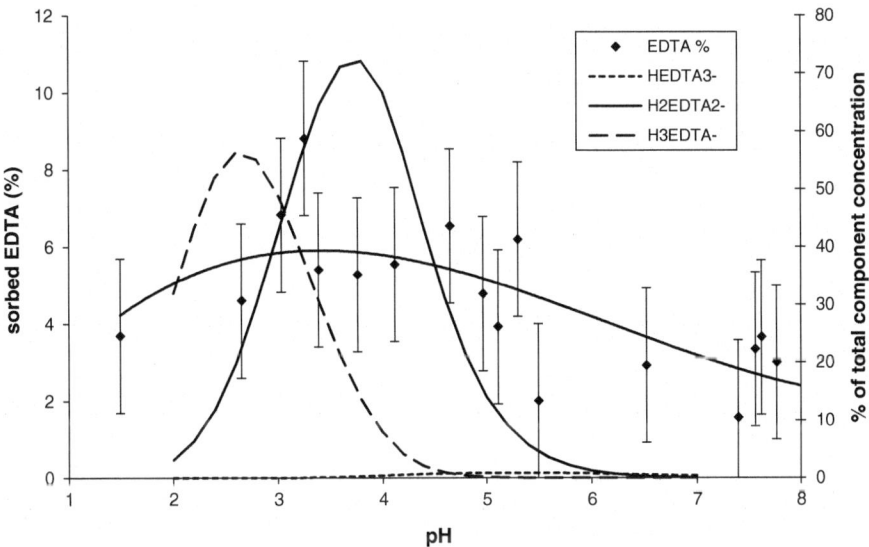

FIGURE 2.15   The ratio of EDTA species in the solution and the relative sorbed quantity of EDTA. $m = 40$ mg Ca-mont, $c_{EDTA} = 5e-4$ mol/dm^3. The error bars show $\pm 5\%$.

charged species, so they cannot enter into the interlayer space. This is confirmed by X-ray diffraction studies where the basal spacing (d_{001}) does not change upon addition of EDTA. The negatively charge EDTA species can be sorbed on the aluminol sites that are protonated in this pH range (Sections 2.4.1, 2.4.3; Figure 2.3). The sorption of the deprotonated, divalent complex (H_2EDTA^{2-}) is obviously greater. The maximum of the H_2EDTA^{2-} concentration in the solution is at the same pH range (3–3.5) as the maximum of EDTA sorption. In the sorption maximum, about 25% of the aluminol sites is covered by EDTA. (Note, that, at pH $\approx$ 4, the formation of $CaEDTA^{2-}$ complex starts; it is not plotted in Figure 2.15 because it has no role in the sorption of EDTA.)

2.9.2 SORPTION OF VALINE ON MONTMORILLONITE

The publications on amino acid adsorption on clay minerals before 1974 are summarized in the book of Theng (1974). It provides information on the adsorbed quantity of different amino acids on cation-exchanged montmorillonites and the characteristic IR bands of amino acid-montmorillonites adsorption compounds. Usually, only the adsorbed quantity of amino acids on montmorillonites is shown; no adsorption mechanism is usually hypothesized (e.g., Friebele et al. 1981; Rak et al. 1982), except Stadler and Schindler (1993a) where the adsorption is evaluated by the surface complexation model, and the possible surface complexes are given for β-alanine.

Here, the adsorption of valine on different cation-exchanged montmorillonites is described (Nagy and Kónya 2004). A discussion of the kinds of interactions that are possible in the ternary system of montmorillonite/valine/metal ions will be presented, and a description of how the metal ions can affect these interactions. The interlayer cations (calcium, zinc, copper ions) were chosen on the basis of the stability constants of their complexes with valine. The adsorption of valine on montmorillonite is interpreted using a surface complexation model.

When studying the sorption of amino acids, their reactions in the solution and on the interface have to be taken into account. In aqueous solution similar to EDTA (Section 2.8.1), these reactions are the protonation and the complex-formation reactions. The protonation reactions of valine are as follows:

$$Val^- + H^+ \rightleftarrows HVal \quad K_{HVal} = \frac{\left[HVal\right]}{\left[Val^-\right]\left[H^+\right]} \tag{2.55}$$

$$Val^- + 2H^+ \rightleftarrows H_2Val^+ \quad K_{H_2Val} = \frac{\left[H_2Val^+\right]}{\left[Val^-\right]\left[H^+\right]^2} \tag{2.56}$$

K's are the stability constants of the protonated valine species (Table 2.11).

When studying the reactions of valine with calcium-, copper-, and zinc-montmorillonite, calcium, copper, and zinc complexes may be formed. The stability constants of the complexes known in the literature (Burger 1990) are listed in Table 2.11.

TABLE 2.11

The Stability Products of Calcium-, Zinc-, and Copper-Valine Complexes, and Protonated Forms (Burger 1990)

Species	lg K
HVal	9.49
H_2Val^+	11.75
$CaVal^+$	1.05
$CaHVal^{2+}$	9.85
$ZnVal^+$	4.45
$ZnVal_2$	8.24
$CuVal^+$	8.11
$CuVal_2$	14.90

Source: Reprinted from Nagy and Kónya 2004, with permission from Elsevier.

At pH = 3–8, only copper complexes are formed in a significant amount, and their quantities depend on pH:

$$Val^- + Cu^{2+} \rightleftarrows CuVal^+ \quad K_{CuVal} = \frac{\left[CuVal^+\right]}{\left[Val^-\right]\left[Cu^{2+}\right]} \quad (2.57)$$

$$2Val^- + Cu^{2+} \rightleftarrows CuVal_2 \quad K_{CuVal_2} = \frac{\left[CuVal_2\right]}{\left[Val^-\right]^2\left[Cu^{2+}\right]} \quad (2.58)$$

In the pH range studied (3 to 8), calcium and zinc ions are practically resent as Ca^{2+} and Zn^{2+} ions.

Valine can be adsorbed on montmorillonite in different ways. The possibilities of valine adsorption have to be studied on the layer charges as well as on the edge sites.

On the Layer Charges

The negative layer charge can be neutralized by protonated valine molecules. The surface complex in the interlayer space is

$$X^- + Val^- + 2H^+ \rightleftarrows H_2ValX \quad K_{H_2ValX} = \frac{\left[H_2ValX\right]}{\left[H^+\right]^2\left[Val^-\right]\left[X^-\right]} \quad (2.59)$$

On the Edge Sites

In an acidic medium, the aluminol sites are mainly present as $AlOH_2^+$ sites (Figure 2.3). Valine molecules are also present in protonated ligands, so sorption can be neglected. The main part of the silanol sites is deprotonated, so valine cannot sorb again. When the pH is close to neutral, aluminol sites and valine can form surface complexes as follows:

$$AlOH + H^+ + Val^- \rightleftarrows -AlOH_2Val \quad K_{AlOHVal} = \frac{[AlOHVal]}{[AlOH][Val^-][H^+]} \quad (2.60)$$

In an alkaline medium, the aluminol sites are deprotonated (AlO^-). Also, the characteristic species of valine is negatively charged and cannot be adsorbed by negative AlO^- sites.

When, however, the system contains a metal ion which can form stable positive complexes with valine (e.g., copper ion), then these complexes may be sorbed on the deprotonated edge sites. Calculations made on the basis of the stability constants show that positively charged $CuVal^+$ complexes form at an acidic pH where the silanol sites can be deprotonated, aluminol sites are protonated (Figure 2.3). As a result, the surface complex can be formed as follows:

$$SiOH-H^+ + Cu^{2+} + Val^- \rightleftarrows SiOCuVal \quad K_{SiOHVal} = \frac{[SiOCuVal]}{[SiOH][Val^-][Cu^{2+}]} \quad (2.61)$$

In addition, valine has another role during the process: the formation of copper-valine complex promotes the desorption of the copper ion from the interlayer space. It means that the sorption of valine produces a copper ion that further consumes another valine molecule, reducing the concentration of valine to undergo sorption. The sorption equilibrium is the result of these two opposing reactions.

When the experimental sorption data of valine on different montmorillonites are to be explained, the aforementioned model can be used. The model suggests that the valine molecules introduce into the interlayer space. This is proved by kinetic studies that show a much longer equilibration period (6 days) than the exchange of inorganic cations in the interlayer space (about 1 hour). This long reaction time shows that the rate-determining step is the diffusion of valine molecules between the layers of montmorillonite, which is much slower than the diffusion of the small cations. Furthermore, the time necessary to establish equilibrium cannot be shortened by stirring.

In Figure 2.16 the sorbed quantity of valine is shown on calcium-, zinc-, and copper-montmorillonite.

As seen in Figure 2.16, the sorbed quantity of valine strongly depends on the quality of the cation in the interlayer space. Calcium-montmorillonite shows a rather low valine adsorption (about 10^{-7} mol valine/g Ca-montmorillonite) compared to the other montmorillonites (10^{-3} mol valine/g Cu-, and Zn-montmorillonites).

For the interpretation of the results using the surface complexation model, reactions 2.47–2.53 have to be taken into account. In addition, the surface acid–base properties and the neutralization reactions of the layer charge have to be included as in Section 2.4.2; the parameters determined there are treated as fixed, input data. In the case of copper- and zinc-montmorillonite, the copper and zinc concentration of the solution and solid also have to be determined, and these data has to be taken

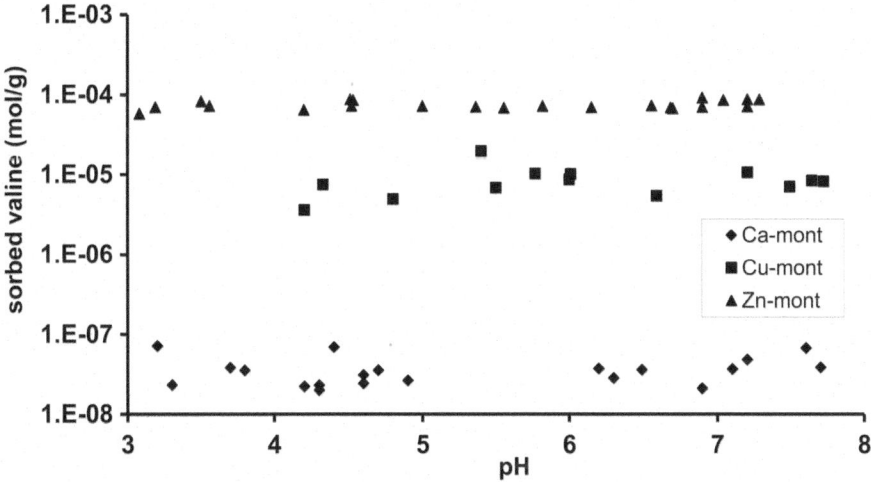

FIGURE 2.16   The sorbed quantity of valine on different cation-exchanged montmorillon-ites. $m = 100$ mg, $V = 20$ cm^3, $c_0 = 1e{-}3$ mol dm^{-3}, $T = 20°C$. (Reprinted from Nagy and Kónya 2004, with permission from Elsevier.)

into consideration. That is, the quantity of the total sorbed valine and the copper or zinc ion concentrations versus pH function can be fitted, and K_{H2ValX}, $K_{AlOH2Val}$, and $K_{SiOCuVal}$ stability constants can be computed. The results of the parameter fit for copper-, and zinc-montmorillonites as well as the obtained stability constants are shown in Figures 2.17 and 2.18, and in Table 2.12, respectively.

As seen in Table 2.12, the stability constants (K_{H2ValX} and $K_{AlOH2Val}$) are very similar in all montmorillonites. The concentration of the different sorbed species, however, strongly depends on the interlayer cation. In the case of copper-montmoril-lonite (Figure 2.17), the total quantity of the sorbed valine consists of two adsorbed species: in the interlayer space, valine is adsorbed as H_2ValX, on the edge sites as SiOCuVal. The amount of $AlOH_2Val$ can be neglected. The concentration of H_2ValX is proportional to the concentration of HX (Figure 2.3) and HVal. The formation of SiOCuVal is related to the concentration of CuVal$^+$ and SiO$^-$ (Figure 2.3), and the concentration of SiOCuVal is maximum when CuVal$^+$ and SiO$^-$ are present in fairly high concentration. The concentration of CuVal$^+$ species is high in the acidic range (there is a maximum at pH ≈ 4).

In the case of zinc-montmorillonite (Figure 2.18), valine is sorbed in the interlayer space and on the aluminol sites. As discussed in Sections 2.5.1.1 and 2.6, zinc ions adsorbed on the deprotonated aluminol sites during the preparation of zinc-montmo-rillonite. These adsorbed zinc ions stimulate the sorption of valine on the edge sites, increasing the quantity of the total sorbed valine. It is shown by the value of $K_{AlOH2Val}$ in Table 2.12, which is in fact $K_{AlOZnVal}$.

In the case of calcium-montmorillonite, the quantity of valine adsorbed is very low, but, similar to zinc-montmorillonite, it is adsorbed in the form of H_2ValX and $AlOH_2Val$.

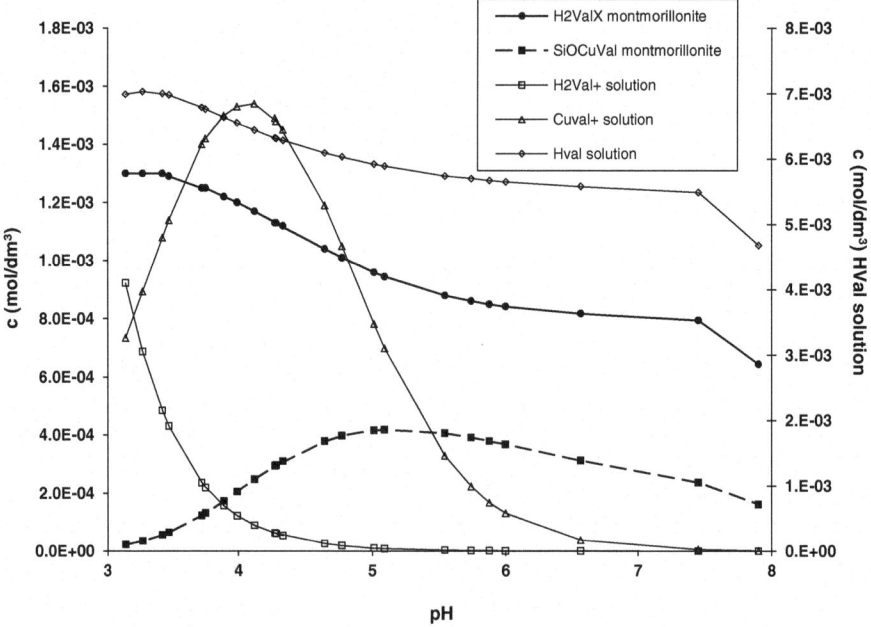

FIGURE 2.17 The quantity of sorbed valine in the interlayer space and on the silanol sites of copper-montmorillonite, and the concentration of valine species in the solution. $m = 100$ mg, $V = 20$ cm^3, $c_0 = $ 1e-3 mol dm^{-3}, $T = 20°C$. The right y-axis shows the concentration of HVal, all other species are labeled on the left y-axis.

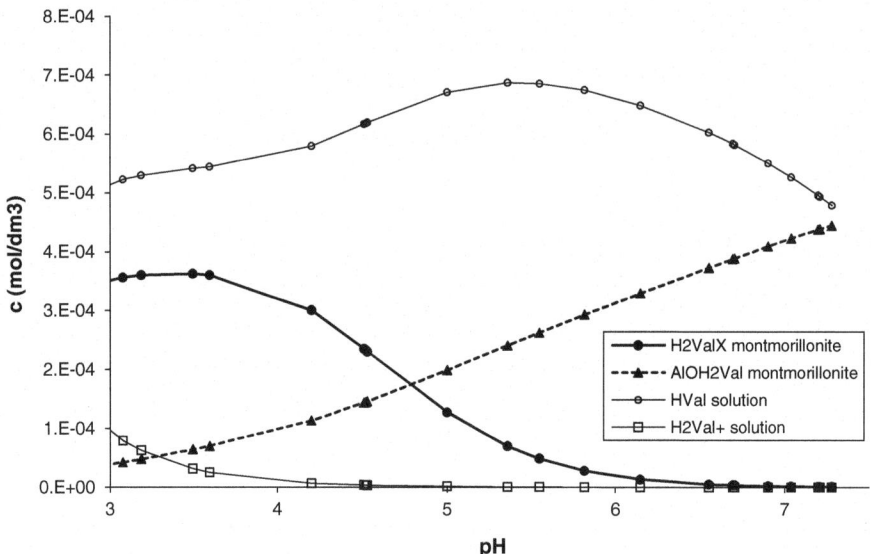

FIGURE 2.18 The quantity of sorbed valine in the interlayer space and on the aluminol sites of zinc-montmorillonite, and the concentration of valine species in the solution. $m = 100$ mg, $V = 20$ cm^3, $c_0 = $ 1e-3 mol dm^{-3}, $T = 20°C$.

TABLE 2.12

The Stability Constants of the Possible Valine Species for Different Cation-Exchanged Montmorillonites

	lg K		
	Ca-mont	**Cu-mont**	**Zn-mont**
$2H^+ + Val^- + X^- \rightleftarrows H_2 ValX$	23.0	22.1	22.7
$AlOH + H^+ + Val^- \rightleftarrows -AlOH_2 Val$	12.1	11.7	15.6
$SiO^- + Cu^{2+} + Val^- \rightleftarrows -SiOCuVal$		6.0	

Source: Reprinted from Nagy and Kónya 2004, with permission from Elsevier.

As a conclusion, we can say that the interlayer cations have an important role in the valine sorption mechanism. The total quantity of the adsorbed valine increases in the following order: calcium-montmorillonite < copper-montmorillonite < zinc-montmorillonite. This order can be explained by the cumulative effect of the affinity of cations to the layer charges (Table 2.2) and the stability constants of metal ion-valine complexes in the solution (Table 2.11).

In order for the valine to be adsorbed on layer charges, the charges have to be protonated. The affinity of calcium ions to the layer charges is the strongest (Table 2.2), and consequently, valine sorption is very low on calcium-montmorillonite. In this case, valine adsorption on the protonated aluminol sites is the dominant process. The affinity of zinc and copper ions to the layer charges is smaller (Table 2.2), so the protonation of layer charges and, consequently, the sorption of valine, is easier. Calcium and zinc ions practically do not form stable complexes with valine in the solution. The copper ions, however, forms stable complexes with valine. Complex formation promotes the desorption of copper ions from the layer charges. Consequently, the protonated layer charges are formed easily, and valine adsorption increases. Theoretically, the positively charged $CuVal^+$ complexes can also be adsorbed on the layer charges, but the surface complexation model did indicate that this adsorption does not take place. $CuVal^+$ complexes are sorbed on the deprotonated silanol sites. In the case of copper-montmorillonite, valine adsorption on protonated aluminol sites is negligible. Valine sorption on aluminol sites, however, is much higher on zinc-montmorillonite. It can be explained by the adsorption of zinc ions on the edge sites during the preparation of zinc-montmorillonite (Sections 2.5.1.1 and 2.6), which can form an AlOZnVal surface complex.

2.10 TRANSFORMATIONS INITIATED BY INTERFACIAL PROCESSES OF MONTMORILLONITE

Metal ions and organic molecules are adsorbed in the interlayer space and on the edges of clay minerals are adsorbed in the interlayer space and on the edges of clay minerals can transform into other chemical species via many different processes. The sorbed metal ions can undergo redox and hydrolytic reactions, which result in the formation of nano- and microparticles in the interlayer space as well as on the outer surfaces. Two-dimensional nanolayer may be formed in the interlayer. Three-dimensional particles are formed on the surfaces initiated by the metal ion adsorption on the edge sites.

The structure of the organic molecules may also undergo changes as a result of adsorption on clay minerals, leading to catalytic transformations (Chapter 1, Section 1.3.5). The cations of the clay minerals may also effect these transformations.

In this section, the redox and hydrolytic processes that result in the formation of nano- and microparticles will be discussed in metal ions (manganese, iron, lead, zinc, and silver ions)/montmorillonite (bentonite) systems. In addition, the catalytic diacetylation reaction of aromatic aldehides will be shown.

2.10.1 OXIDATION OF MN(II) ION AND FORMATION OF A NANOLAYER IN THE INTERLAYER SPACE OF MONTMORILLONITE

Manganese(II) ions can be sorbed in the interlayer space of montmorillonite using a calcium-manganese(II) cation-exchange reaction (Section 2.2). The equilibrium constant of calcium-manganese(II) cation exchange on montmorillonite is about 1 (Table 2.2). As a result of the exchange reaction, all the calcium ions can be exchanged for manganese(II) ions, and in this way it is possible to prepare manganese(II)-montmorillonite (Section 2.3). When this substance is kept for a long time (1–2 years) under atmospheric conditions (at ambient temperature, in the presence of air and humidity), its color, physical, and chemical properties change. The freshly prepared Mn(II)-montmorillonite is light, while the old Mn-montmorillonite is dark brown, similar to the color of manganese dioxide (Figure 2.19). The change in color indicates oxidation (Nagy et al. 2004; Kónya et al. 2005).

At the same time, the concentration of manganese ions remains the same. The basal spacing (d_{001}) of montmorillonite determined by X-ray diffraction is very similar for the newly prepared (1.51 nm) and old (1.48 nm) samples. The distribution of

FIGURE 2.19 Color of the fresh Mn–bentonite (left) and that of the old Mn–bentonite (right). (Reprinted from Nagy et al. 2004, with permission from Elsevier.)

manganese determined by scanning electron microscopy is also uniform in the old samples, similar to that in the fresh samples.

However, redox potential measurements, ESCA scans, and thermoanalytical studies indicate that manganese(II) has, in fact, undergo oxidation. The redox potential of the old manganese-montmorillonite is more positive than that of the fresh sample, showing an oxidation process. According to the results of ESCA studies, the oxidation state of the fresh manganese-montmorillonite is II, whereas the old samples refer to III, IV oxidation states.

In Figure 2.20, TG, DTG, and DTA curves of fresh and old manganese-montmorillonites are shown.

The thermoanalytical curves of clays can usually be divided into two regions:

1. The region of dehydration of clay (below 200–250°C) (Yermiyahu et al. 2003), and

2. The endothermic dehydroxylation of clay (above 500°C), which is followed by a small exothermic peak of the recrystallization. Both regions can be observed in Figure 2.20, namely, the elimination of interlayer water (at 120°C) and the endothermic dehydroxylation (above 500°C). In addition, the elimination of water coordinated to manganese (II) ion (at 260°C) is well distinguished in the DTA and DTG curves (Földvári et al. 1998). This peak of the dehydration of manganese (II) is more intense in the DTG curve for the fresh manganese-montmorillonite than for the old manganese-montmorillonite where this peak cannot readily be observed. The exothermal oxidation of manganese (II) into manganese (IV) is also more significant in the DTA curve for the fresh sample where the endothermic dehydroxylation of montmorillonite at high temperatures is compensated by the exothermic oxidation of manganese (II) into manganese (IV). This compensation is not observed on the DTA curve for the old manganese-montmorillonite where the major part of manganese has been oxidized before undergoing the thermal treatment during the analysis.

The oxidation of manganese affects the adsorption and catalytic properties of clay. For example, valine amino acid does not adsorb on fresh manganese-montmorillonite, while it is adsorbed on old manganese-montmorillonite (Nagy and Kónya 2004). The old manganese-montmorillonite catalyzes the destruction of hydrogen peroxide, fresh manganese-montmorillonite, however, has no catalytic effect on the destruction of hydrogen peroxide. The surface acid–base properties are also different (Nagy et al. 2004).

These results indicate that a two-dimensional nanolayer is formed in the interlayer space where manganese ion is bonded to the clay layers by two positive charges, and to oxides or hydroxides by the other two positive charges. The presence of a two-dimensional layer of manganese in the interlayer space of montmorillonite is understandable because the montmorillonite has a layered structure and manganese ions are introduced into the interlayer space by cation exchange. For this reason, the distribution of manganese ions is uniform (as also seen by scanning electron microscope [SEM]), and it does not change under oxidation.

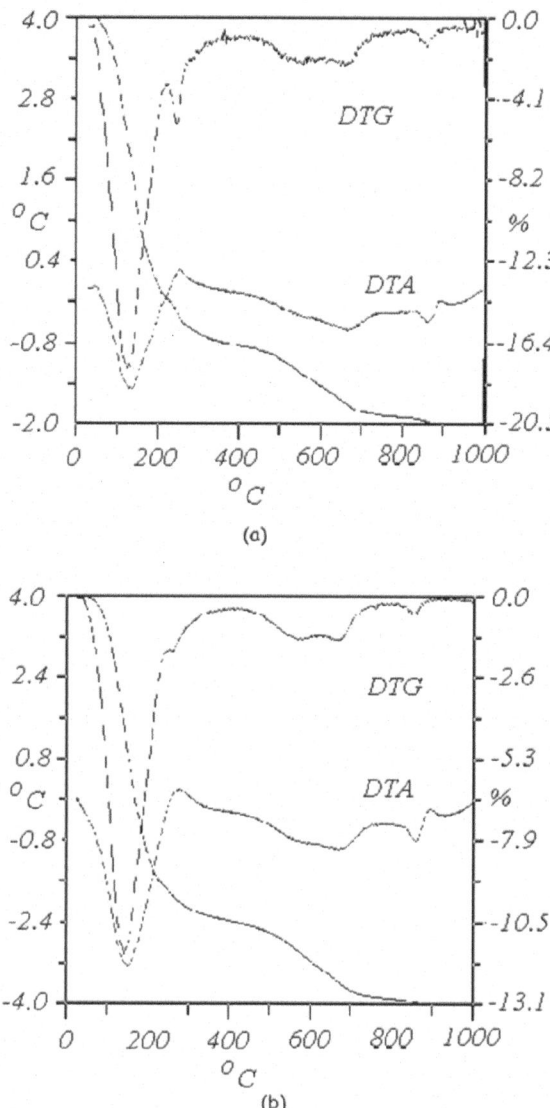

FIGURE 2.20 Thermoanalytical curves of fresh (a) and old (b) Mn–bentonite (x-axis: heating temperature, left y-axis: temperature change for *DTA* curve, right y-axis: loss of mass in percent for *TG* curve). (Reprinted from Kónya et al. 2005, with permission from Springer.)

Another example of the formation of nanoparticles in the interlayer space is the solid-state reaction of manganese(II)-montmorillonite and sodium sulfide. The resulting MnS nanoparticles can be used as a phosphor in thin-film electroluminescence devices (Khaorapapong 2009). The incorporation of CdS, ZnS and/or PbS into mesoporous silica (Chen et al. 1998; Zhang et al. 2001; Gao et al. 2001), layered metal oxides (Shangguan and Yoshida 2002), and layered metal phosphate (Cao et al. 1991) has also been reported.

A similar oxidation process occurs during the preparation of tin-montmorillonite. The calcium ions in the interlayer are exchanged to tin(II) ions in some minutes (about 5 min) because of the acidic pH of tin(II) chloride solution. This short time is enough to result in the oxidation of tin(II) into tin(IV) as shown by Mössbauer and electron spectra. At the same time, the d(001) basal spacing of montmorillonite decreases to 1.351 nm (Buzetzky et al. 2019b).

2.10.2 FORMATION OF AN IRON(III) OXIDHYDROXIDE NANOLAYER IN THE INTERLAYER SPACE OF MONTMORILLONITE

As seen in Figure 2.1, the substitution of the aluminum ions of the octahedral layer with iron(II) ions leads to the formation of the negative layer charge of montmorillonite. In addition, as a cation, iron ions can be present in the interlayer space also. In both places, the oxidation state of iron can be Fe(II) and Fe(III). Reactions involving the back-and-forth transition between the two oxidation states play a significant role in the redox processes of rocks and soils as well as in the catalysis of redox reactions ('Choudary et al., 1994; Stucki et al. 2002; Stucki 2008).

It is well known that Fe(II) ions are easily oxidized and hydrolyzed in the presence of oxygen and water, so the preparation of iron-montmorillonite is problematic. In aqueous suspension or in any Fe-salt (either Fe(II)– or Fe(III)–salt) in the presence of air, iron(III) oxyhydroxides or oxides are formed (e.g., Berry et al. 1986; Drame 2005; Green-Pedersen and Pind 2000; Izumi et al. 2005; Kong et al. 2005; Kozai et al. 2001; Oliviera et al. 2003; Pillai and Sahle-Demessie 2003; Shrigadi et al. 2003).

To avoid the hydrolysis of iron cations, iron(III)-montmorillonite can be prepared with $FeCl_3$ dissolved in acetone (Li et al. 1998). When studying the preparation process and properties of the product, the effects of acetone as well as $FeCl_3$ have to be investigated (Komlósi et al. 2006, 2007). When sodium-montmorillonite is suspended in acetone, the iron content decreases. X-ray diffraction studies show that the basal spacing of the montmorillonite does not change under the effect of acetone, but there is some broadening of d_{001} peak and slight changes in the relative intensities of certain peaks (mainly at $2\Theta = 30–32°$, related to montmorillonite reflections). This may be connected with some degradation in the structure of the montmorillonite, namely, acetone may cause the dissolution of some of the iron from the montmorillonite. Similar changes were observed in the X-ray diffraction analysis (XRD) patterns and were also found by Vlasova et al. (2003) when bentonite was dehydrated by an acidic treatment. The changes were explained with the destruction of Si-O-Al bonds and with the increase of the number of Si-O bonds due to the destruction of montmorillonite.

When montmorillonite is suspended in the acetoneous solution of $FeCl_3$, the iron content of the clay obviously increases. The increase of iron concentration is proportional to the quantity of iron(III) added to montmorillonite, and it can even surpass the cation exchange capacity. However, no excess of iron salt can be observed by Mössbauer spectra or X-ray diffraction.

The basal spacing (d_{001}) of montmorillonite increases during treatment with $FeCl_3$; it is 1.6 nm (Table 2.3), compared to the basal spacing of the original sodium-montmorillonite, which is 1.26 nm. It can be explained by the incorporation of Fe^{3+} ions into the interlayer space (e.g., Izumi et al. 2005; Kong et al. 2005). The relative intensities of the main peaks do not change after this treatment, indicating that the structure of montmorillonite does not undergo similar changes as observed in the presence of acetone alone.

The microenvironment of iron can well be studied by Mössbauer spectroscopy (Kuzmann et al. 1998). The Mössbauer spectra (taken at room temperature) of sodium-bentonite, treated with pure acetone and acetoneous $FeCl_3$ solution, are shown in Figure 2.21 (upper figure).

As seen in Figure 2.21, there are three main peaks in the 0 to 2.5 mm/s range that can be decomposed into two doublets (the corresponding parameters are given in Table 2.13).

The Mössbauer spectrum of the original sodium-bentonite shows the isomer shift and quadrupole splitting values as usual for montmorillonite (Stevens et al. 1983). The values δ_1 and Δ_1 exhibit Fe^{3+}, while the values of δ_2 and Δ_2 reveal Fe^{2+} microenvironments. These values are typical for Fe^{3+} and Fe^{2+} ions that are in the central positions of octahedrons of aluminosilicates (Kuzmann et al. 1998; Stevens 1958-2002). The relative areas (A's) show that the Fe^{3+} species is dominant as usual for Na-bentonites (Stevens et al. 1983). The spectra at the temperature of liquid nitrogen show no magnetically split components.

The treatment of sodium-bentonite with acetone causes no significant changes in the Mössbauer spectrum. The Fe^{2+}: Fe^{3+} ratio practically remains approximately the same. It is in agreement with the results of X-ray diffraction studies showing no change in the lattice parameters.

The room-temperature Mössbauer spectra of the samples treated with acetoneous $FeCl_3$ solution, however, exhibit changes. The relative occurrence of Fe^{2+} decreases considerably; it is clearly the result of the treatment with $FeCl_3$. The decrease of the ratio of Fe^{2+}:Fe^{3+} is proportional to the quantity of added $FeCl_3$. The Mössbauer pattern of Fe^{3+} microenvironments, freshly introduced by the treatment, can hardly be distinguished from those in the original montmorillonite on the basis of the room temperature Mössbauer spectra. However, the Mössbauer spectrum recorded at the temperature of liquid nitrogen displays significant changes (Figure 2.21, lower): new magnetically split subspectra, two sextets, have appeared. The Mössbauer parameters of the sextets (Table 2.13) are similar to data of Fe^{3+} species intercalated between the layers in iron-exchanged montmorillonite (Berry et al. 1986). The relative areas (A's) of the subspectra (Table 2.13) indicate that the iron introduced by the treatment is dominantly in microenvironments producing the sextets. Treatment with acetoneous $FeCl_3$ is a successful method to introduce Fe^{3+} ions into the interlayer space of montmorillonite, as was also confirmed by the increase of d_{001} basal spacing and the catalytic activity (Section 2.10.5). The Mössbauer parameters of Fe^{3+} in the interlayer are close to those characteristics for Fe^{3+} microenvironments in oxyhydroxides. It can indicate that the ligand environment of Fe^{3+} in the interlayer is similar to iron-oxyhydroxides. On the other hand, the presence of $\alpha\text{-}Fe_2O_3$ can be excluded based

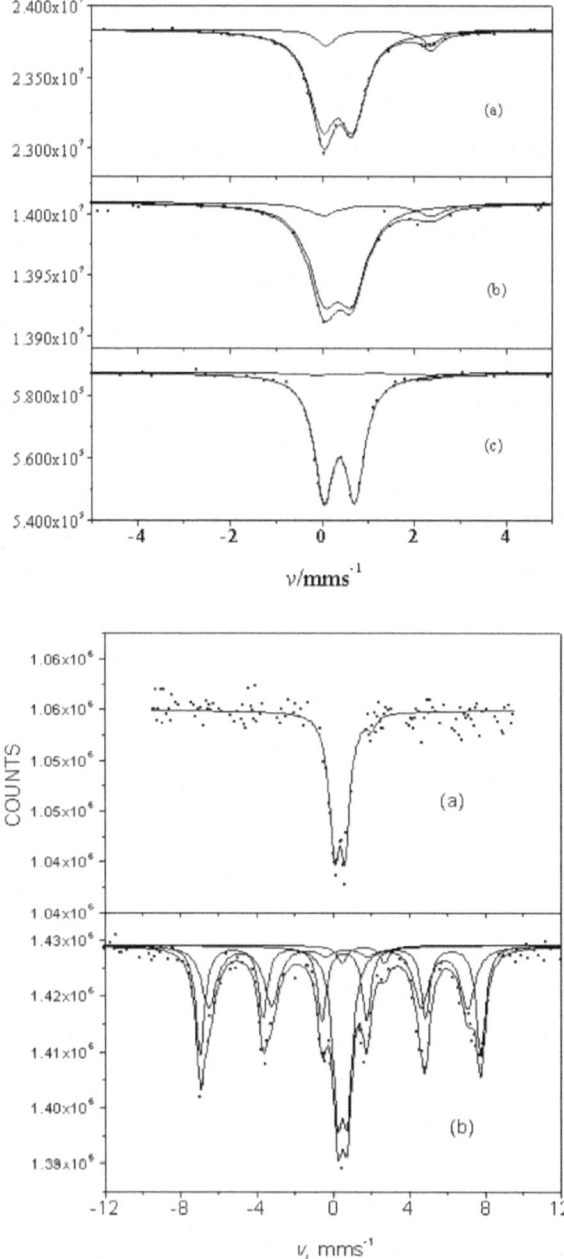

FIGURE 2.21 ^{57}Fe Mössbauer spectra of sodium-bentonite (a), after treatment with acetone (b), and acetoneous solution of $FeCl_3$ (c), taken at room temperature (upper). Mössbauer spectra of bentonite before (a) and after (b) treatment with $FeCl_3$ in acetone, recorded at 74 K (lower). (Reprinted from Komlósi et al. 2007, with permission from Clay Mineral Society.)

TABLE 2.13

Mössbauer Parameters of Bentonites.

Sample	Na-bentonite	Na-bentonite	Na-bentonite treated with acetone	Na-bentonite treated with FeCl$_3$ in acetone	Na-bentonite treated with FeCl$_3$ in acetone	Na-bentonite treated with FeCl$_3$ in acetone	Na-bentonite treated with FeCl$_3$ in acetone
Mass ratio Na-bent:FeCl$_3$				1.5:1	3:1	6:1	3:1
T (K)	293	88	293	293	293	293	74
δ_1 (mms^{-1})	0.34	0.44	0.34	0.35	0.36	0.36	0.47
Δ_1 (mms^{-1})	0.64	0.70	0.63	0.63	0.66	0.69	0.53
A_1 (%)	88.4	89.7	87.0	96.7	97.3	97.8	26.3
δ_2 (mms^{-1})	1.21	1.34	1.20	1.23	1.22	1.24	1.40
Δ_2 (mms^{-1})	2.28	2.35	2.31	2.31	2.32	2.31	2.41
A_2 (%)	11.6	10.3	13.0	3.3	2.7	2.2	2.7
δ_3 (mms^{-1})							0.48
B_3 (T)							41
A_3 (%)							32.1
δ_4 (mms^{-1})							0.47
B_4 (T)							46
A_4 (%)							37.9

Note: δ denotes the isomer shift, Δ denotes the quadrupole splitting, B is the magnetic induction, and A denotes the relative spectral area. The indices 1, 2, 3, and 4 are related to the spectral components of Fe^{3+} doublet, Fe^{2+} doublet, and two Fe^{3+} sextets, respectively.

Source: Reprinted from Komlósi et al. 2007, with permission from Springer.

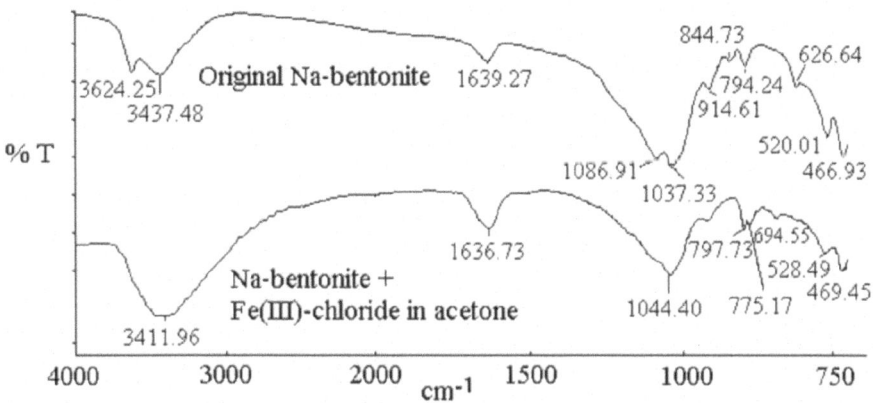

FIGURE 2.22 IR spectra of Na-bentonite, and after its treatment with FeCl₃ in acetone. (Reprinted from Komlósi et al. 2007, with permission from Clay Mineral Society.)

on the Mössbauer measurement. So, the results show that some kind of nanoparticles may be present in the interlayer space of montmorillonite. Some authors suggested that Fe-oxyhydroxides are formed in the interlayer space (e.g., Izumi et al. 2005; Pillai and Sahle-Demessie 2003).

Infrared spectra also show some changes in the montmorillonite structure (Figure 2.22) during the treatment with FeCl₃ in acetone.

The bands characteristic of montmorillonite lattice (e.g., Si-O stretching near 1040 cm⁻¹, H-O-H bending near 1630 cm⁻¹, and OH-stretching in the range of 3400–3600 cm⁻¹) remain the same. However, the intensities of the bands of the stretching and bending of H-O-H (3624 and 1630 cm⁻¹) decrease. This can be associated with the dehydration of montmorillonite. It is further confirmed by the intensity decrease and shift of Si-O bending vibrations at 520 cm⁻¹ to a higher frequency (775 cm⁻¹). At the same time, a new absorption band at 775 cm⁻¹ due to a dehydroxylated phase is observed. Furthermore, the hydroxyl bending absorption bands (840–940 cm⁻¹) are less intense, and a broad shoulder appears between 1070 and 1150 cm⁻¹. These changes are associated with the dehydration and dehydroxylation of montmorillonite (Bray and Redfern 2000).

The decrease in the intensity of the band at 3640 cm⁻¹ is especially interesting because this peak belongs to the hydrate water of the interlayer cation (Section 2.1.2). It means that the Fe^{3+} cation in the interlayer has no hydrate water. This shows in the direction of a layer similar to iron oxidhydroxide in the interlayer space. This layer can be formed by the hydrolysis of Fe^{3+} ions in the interlayer water: even when the hydrolysis is avoided in the solvent (e.g., by the application of acetone here), it can occur with the water molecules in the interlayer of montmorillonite.

It is interesting to note here that the suspension of iron-montmorillonite prepared from acetoneous solution of FeCl₃ has pH ≈ 2. By gradually washing it with water, that is, by the increase of pH of the suspension, hydrogen and chloride ions in equivalent quantities, namely hydrochloric acid, leaves iron-montmorillonite. At the same time, the basal spacing of montmorillonite and the relative ratio of the sextet in

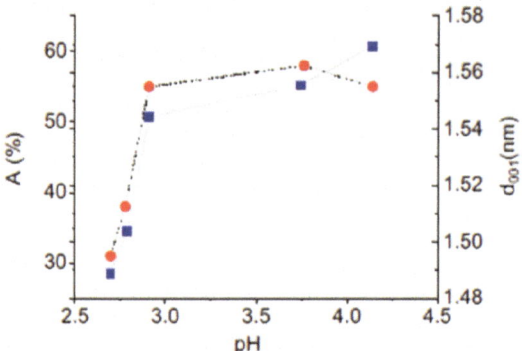

FIGURE 2.23 pH dependence of relative Mössbauer spectral area of sextets to iron intercalated (red circles) and basal spacing d(001) (blue squares). (Reprinted from Kuzmann et al.
2016b, with permission from Springer.)

Mössbauer spectra increase. In the range of pH = 2.7–2.9, an abrupt increase can be
observed in these values, associated with the hydrolysis of Fe^{3+} ion (Figure 2.23). In
aqueous solution, this is the characteristic range of the hydrolysis of Fe^{3+}. The hydrolyzation is stabilized at higher pH values, where iron oxide hydroxide stalilizes the
H-montmorillonite structure.

2.10.3 REDUCTION OF IONS

2.10.3.1 Reduction of Silver and Palladium Ion

Naturally, the chemical form that the cation will have in solution, in the interlayer,
and on the outer surfaces of montmorillonite, is governed by thermodynamics.
As seen in Section 2.10.2 iron(III) ions can be hydrolyzed in the water of the
interlayer even if the hydrolysis is avoided through the application of an organic
solvent. Similarly, manganese(II) ions are oxidized in the interlayer space of montmorillonite under the effect of the oxygen of the air and the interlayer water. The
size of the interlayer space limits the size of the oxide and oxidhydroxide layers.
As seen in the case of iron- and manganese-montmorillonite, the result is a two-
dimensional nanolayer.

Besides oxidation, reduction processes can also take place spontaneously or by
chemical treatments when the redox potential of the sorbed cation makes it possible.
As an example, the spontaneous reduction of silver ions in silver-montmorillonite is
shown here (Kónya et al. 2005).

(The possible application of Ag-bentonite in anion sorption will be discussed in
Section 2.11.)

Silver-montmorillonite can be produced from sodium- or calcium-montmorillonite. To avoid the hydrolysis of silver ion in the solution, the pH has to be adjusted
at 4 so silver-hydrogen-sodium/calcium ions are present in the interlayer. The SEM
picture (Figure 2.24) and thermal analytical studies (Figure 2.25) of this sample
show the following features.

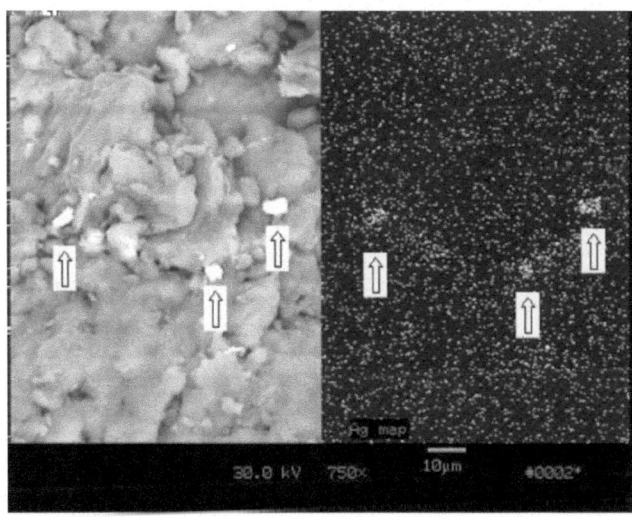

FIGURE 2.24 Scanning electron microscopic picture of silver-montmorillonite. Left side: morphology of the sample made by backscattered electrons. Right side: silver map made by characteristic X-ray photons. Silver concentration is proportional to the density of the light spots. The arrows show spots with 100% silver concentration. (Reprinted from Kónya et al. 2005, with permission from Springer.)

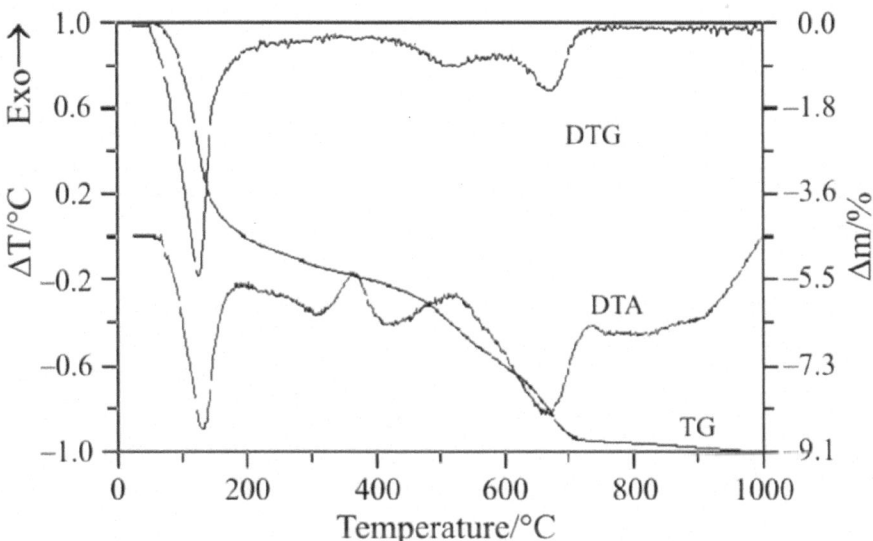

FIGURE 2.25 Thermoanalytical curves of silver-montmorillonite. (Reprinted from Kónya et al. 2005, with permission from Springer.)

The horizontal resolution of SEM is about 0.5–1 µm, and the density of light spots in a SEM picture is proportional to the concentration. Figure 2.24 shows that the distribution of silver ions on montmorillonite is usually uniform, but there are places where the concentration of silver is much higher, which refers to silver enrichments. The horizontal sizes of the silver enrichments are some micrometers in diameter. The vertical size cannot be determined from the SEM pictures. The silver content can reach 100% (marked with arrows in Figure 2.24). It shows that the reduction of silver (I) to metallic silver also takes place.

The presence of the metallic silver is supported by the thermal analytical studies as well (Figure 2.25).

The thermal analytical curves of silver-montmorillonite (Figure 2.24) show the usual reactions of monovalent montnmorillonites (Section 2.1.2). An additional exothermal reaction appears at 361°C. It corresponds to the oxidation of metallic silver, and this correspondence is further confirmed by the change of the color of silver-montmorillonite, which is originally dark gray and becomes light after heat treatment.

The application of silver bentonite for the sorption of (radioactive) halogenide ions will be discussed in Section 2.11.

Palladium ions sorbed on the edge sites of montmorillonite (Section 2.5.1.3) can also be reduced to metallic palladium with ethanol. As a result of the reduction, palladium enrichments, similar to silver enrichments in the previous example, can be detected by SEM.

Thus, the reduction of both silver and palladium ions results in the formation of metallic microparticles on montmorillonite.

2.10.3.2   Reduction of Fe(III) Ions

Fe(III) ions of Fe(III)-montmorillonite can be reduced to Fe(II) ions by ascorbic acid (Manjanna 2008), so Fe(II)-montmorillonite can be produced. This product, however, is not stable in the presence of oxygen and water. Fe(II) ions are oxidized again to Fe(III) ions.

2.10.4   HETEROGENEOUS NUCLEATION ON EDGE SITE: FORMATION OF LEAD OXIDE FINE PARTICLES ON THE EDGES OF MONTMORILLONITE

As mentioned in Section 2.5.1.4, lead ions can sorb on Ca-montmorillonite through other than simple ion-exchange reaction (Nagy and Kónya 1998; Nagy et al. 1998b, 2001, 2003a). The main sorption mechanism is obviously still the cation exchange between lead ions and calcium ions in the interlayer space of montmorillonite. However, there is also the sorption of lead ions on the edge sites where Pb-O bonds can be formed. This process serves as an initiator of the formation of a heterogeneous nucleation on particle surface, followed by crystal growth. As a result, areas with lead enrichment form over the surface of the clay particles. These can be detected using different analytical methods. The third possibility of lead sorption is on the impurities of montmorillonite. Some of these impurities, namely, other aluminosilicates, have the same characteristic edge sites as montmorillonite; thus, in this case, the reaction is also the same. Other minerals (e.g., iron containing minerals) can also form co-precipitate with lead ions. In this section, the mechanism of these different lead sorption processes will be discussed in detail.

Cation exchange in the interlayer space can be studied by comparing the characteristic properties of the interlayer as well as by measuring the quantity and distribution of sorbed lead ion. The characteristic properties of the interlayer are the basal spacing (d_{001}) and the water content. The basal spacing of original Ca-montmorillonite is 1.533 nm, and that of lead-montmorillonite is 1.254 nm (Table 2.3; Section 2.3). The decrease of basal spacing proves that cation exchange has occurred.

Cation exchange has an effect on the quantity of the interlayer water, especially on the hydrate water of cations. The transmittance of the adsorbed water band (3600 cm^{-1}) and the hydrate water band (3450 cm^{-1}) as a function pH is shown in Figure 2.26.

As seen in Figure 2.26, the transmittance of the adsorbed water of the calcium as well as lead-montmorillonite varies slightly with the pH. The higher transmittance of lead-montmorillonite shows a slight decrease of the adsorbed water of the interlayer space. However, the transmittance of the hydrated water of lead montmorillonite also increases, indicating that the quantity of hydrate water decreases during the exchange of calcium ions to lead ions. The value of basal spacing (d001) of lead-montmorillonite is in agreement with the decrease of water content of the interlayer space.

It can be shown that the quantity of lead ions in montmorillonite in atomic percent is equal to the quantity of calcium ions in calcium-montmorillonite (Table 2.14). Since the quantity of calcium ions is equal to the CEC, it means total cation exchange.

The lead ion sorption of the edge sites of montmorillonite is shown in the IR spectra by the disappearance of the band at 2920 cm^{-1}. This band has been assigned to the OH vibrations at the edges of the montmorillonite sheets (Peker et al. 1995), so its disappearance is due to proton-lead exchange at the edge sites.

Similar results have been obtained by Strawn and Sparks (1999) who used X-ray absorption fine structure (XAFS) spectroscopy. They studied the adsorption of lead ion by montmorillonite and found there were two different lead adsorption mechanisms depending on ionic strength. At low ionic strength, the mechanism was to be found to be pH independent and consistent with an outer-sphere complexation, that is, ion exchange. At high ionic strength, the mechanism was pH dependent and suggests inner-sphere complexation where lead forms covalent bonds with the oxygen of the edge sites.

The two different species of sorbed lead ions show different desorption properties. The desorption behavior in solutions of complex-forming agents was discussed in Section 2.8.4.

In Figure 2.27, the SEM picture of lead-montmorillonite is shown.

The back scattered image is shown on the lower left side of the figure, and a lead distribution map of the same surface obtained by characteristic X-ray photons is shown on the lower right side. On the top figure, the environment of the spot indicated with No. 2 is enlarged. The bright spots on the backscattered image indicate concentrations of elements with high atomic numbers, and they are in agreement with areas of higher lead concentration observed in the distribution map.

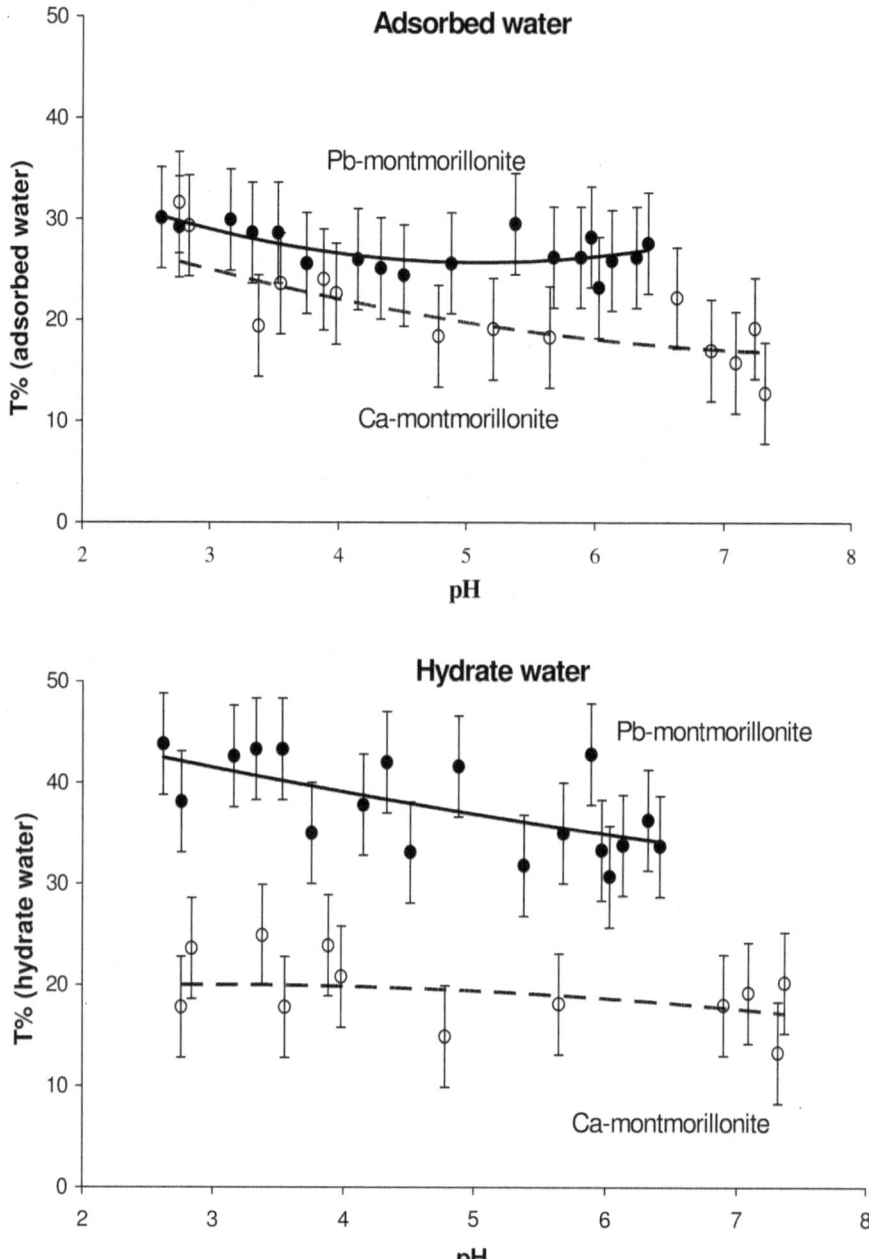

FIGURE 2.26 The transmittance values of adsorbed (upper) and hydrate (lower) water of Ca-montmorillonite and lead-montmorillonite as a function of pH, using the 3600 and 3450 cm^{-1} infrared bands. (Reprinted from Nagy et al. 2003a, with permission from Elsevier.)

TABLE 2.14

The Composition of Montmorillonites in Atomic Percent and the Selectivity Coefficient of Calcium-Lead Cation Exchange

Sample	O (%)	Al (%)	Si (%)	K (%)	Ca (%)	Fe (%)	Mg (%)	Pb (%)	K_{sel}
Ca-mont(exp)	63.3	6.6	23	0.3	1.3	4	1.5		
Ca-mont(calc)	65.8	8	21.2		1.2	2.2	1.6		
Pb-mont.	75.3 ± 2.4	5.6 ± 0.5	15.5 ± 1.6			1.5 ± 0.2	0.9 ± 0.1	1.2 ± 0.2	
Ca-Pb-mont.									
pH = 3.33, 5.10^{-4} mol/dm³ Pb²⁺	75.9	5.1	17	0.2	0.43	0.6	0.8	0.36	0.85
	74.8	5.2	18	0.2	0.41	0.6	0.8	0.36	0.85
	70.9	5.8	21.1	0.3	0.42	0.6	0.8	0.36	0.85
	66.5	6.7	24.4	0.3	0.45	0.7	0.9	0.36	0.85
pHh = 4.33, 5.10^{-4} mol/dm³ Pb²⁺	75.2	4.9	17.6	0.3	0.53	0.7	0.7	0.46	1.3
	74.9	5	17.8	0.4	0.55	0.6	0.7	0.46	1.31
	74.2	5	18.9	0.3	0.32	0.8	0.4	0.46	1.21
	68.6	5.7	23.5	0.1	0.56	0.6	0.9	0.46	1.32
	74.2	3.1	21.3	0.2	0.37	0.4	0.3	0.46	1.23
	62.8	7.7	26.3	0.6	0.54	0.9	1	0.46	1.31
	61.1	8.5	27	0.6	0.61	0.7	1.3	0.46	1.36
	72.8	5.8	18.5	0.2	0.67	1	1	0.46	1.41
pH=6.13, 5.10^{-4} mol/dm³ Pb²⁺	64.2	7	25.7	0.5	0.69	0.8	0.8	0.54	2.55
	63.4	6.2	27.9	0.2	0.65	0.7	0.8	0.54	2.3
	57.6	8.5	30.8	0.2	0.73	0.7	1.3	0.54	2.9
Mean	69.1 ± 6.1	6 ± 1.5	22.4 ± 4.4	0.3 ± 0.2		0.7 ± 0.1	0.8 ± 0.3		
pH = 3.55, 5.10^{-3} mol/dm³ Pb²⁺	69.4	6.1	22	0.2	0.28	0.8	1	0.8	2.87
	68	6.3	23.2	0.2	0.33	0.7	0.9	0.8	3.18
	67	6.4	24.1	0.2	0.35	0.6	1	0.8	3.41
	67.9	6.7	21.9	0.5	0.3	1.3	1	0.8	2.99
	75.1	5.1	17.5	0.1	0.28	0.7	0.9	0.8	2.87
	67.9	6.2	23.2	0.4	0.45	0.7	0.9	0.8	4.91

Note: The reaction of calcium-montmorillonite-lead ion was studied with 20 mg Ca-montmorillonite in 20 cm³ solution, $T = 25°C$ (Nagy et al. 2003).

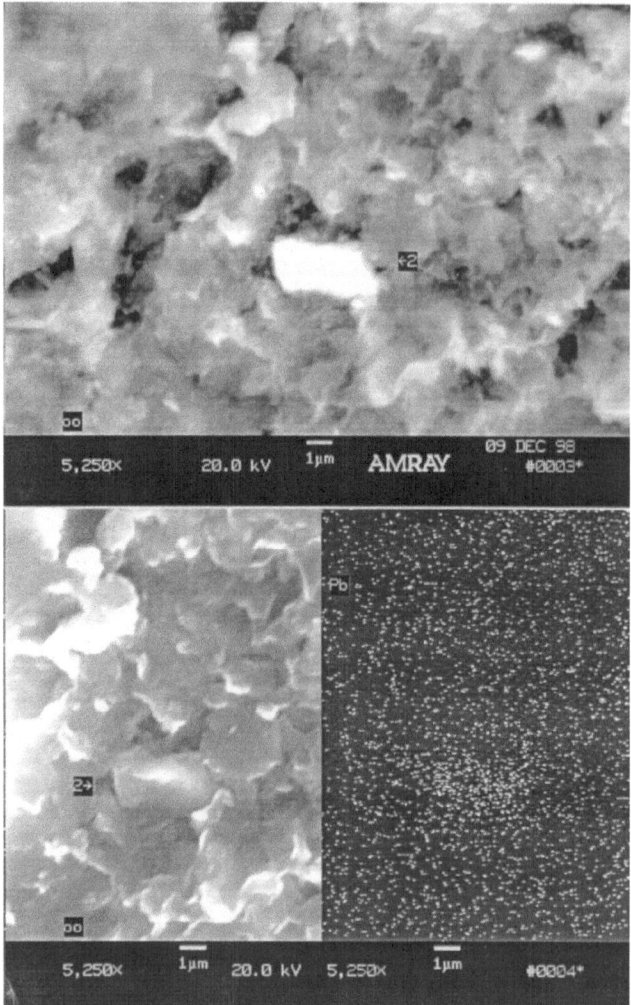

FIGURE 2.27   SEM picture of montmorillonite treated with 5e-4 mol/dm³ lead perchlorate solution. Lower left side: morphology of the sample made by backscattered electrons. Lower right side: lead map made by characteristic X-ray photons. Upper: morphology of the sample made by backscattered electrons, enlarged from site No. 2.

As seen in Figure 2.27, lead ions are fairly evenly distributed across the surface, proving cation exchange. However, there are spots where the concentration of lead is larger than average. In other words, lead enrichments can be observed.

The results of the quantitative analysis of the montmorillonite samples obtained with different Ca-montmorillonit/lead ion ratio and pH are shown in Table 2.14 for the surfaces with even lead distribution. The chemical composition of lead enrichments is shown in Table 2.15.

TABLE 2.15

The Composition of Lead Enrichments on Montmorillonites

	O (%)	Al (%)	Si (%)	K (%)	Ca (%)	Fe (%)	Mg (%)	Pb(%) in enrichment	Pb(%) mean
Mean Ca-mont.	56.1	8.3	28.2	0.6	1.0	4.1	1.2		
St. deviation	1.8	0.6	2.0	0.5	0.3	0.3	0.3		
Lead enrichments									
Sample 1	75.4	4.7	16.5		0.7	1.2	0.8	0.6	0.3
Sample 2	65.0	5.3	18.3		9.8	0.6	0.8	0.9	0.8
Sample 3	57.5	7.2	24.4		1.2	0.5	1.3	1.2	0.8
Sample 4	48.0	6.3	23.8		1.6	14.8	0.0	5.5	0.4

Source: Reprinted from Nagy et al. 2003a, with permission from Elsevier.

In Tables 2.14 and 2.15, the average compositions of calcium-, lead-montmoril-lonite, and calcium-lead montmorillonites are listed as atomic percent. (The data was determined by energy- dispersive X-ray spectroscopy, which cannot determine hydrogen content. Note that the hydrogen content is not indicated.)

In the first row of Table 2.14, the average composition of calcium-montmorillonite is given. In the second row, the mean composition of lead-montmorillonite, where lead concentration is even (no enrichments), is provided. The atomic percent of lead in the lead-montmorillonite is about equal, within the experimental error of 5 to 10%, of the atomic percent of calcium in calcium-montmorillonite. Since the inter-layer cation of the original montmorillonite is calcium ion, lead ions can completely exchange calcium ions.

The lead concentration of the calcium-lead montmorillonites increases with pH and with the initial lead concentrations of the suspensions.

As discussed previously in this chapter, there are spots where the lead content (and the elementary composition) is higher and so-called lead enrichments are formed. Table 2.15 shows that the composition of the lead enrichments can be rather different in the different enrichments. Besides the increase in lead con-centration, other changes in the elementary composition may also occur. There are examples of lead enrichment (Figure 2.28; Sample 3 in Table 4) where the average composition remains unchanged even though the lead concentration is elevated. However, there are lead enrichments where the increase of lead con-centration is accompanied by a simultaneous increase in iron (Sample 4 in Table 2.15), or a simultaneous increase of calcium and lead (Sample 2). Lead enrich-ment in these cases is likely due to lead coprecipitation reactions with other minerals (iron oxide and other calcium silicate phases) that were present in the sample.

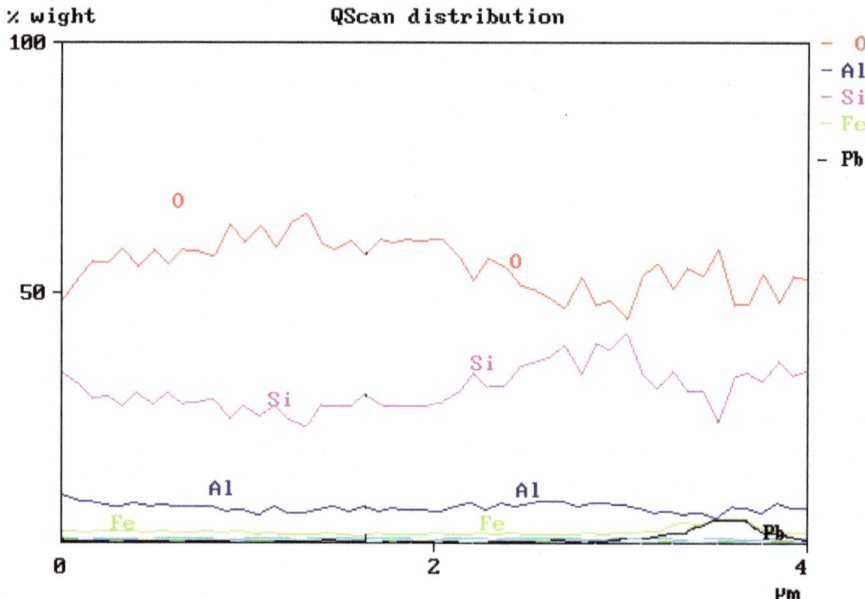

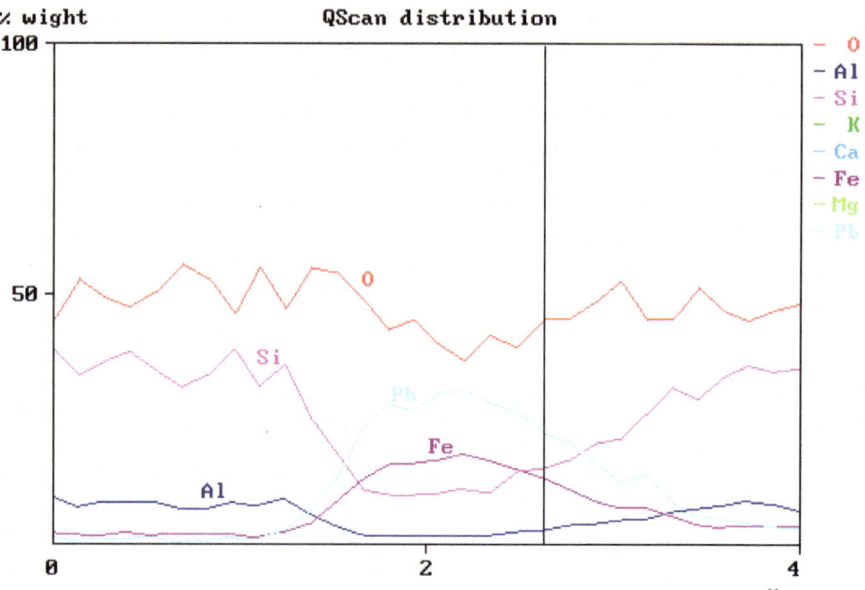

FIGURE 2.28 The concentration profiles of different elements of lead-montmorillonite. Upper: only the concentration of lead increases; the other elements show concentrations characteristic of montmorillonite. Lower: the increase of lead concentration is parallel to the increase of iron concentration. (Reprinted from Nagy et al. 2003a, with permission from Elsevier.)

The selectivity coefficients (Equation 1.79) for the different pH's and lead concentrations are listed in Table 2.14. At the most dilute concentration (5.10^{-4} mol dm^{-3}), the values of the selectivity coefficients increase as the pH increases. At similar pH values (3.33 and 3.55), the selectivity coefficients increase when increasing the lead concentration with an order of magnitude (from 5.10^{-4} mol dm^{-3} to 5.10^{-3} mol dm^{-3}). The standard deviation of selectivity coefficients is also greater at higher pH values and higher lead concentration. The increase of selectivity coefficients and its deviation with the increase of pH and lead concentration is due to the formation of lead enrichment spots. As a result, the equilibrium constant of the lead-calcium cation exchange reaction (Table 2.2) cannot be determined correctly (Chapter 1, Equation 1.81, Section 1.3.4.2.1) because of the two parallel processes and the formation of a new phase (lead enrichment phase). This limitation stands for ion-exchange isotherms, too (Chapter 1, Equation 1.94, Section 1.3.4.2.2).

As mentioned in Section 2.10.3.1, the resolution of SEM (Figure 2.27) is about 1 μm. Thus, the size of enrichments greater than 1 μm can be measured by SEM. When smaller spots of lead enrichments are found, their size should be determined by another method below 1 μm resolution. For example, atomic force microscopy (AFM) can be a suitable method since it has nanometer resolution. This technique, however, requires extremely smooth surfaces. The surface of freshly cleaved mica is smooth in molecular dimensions because the cleavage takes place along the plane of the crystal. Those lead enrichments that are weakly adsorbed on the surface of montmorillonite may be transferred to the surface of mica: the freshly cleaved mica plate is immersed into the suspension of calcium-montmorillonite in lead ion solution. The lead enrichments weakly adsorbed on montmorillonite can be absorbed onto the freshly cleaved flakes of mica. An AFM picture of a mica surface prepared in this way is shown in Figure 2.29.

As seen in Figure 2.29, the surfaces of mica flakes had 170 to 200 nm diameter objects on them. These objects are not layered (not montmorillonite), are likely to be amorphous, and it is likely that these particles that were weakly absorbed on the montmorillonite and transferred to the mica surfaces. However, AFM cannot provide analytical information on the samples; only morphological information can be obtained. In order to have some analytical information, the prepared mica surfaces can be examined by SEM (Figure 2.30).

The SEM picture (Figure 2.30) shows two types of particles: (1) lead enrichments without montmorillonite and (2) montmorillonite particles with lead enrichment on it. The montmorillonite particles are much larger, their size can reach 10 μm, and their elemental SEM maps show even distributions of Al, Si, Mg, Fe, Ca, and O; only the distribution of Pb shows spots with higher concentration (white spots in Figure 2.30).

Lead enrichments can be detected only when the system contains both lead ions and montmorillonite. However, they cannot be observed on mica flakes immersed into the solution of lead ions, or suspensions of calcium-montmorillonite. Similar studies (SEM, AFM) with calcium-montmorillonite (without lead ions) show even distribution of all elements; no enrichments were found.

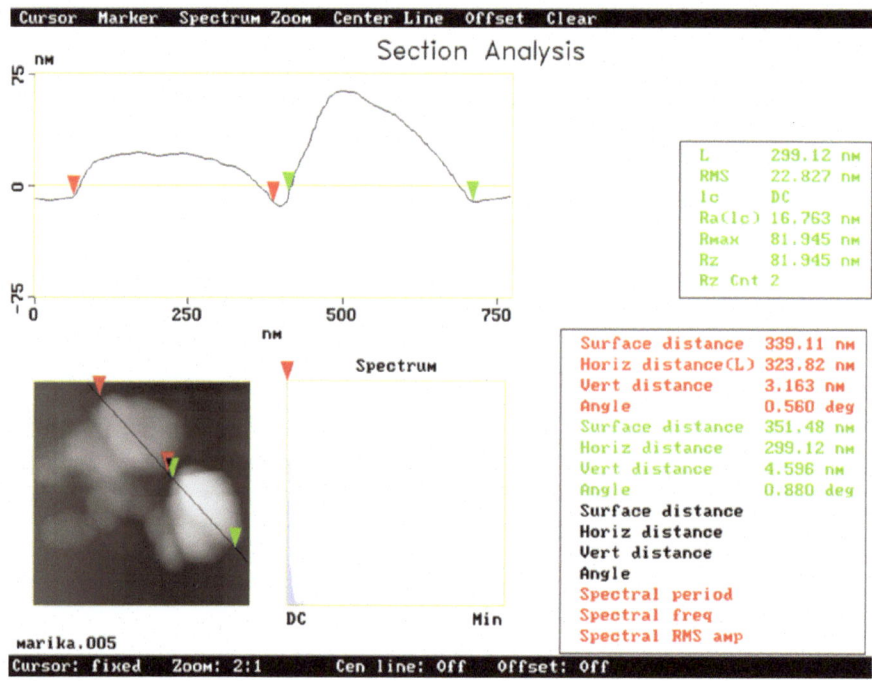

FIGURE 2.29 Atomic force microscopic (AFM) picture of a mica plate immersed into the suspension of Ca-montmorillonite and 5.10^{-4} mol dm^{-3} lead perchlorate solution.

Of course, only the weakly adsorbed lead enrichments can be transferred onto the surface of mica. The presence of other, more strongly adsorbed, lead enrichments on montmorillonite cannot be excluded.

As a conclusion, lead ions can interact with calcium-montmorillonite in different ways. The first is the usual lead-calcium cation exchange in the interlayer space, that is, outer-sphere complexation takes place. As a result, lead is distributed evenly on the internal surfaces. The second way is the reaction of lead ions with the edge sites of montmorillonite, or mineral impurities containing similar edge sites as montmorillonite (e.g., kaolinite). In this reaction, Pb-O bonds are formed, and an inner-sphere complexation takes place. This process is the initial step of heterogeneous nucleation on particle surface, followed by crystal growth. The third possibility is coprecipitation of lead ions with mineral impurities (e.g., iron-containing minerals; Figure 2.28). As a result of the latter two processes, nano- and microparticles (lead enrichments) are formed, and they can be detected by SEM and AFM.

The precipitation and colloid formation of different metal oxide hydroxides is known in soils when the concentration of the ions reaches the value of stability products. In this case, the precipitation can be explained by the thermodynamic properties of the bulk solution. In the lead ion/calcium-montmorillonite system, however, the production of lead enrichments cannot be explained by the thermodynamic constants of the solution of lead ions (Pourbaix 1966) and cannot be observed in the

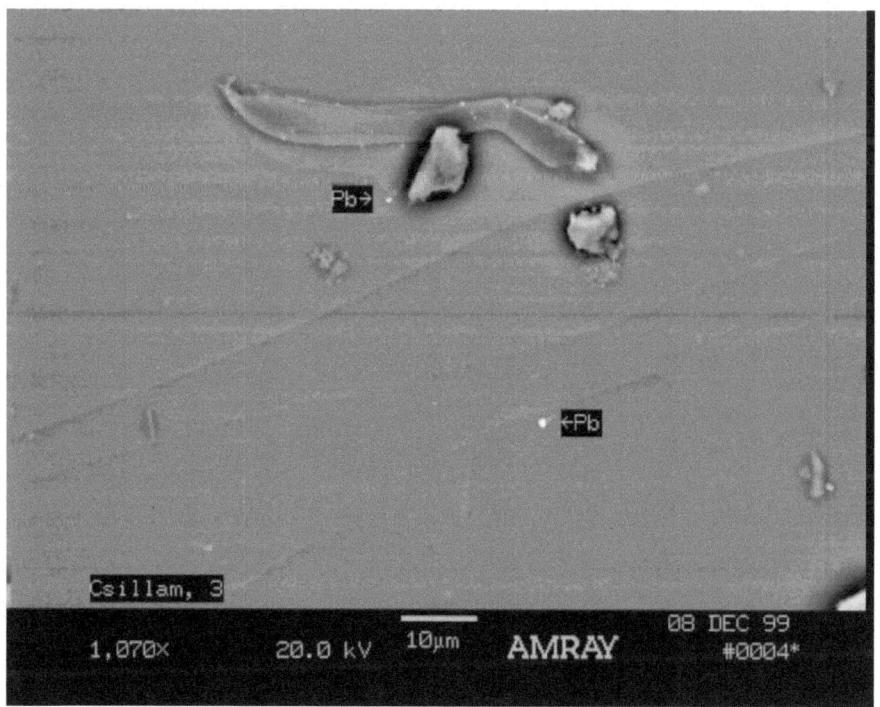

FIGURE 2.30 SEM picture of a mica plate immersed into the suspension of Ca-montmorillonite and 5.10^{-4} mol dm^{-3} lead perchlorate solution.

absence of montmorillonite. On the basis of the solubility product of Pb(OH)$_2$ (L = $6.8*10^{-13}$), lead is present as Pb^{2+} at pH values and lead ion concentrations of the experiments. The formation of lead enrichments requires the simultaneous presence of lead ions and montmorillonite.

The formation of lead enrichments can also be detected in the presence of complex-forming agents. As seen in Section 2.8.4, a part of lead ions cannot be desorbed from montmorillonite with complex-forming agents. Since the interlayer cations can be desorbed by complex-forming agents, provided that the stability of complexes is great enough, we can suppose that the lead ions remaining on montmorillonite are bonded in the lead enrichments.

2.10.5 LANTHANIDE ION EXCHANGE: STRUCTURAL MODIFICATION DUE TO THE EXCHANGE WITH LIGHT LANTHANIDE AND ITTRIUM IONS

Lanthanide cation-exchanged montmorillonite (La, Ce, Pr, Nd, Sm, Eu, Gd, Tb, Dy, Ho, Er, Tm, Yb, and Lu; except for radioactive promethium) and Y-exchanged bentonites were prepared from Ca-bentonite (Istenmezeje, Hungary) by ion exchange in three consecutive exchanges with lanthanide ion solutions. Their composition and structure were studied by different analytical techniques.

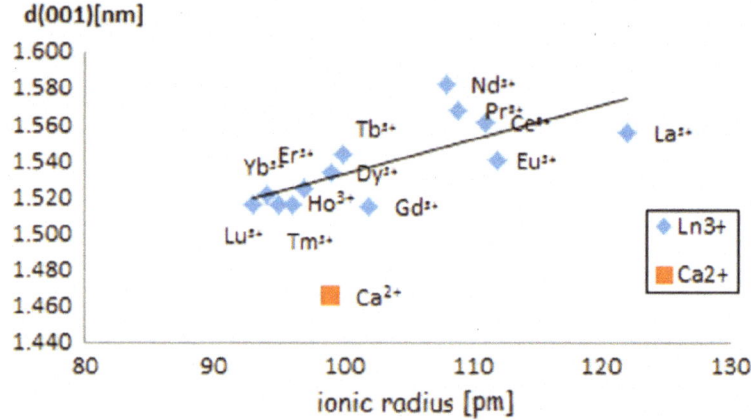

FIGURE 2.31 d(001) basal spacing of calcium- and lanthanide-montorillonites as a function of ion radius of interlayer cation. (Adapted from Kovács et al. 2017, with permission from Elsevier.)

XRD patterns of all Ln-bentonites and the initial original Ca-bentonite are quite similar to each other, reflecting that no essential changes occur in the crystal structure and in mineral composition upon the lanthanides incorporation in the montmorillonite interlayer space (Stefani et al. 2014). However, differences can be observed in the position of the peaks relating to the d_{001} basal spacing of lanthanide montmorillonites (Figure 2.31). The difference in the basal spacing between La and Ca montmorillonites is similar as reported by Stefani et al. 2014. Trivalent lanthanide ions' basal spacing d_{001} is greater than that of divalent calcium ion. The d_{001} basal spacing decreases from La (1.577 nm) to Lu (1.479 nm), in connection with the changes is ionic radii as mentioned in Section 2.3.

X-ray fluorescence spectroscopic and SEM-EDX measurements showed that some light lanthanide, such as La, Ce, and Gd, and ittrium cation concentrations exceed the CEC. At the same time, the iron concentration of bentonite significantly decreases. The sorption is the highest for ittrium (Kovács et al. 2019), as much as 130% of CEC at 25°C. The sorbed quantity increases as the initial lanthanum concentration of the solution and the temperature increases. In the case of lanthanum, as high as 136% of CEC sorbed quantity is obtained at 40°C (Kovács et al. 2017). This observation suggests that light lanthanide and ittrium ions may be also bounded in another way than simple cation exchange in montmorillonites interlayer. The question arises: how is it possible and is there any relationship between the sorption over CEC and the decrease in iron content of bentonite? A possibility is that iron ions in the octahedral positions of montmorillonite move into the interlayer space and the external surfaces. The emission of the positively charged iron ions from the crystal lattice increases the negative layer charge, and consequently, the CEC. Iron ions are hydrolyzed in the interlayer and the solvent water, forming iron oxide hydroxide species as shown by Mössbauer spectrum (Figure 2.32; Kuzmann et al. 2016a, 2016b).

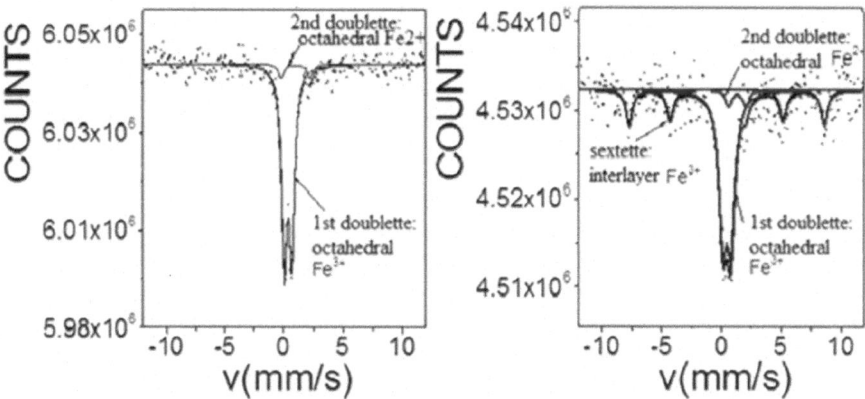

FIGURE 2.32 The ^{57}Fe Mössbauer spectra of Ca-bentonite (on the left) and La-bentonite (on the right) measured at 78 K. (Reprinted from Kovács et al. 2017, with permission from Elsevier.)

The 78 K Mössbauer spectrum (Figure 2.32, left side) of initial Ca-bentonite consists of the two doublets as usual in bentonites. These doublets show the octahedral iron ions. However, in the Mössbauer spectra of some lanthanum montmorillonite (and in other lanthanide montmorillonites also), a magnetically split component is observed, proving that during the preparation of lanthanide-bentonites a part of the octahedral iron atoms can move into the interlayer space.

2.10.6 Structural Change upon Heating Lanthanide-Bentonites

Structural changes in lanthanide (La, Ce) and ittrium-montmorillonites can be observed upon heating. Their ^{57}Fe Mössbauer spectra (without heat treatment) have a doublet envelop at room temperature, showing dominantly Fe^{3+} assigned to (cis) octahedral sites in the montmorillonite. 80 K spectra, however, show an additional magnetically slit component, associated with iron atoms intercalated into the interlayer space of montmorillonite (Section 2.10.5). Heating these samples to 250, 360, and 500°C, respectively, both room temperature and 80 K Mössbauer spectra reveal a new doublet, associated with Fe^{3+} at trans octahedral sites. This is illustrated in the case of Ce-bentonite in Figure 2.33. At the same time, X-ray diffractograms of the heat treated ittrium and lanthanide (La, Ce) montmorillonites show that a gradual mineral phase transformation of montmorillonite to muscovite happens; see Figure 2.34. This is in agreement with the fact that the characteristic position of iron in montmorillonite is cis, while that in muscovite is trans. Electron paramagnetic resonance (EPR) spectra also support this transformation, that is the gradual change (narrowing) in the shape of the signal around 1580 G as the temperature of heat treatment increases (Figure 2.35; Kuzmann et al. 2019; Lück et al. 1993).

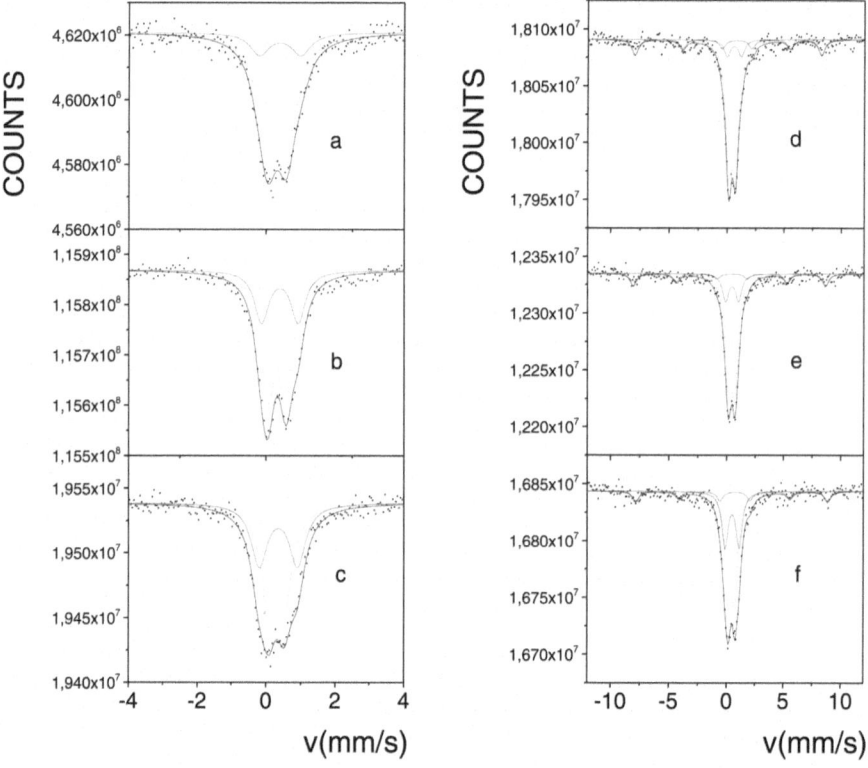

FIGURE 2.33 ^{57}Fe Mössbauer spectra, recorded 293K (on the left) and at 80K (on the right) of Ce-bentonites heat treated at 250°C (a, d), 360°C (b, e) and 500°C (c, f). (Reprinted from Kuzmann et al. 2019, with permission from Springer.)

2.11 EFFECT OF CHEMICAL MODIFICATION OF BENTONITE ON SORPTION PROPERTIES: SORPTION OF ANIONS

As mentioned several times in Chapters 1 and 2, natural clays, including bentonite clay rock and montmorillonite clay mineral, can uptake cation species, mainly by cation exchange processes in the interlayer space of montmorillonite. The sorption of anions is possible on the edge sites, at pH values lower than pzc; its degree is much less than the sorption of cations (Section 2.4). During the environmental applications, however, the sorption of anionic species should be desirable since the anionic pollutants, not sorbing to a significant extent, can relatively quickly migrate with water in the environment. By chemical modifications, sorbing sites can be created in the clays where anions can also be sorbed or precipitated as very insoluble salts. As a result, the migration rate of anions and the concentration of the anionic species is also reduced. In this section, the application of modified bentonites will be presented. The possibilities for the reduction of migration rate will be shown in the case of some radioactive anions (^{36}Cl$^-$, ^{129}I$^-$, ^{99}Tc$^-$ isotopes as pertechnetate ions, TcO$_4$$^-$).

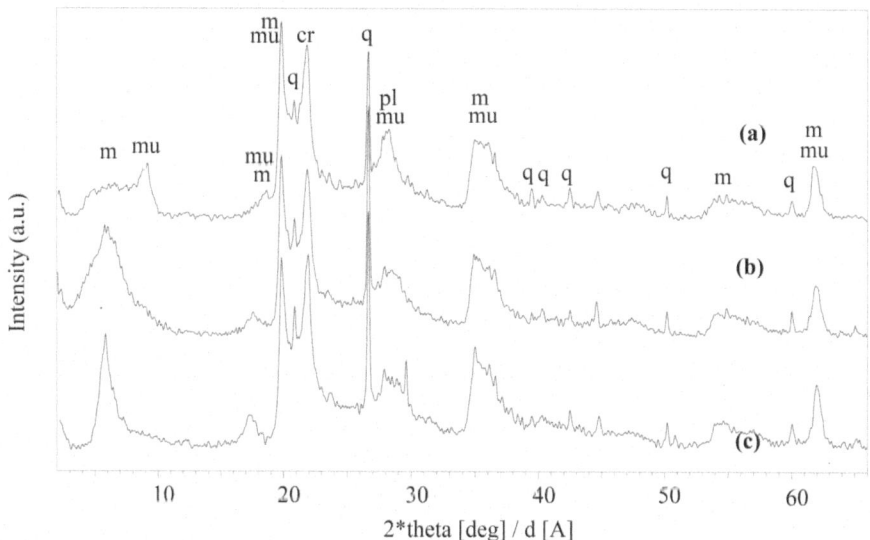

FIGURE 2.34 XRD patterns of Ce-montmorillonite heat treated at 250°C (a), 360°C (b) and 500°C (c), (m: monmorillonite, m: montmorillonite, mu: muscovite, q: quartz, cr: crystoballite and pl: plagioclase. (Reprinted from Kuzmann et al. 2019, with permission from Springer.

The sorption of phosphate and arsenite ions by modified bentonites will be presented as examples of concentration decrease of anionic pollutants. The decrease of the concentration of phosphate and arsenite ions is important to preventing eutrophication of waters and to providing clear drinking water, respectively.

The idea of anion uptake by bentonite was behind the preparation and application of cation-exchanged bentonite in which the cations in the interlayer can precipitate with the anionic species (Figure 2.36).

Considering the solubility of precipitates, $^{36}Cl^-$, $^{131}I^-$ isotopes can be precipitated on silver bentonite (Buzetzky et al. 2020), while phosphate (Buzetzky et al. 2017) and arsenite (Buzetzky et al. 2019a) ions can be precipitated on rare earth (REE) and Fe(III)-bentonites.The sorption of ^{99m}Tc isotopes as pertechnetate ions (TcO_4^-) was studied on Mn−, Cr−, Sn-bentonites (Buzetzky et al. 2019b).

As seen in the case of iron-bentonite (Section 2.10.2), the precipitation reaction occurs similarly to those in a bulk aqueous solution. The reaction was fast with small activation energy (some 10 kJ/mol). The exothermic or endothermic character of the precipitation is similar to the precipitates formed in bulk solutions. As an example, the uptake of halogenide ions by silver bentonite is mentioned: the precipitation of chloride ion is exothermic, while that of iodide ion is endothermic, as it is the case for the silver halogenide precipitates. In addition, the precipitation processes are influenced by the same factors. The effect of influencing factors is shown in the case of the sorption of iodide ion on silver bentonite. The application of inactive iodide carrier solution reduces the sorption

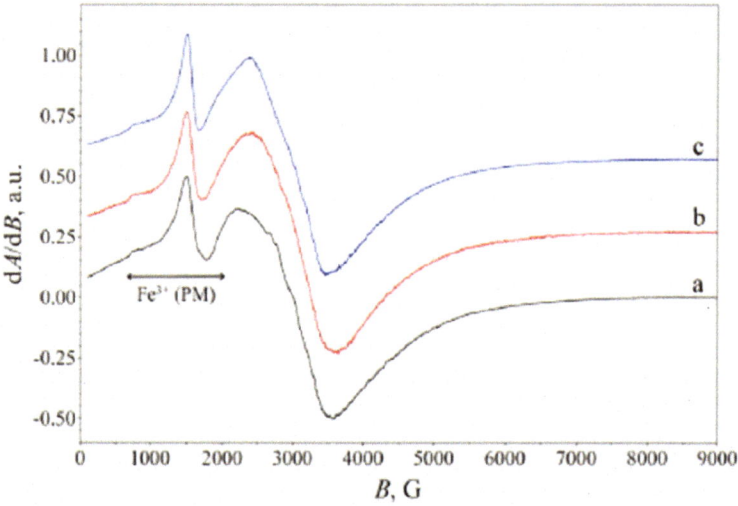

FIGURE 2.35 Electron magnetic resonance spectra of Ce-bentonite heat treated at at 250°C (a), 360°C (b) and 500°C (c). Apart from an arbitrary shift along the vertical axis, the spectra are proportional to dA/dB, i.e., the first derivative of the power of microwave absorption with respect to the applied magnetic field. The signal observed in the magnetic field range senoted with "Fe^{3+} (PM)" refers to the presence of paramagnetic Fe^{3+} ions in a variety of rhombiv crystal field environments. (Reprinted from Kuzmann et al. 2019, with permission from Springer.

of the radioactive iodide ion due to the formation of a soluble silver diiodide complex, as is well known in analytical chemistry:

$$Ag^+ + I^- = AgI \quad \left(\text{solubility product of } AgI(s) \text{ is } lgK = -16.08\right) \qquad (2.62)$$

$$AgI + I^- = [AgI_2]^- \quad (\text{stability constant is } lgK = 11.7) \qquad (2.63)$$

The relative ratios of Ag^+, AgI, and AgI_2^- can be calculated by using the thermodynamical equilibrium model (Figure 2.37).

Figure 2.37 shows the relative ratios of the chemical species expressed as a percentage of silver ion of bentonite as a function of the total equilibrium concentration of the iodide ion as well as the sorbed amount of iodide ions on Ag-bentonite. As seen, the distribution of the species is in a good agreement with the sorption data, namely the decrease in the sorbed quantity starts when the formation of the soluble AgI_2^- species begins. In the case of chloride ion sorption, the phenomenon is the same as in the case of the iodide ion sorption. However, the sorption quantity differs because of the difference in the stability constant of the dihalogenide complexes (the solubility product and the stability constant of $AgCl(s)$ and $AgCl_2^-$ are $lgK = -9.75$ and 5.5, respectively (Martell and Smith, 1976; Buzetzky et al. 2020).

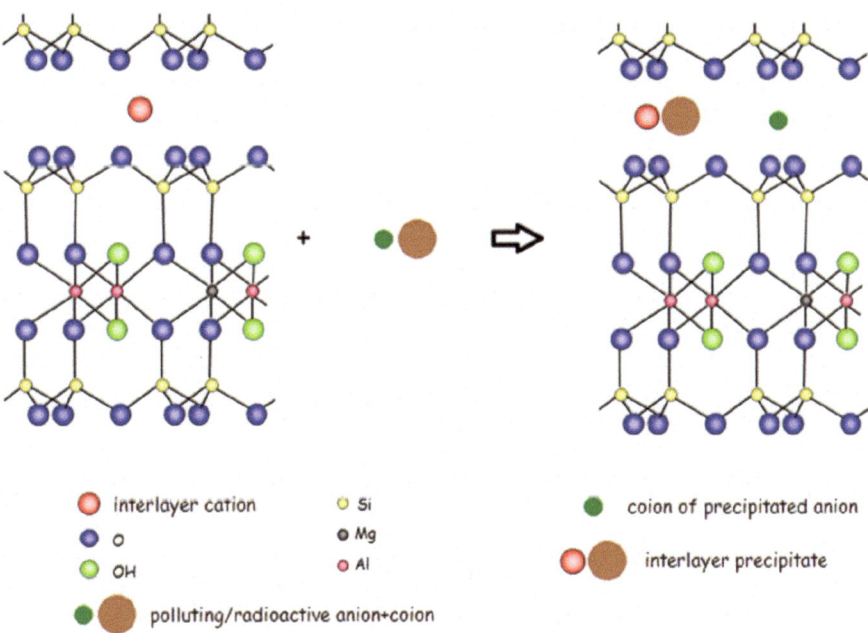

FIGURE 2.36 The scheme of precipitation of anions and cations in the interlayer space of montmorillonite.

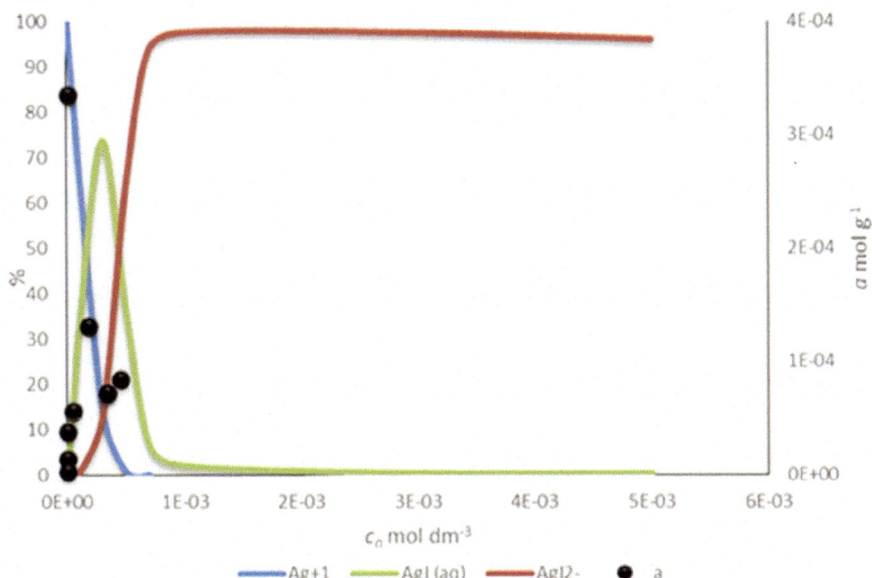

FIGURE 2.37 Relative ratios of Ag[+], AgI, and AgI_2^- as a percentage of silver ion and the sorbed amount of iodide ions (a, mol g[−1]) at 25°C, c_o is the concentration of iodide solution (mol dm[−3]). (Reprinted from Buzetzky et al. 2020, with permission from Springer.)

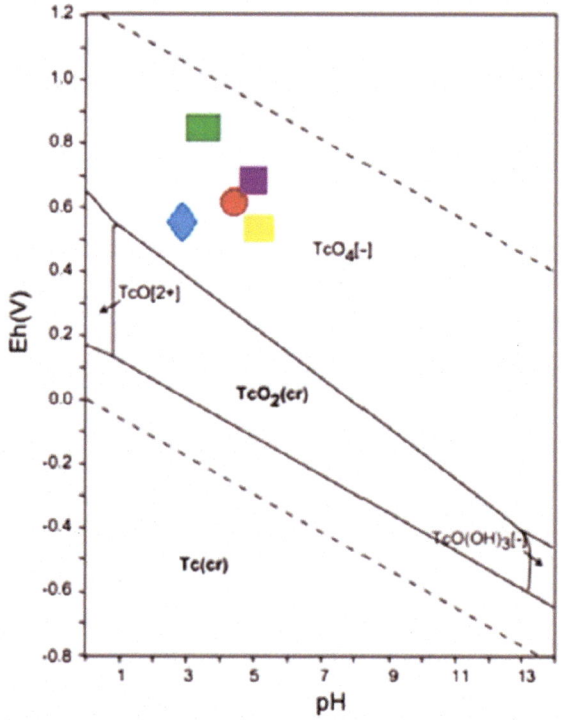

FIGURE 2.38 The pH-Eh predominance diagrams for the the sorption of technetium. Mn–bentonite: green square pH = 3.5–3.6, sorption is violet square pH = 4.6–4.7, yellow square pH = 5; Cr-bentonite: red circle; Sn-bentonite: blue diamond). The related sorption values are 15, 25, and 35%, respectively, for manganese-bentonite at pH 3.5–3.6, 4.6–4.7, and 5, respective. On Cr- and Sn-bentonite about 100% technetium sorption can be reached. (Reprinted from Bużetzky et al. 2019b, with permission from Springer.)

Redox reactions can also occur in the interlayer space, as shown in Sections 2.10.1 and 2.10.3. The sorption of technetium on different cation exchange bentonite (manganese-, chromium-, and tin-bentonites) increases when the redox potential gets closer to the +4-oxidation state of technetium (Figure 2.38).

This experience can be interpreted by the Hahn's adsorption rule (Chapter 1, Section 1.2.4) which says that an ion in extremely low concentration adsorbs well on the surface of a solid crystal when the adsorbed compound is very insoluble which is the case when technetium has a +4 oxidation state (technetium at a +7 oxidation state dissolves in water as pertechnetate anion).

An interesting observation during the sorption of anions on cation exchange bentonite is that the equilibrium sorption can be evaluated by using the simple Langmuir representation (Chapter 1, Equation 1.65). The term "representation" is used to emphasize that the process is obviously not a simple adsorption process. This is valid for all studied systems (sorption of phosphate and arsenite ion on rare earth bentonite, sorption of halogenide ion on silver bentonite), except technetium where no isotherm can be constructed because there is no stable isotope of technetium. The four basic assumptions of the Langmuir model have been discussed in Chapter 1, Section 1.3.4.1.2. Now we consider how they are fulfilled during the uptake of anions by the precipitation with cations in the interlayer space of montmorillonite.

1. The adsorption takes place on the free sites of the adsorbent: the sorption of anions takes place on free sites produced by the exchanged cations. It is supposed that in the case of natural calcium-bentonite, no sorption of anion can be observed.
2. One adsorption site can adsorb one molecule; monolayer coverage is maximum. The anions precipitate or adsorb in the interlayer space on the exchanged cations.
3. The adsorption sites have the same energy (homogeneous surface), no interaction between the adsorbed molecules. This means that the adsorption energy is independent of coverage. As a result of the precipitation reaction, the produced species are neutral, thus no electrostatic interaction occurs between particles.
4. There is adsorption equilibrium between the phases. The results of anion uptake proved this fact.

REFERENCES

Abollino, O., M. Aceto, M. Malandrino, C. Sarzanini, and E. Mentasti. 2003. Adsorption of heavy metals on Na-montmorillonite. Effect of pH and organic substances. *Water Research* 37: 1619–1627.

Aceman, S., N. Lahav, and S. Yariv. 2000. A thermo-XRD study of Al-pillared smectites differing in source of charge, obtained in dialyzed, non-dialyzed and washed systems. *Applied Clay Science* 17: 99–126.

Adams, J.M. 1987. Synthetic organic chemistry using pillared, cation-exchanged and acid-treated montmorillonite catalysts-a review. *Applied Clay Science* 2: 309–342.

Adams, J.M., T.V.Clapp, and D.E. Clement. 1983. Catalysis by montmorillonite. *Clay Minerals* 18: 411–421.

Adams, J.M., S. Dyer, K. Martin, W.A. Matear, and R.W. McCabe. 1994. Diels–Alder reactions catalysed by cation-exchanged clay minerals. *Journal of the Chemical Society, Perkin Transactions* 1: 761–765.

Barakan, S., and V. Aghazadeh. 2019. Synthesis and characterization of hierarchical porous clay heterostructure from Al, Fe -pillared nano-bentonite using microwave and ultrasonic techniques. *Microporous and Mesoporous Materials* 278: 138–148.

Barbier, F., G. Duc and M. Petit-Ramel. 2000. Adsorption of lead and cadmium ions from aqueous solution to the montmorillonite/water interface. *Colloids and Surfaces* 166: 153–159.

Barshad, I. 1969. Preparation of H saturated montmorillonites. *Soil Science* 108: 38–42.

Bergaya, F., G. Lagaly, and M. Vayer. 2008. Cation and anion exchange. In *Handbook of clay science*, 3rd ed., eds. F.B.K.G. Theng, and G. Lagaly, 979–1001. Amsterdam: Elsevier.

Berry, F.J., M.H.B. Hayes, and S.L. Jones. 1986. Investigations of intercalation in inorganic solids with layered structures: Iron-57 Mössbauer spectroscopy studies of size fractionated and iron-exchanged montmorillonite clays. *Inorganica Chimica Acta* 122: 19–24.

Bhattacharyya, K.G., and S.S. Gupta. 2008. Kaolinite and montmorillonite as adsorbents for Fe(III), Co(II) and Ni(II) in aqueous medium. *Applied Clay Science* 41:1–9.

Bishop, J.L., C.M. Pieters, and J.O. Edwards. 1994. Infrared spectroscopic analyses on the nature of water in montmorillonite. *Clays Clay Minerals* 42: 702–716.

Bouabid, R., M. Badraoui, and P. Bloom. 1991. Potassium fixation and charge characteristics of soil clays. *Soil Science Society of America Journal* 55: 1493–1498.

Boyd, G.E., J. Schubert, and A.W. Adamson. 1947. The exchange adsorption of ions from aqueous solutions by organic zeolites. I. Ion-exchange equilibria. *Journal of the American Chemical Society* 69: 2818–2829.

Bray, H.J., and S.A.T. Redfern. 2000. Influence of counterion species on the dehydroxylation of Ca^{2+}-, Mg^{2+}-, Na^+- and K^+-exchanged Wyoming montmorillonite. *Mineralogical Magazine* 64: 337–346.

Burger, K. 1990. *Biocoordination chemistry*. New York: Ellis Horwood.

Buzetzky, D., E.M. Kovács, N.M. Nagy, and J. Kónya. 2019b. Sorption of pertechnetate anion by cation modified bentonites. *Journal of Radioanalytical and Nuclear Chemistry* 322: 1771–1776.

Buzetzky, D., N.M. Nagy, J. Kónya. 2017. Use of La-, Ce, Y-, and Fe-bentonites for removal phosphate ion from aqueous media. *Periodica Polytechnica Chemical Engineering* 61: 27–32.

Buzetzky, D., N.M. Nagy, and J. Kónya. 2020. Use of silver-bentonite in sorption of chloride and iodide ions, *Journal of Radioanalytical and Nuclear Chemistry*. 326: 1795–1804. doi: 10.1007/s10967-020-07457-2.

Buzetzky, D., N.C. Tóth, N.M. Nagy, and J. Kónya. 2019a. Application of modified bentonites for arsenite(III) removal from drinking water. *Periodica Polytechnica Chemical Engineering* 63: 113–121. doi: 10.3322/PPch 12.197

Cao, G., L.K. Rabenberg, C.M. Nunn, and T.E. Mallouk. 1991. Formation of quantum-size semiconductor particles in a layered metal phosphonate host lattice. *Chemistry of Materials* 3: 149–156.

Chen, W., Y. Xu, Z. Lin, Z. Wang, and L. Lin. 1998. Formation, structure and fluorescence of CdS clusters in a mesoporous zeolite. *Solid State Communications* 105: 129–134.

Chen, Y-G., X-M. Liu, X. Mu, W-Min Ye, Y-J. Cui, B. Chen, and D-Bei Wu. 2018. Thermal conductivity of compacted GO-GMZ bentonite used as buffer material for a high-level radioactive waste repository. *Advances in Civil Engineering* Article ID 9530813. doi: 10.1155/2018/9530813.

Chmielarz, L., A. Kowalczyk, M. Skoczek et al. 2017. Porous clay heterostructures intercalated with multicomponent pillars as catalysts for dehydration of alcohols. *Applied Clay Science* 160: 116–125.

Choudary, B.M., R.M. Sarma, and K.V. Kumar. 1994. Fe^{3+}-Montmorillonite catalyst for selective nitration of chlorobenzene. *Journal of Molecular Catalysis A* 87: 33–38.

Criscenti, L.J., and D.A. Sverjensky. 1999. The role of electrolyte anions (ClO_4^-, NO_3^- and Cl^-) in divalent metal (M^{2+}) adsorption on oxide and hydroxide surfaces in salt solutions. *American Journal of Science* 299: 828–899.

Davis, A.J., R.O. James, and J.O. Leckie. 1978. Surface ionization and complexation at the oxide-water interface, I. Computation of electrical double layer properties in simple electrolytes. *Journal of Colloid and Interface Science* 63: 480–499.

Delgado, A., F. Gonzales-Cabellero, and J.M. Bruque. 1986. On the zeta potential and surface charge density of montmorillonite in aqueous electrolyte solutions. *Journal of Colloid and Interface Science* 113: 203–211.

Drame, H. 2005. Cation exchange and pillaring of smectites by aqueous Fe nitrate solutions. *Clays and Clay Minerals* 53: 335–347.

Eberl, D.D. 1980. Alkali cation selectivity and fixation by clay minerals. *Clays and Clay Minerals* 28: 161–172.

Eltantawy, I.M., and P.W. Arnold. 1993. Reappraisal of the ethylene glycol mono-ethyl ether for surface area estimation of clays. *Journal of Soil Science* 24: 232–238.

Fernandez, A.M., B. Baeyens, M. Bradbury, and P. Rivas. 2004. Analysis of the porewater chemical composition of a Spanish compacted bentonite used in an engineered barrier. *Physics and Chemistry of the Earth* 29: 105–118.

Fetter, G., P. Salas, L.A. Velazquez, and P. Bosch. 2000. Ce-Al-pillared clays: Synthesis, characterization and catalytic performance. *Industrial & Engineering Chemistry Research* 39: 1944–1949.

Filep, Gy. 1999. *Soil chemistry*. Budapest: Akadémiai Kiadó.

Földvári, M. 2008. Unpublished data.

Földvári, M., and P. Kovács-Pálffy. 2005. Thermoanalytical investigation of mono- and bivalent interlayer cations in montmorillonite (in Hungarian language). *Annual Report of Hungarian Geological Survey, Budapest* pp.167–176.

Földvári, M., P. Kovács-Pálffy, N.M. Nagy, and J. Kónya. 1998. Use of the second derivative of TG curves for investigation of the exchanged interlayer cation in montmorillonite. *Journal of Thermal Analysis* 53: 547–558.

Friebele, E., A. Shimoyana, P.E. Hare, and C. Ponnamperuma. 1981. Adsorption of amino acid enantiomers by sodium-montmorillonite. *Origins of Life and Evolution of the Biosphere* 11: 173–184.

Fudala, A., I. Pálinkó, and I. Kiricsi. 1999. Amino acids, precursors for cationic and anionic intercalation synthesis ad characterization of amino acid pillared materials. *Journal of Molecular Structure* 482–483: 33–37.

Fujikawa, Y., and M. Fukui. 1997. Radionuclide sorption to rocks and minerals: Effects of pH and inorganic anions. Part 1. Sorption of cesium, cobalt, strontium and manganese. *Radiochimica Acta* 76: 153–162. doi.org/10.1524/ract.1997.76.3.153

Gaines, G.L., and H.C. Thomas. 1953. Adsorption studies on clay minerals II. A formation of the thermodynamics of exchange adsorption. *The Journal of Physical Chemistry* 21: 714–718.

Gao, F., Q. Lu, X. Liu, Y. Yan, and D. Zhao. 2001. Controlled synthesis of semiconductor PbS nanocrystals and nanowires inside mesoporous silica SBA-15 phase. *Nano Letters* 1: 743–748.

Gaskova, O.L., and M.B. Bukaty. 2008. Sorption of different cations onto clay minerals: Modelling approach with ion exchange and surface complexation. *Physics and Chemistry of the Earth, Parts A/B/C* 33: 1050–1055.

Gast, R.G. 1972. Alkali metal exchange on chambers montmorillonite. *Soil Science Society of America Proceeding* 36: 14–19.

Goulding, K.W.T., and O. Talibudeen. 1980. Heterogeneity of cation exchange sites for K-Ca exchange in aluminosilicates. *Journal of Colloid and Interface Science* 78: 15–24.

Gratzel, M., and K. Kalyanasundaram. 1991. *Kinetics and catalysis in microheterogeneous systems*. New York: Marcel Dekker, Inc.

Green-Pedersen, H., and N.Plnd. 2000. Preparation, characterization, and sorption properties for Ni[II] of iron oxyhydroxide-montmorillonite. *Colloids and Surfaces A* 168: 133–145.

Green-Ruiz, C. 2009. Effect of salinity and temperature on the adsorption of Hg(II) from aqueous solutions by a Ca-montmorillonite. *Environmental Technology* 30: 63–68.

Gustafsson, J.P. 2013. VisualMinteq, Database. https://vminteq.lwr.kth.se/ (accessed June 19, 2021)

Hachiya, K., M. Ashida, M. Sasaki, H. Kan, T. Inoue, and T. Yasunaga. 1979. Study of the kinetics of adsorption-desorption of lead(2+) ion on a gamma-aluminum oxide surface by means of relaxation techniques. *The Journal of Physical Chemistry* 83: 1866–1871.

Han, Z., Q. Yu, H. Xie, et al. 2018. Fabrication of manganese-based Zr-Fe polymeric pillared interlayered montmorillonite for low-temperature selective catalytic reduction of NOx by NH3 in the metallurgical sintering flue gas. *Environmental Science and Pollution Research* 25: 32122–32129.

Heller-Kallai, L., and C. Mosser. 1995. Migration of Cu ions in Cu montmorillonite heated with and without alkali halides. *Clays and Clay Minerals* 43: 738–743.

Helmy, A.K. 1998. The limited swelling of montmorillonite. *Journal of Colloid and Interface Science* 207: 128–129.

Hlavay, J., and K. Polyák. 2005. Determination of surface properties of iron hydroxide-coated alumina adsorbent prepared for removal of arsenic from drinking water. *Journal of Colloid and Interface Science* 284: 71–77.

Horne, R.A., R.H. Holm, and M.D. Meyers. 1957. The adsorption of zinc(II) on anion-exchange resins. II. Stoichiometry, thermodynamics, loading studies, dowex-2 adsorption and factors influencing the rate of the adsorption process. *The Journal of Physical Chemistry* 61: 1655–1661.

Howery, D.G., and H.C. Thomas. 1965. Ion exchange on mineral clinoptilolite. *The Journal of Physical Chemistry* 69: 531–537.

Huertas, F.J., E. Caballero, C. JimeÂnez de Cisneros, F. Huertas, and J. Linares. 2001. Kinetics of montmorillonite dissolution in granitic solutions. *Applied Geochemistry* 16: 397–407.

Izumi, Y., D. Masih, K. Aika, and Y. Seida. 2005. Characterization of intercalated iron(III) nanoparticles and oxidative adsorption of arsenite on them monitored by X-ray absorption fine structure combined with fluorescence spectrometry. *Journal of Physical Chemistry B* 109: 3227–3232.

Janssen, M.J.G., and H.N. Stein. 1986. The TiO2/electrolyte solution interface II. Calculations by means of the site binding model. *Journal of Colloid and Interface Science* 111: 112–118.

Jeon, I., and K. Nam. 2019. Change in the site density and surface acidity of clay minerals by acid or alkali spills and its effect on pH buffering capacity. *Scientific Reports* 9: 9878.

Johnson, J.W., and J.F. Brody. 1988. Pillared clays and micas. *Materials Research Society Symposium Proceedings* 111: 257.

Kabata-Pendias, A., and H. Pendias. 1985. *Trace elements in soils and plants*. Boca Raton: CRC Press, Inc.

Kang, L., H. Liu, H. He, et al. 2018. Oxidative desulfurization of dibenzothiophene using molybdenum catalyst supported on Ti-pillared montmorillonite and separation of sulfones by filtration. *Fuel* 234 :1229–1237.

Kannan, P., K. Pitchumani, S. Rajagapal, and C. Srinivasan. 1997. Sheet silicate catalysed demethylation and Fischer-Hepp rearrangement of *N*-methyl-*N*-nitrosoaniline. *Journal of Molecular Catalysis A* 118: 189–193.

Khaorapapong, N., A. Ontam, J. Khemprasit, and M. Ogawa. 2009. Formation of MnS- and NiS-montmorillonites by solid-solid reactions. *Applied Clay Science* 43: 238–242.

Komadel, P., J. Madejova, J. Hrobarikova, M. Janek, and J. Bujdak. 2003. Fixation of Li⁺ cations in montmorillonite upon heating. *Journal of Solid State Chemistry* 90–91: 497–502.

Komlósi, A., E. Kuzmann, Z. Homonnay, N.M. Nagy, S. Kubuki, and J. Kónya. 2006. Effect of FeCl₃ and acetone on the structure of Na-montmorillonite studied by Mössbauer and XRD measurements. *Hyperfine Interactions* 166: 643–649.

Komlósi, A., E. Kuzmann, N.M. Nagy, Z. Homonnay, S. Kubuki, and J. Kónya. 2007. Interlayer incorporation of iron into Na-bentonite via treatment with FeCl₃ in acetone. *Clays and Clay Minerals* 55: 91–97.

Kong, Q., Y. Hu, H. Lu, Z. Chen, and W. Fan. 2005. Synthesis and properties of polystyrene/Fe-montmorillonite nanocomposites using synthetic Fe-montmorillonite by bulk polymerization. *Journal of Materials Science* 40: 4505–4509.

Kónya, J., and N.M. Nagy. 1998. The effect of complex-forming agent (EDTA) on the exchange of manganese ions on calcium-montmorillonite, I. Reaction scheme and Ca-montmorillonite/Na$_2$EDTA system. *Colloids and Surfaces* 136: 297–308.

Kónya, J., and N.M. Nagy. 2011 Sorption of dissolved mercury (II) species on calcium-montmorillonite: An unusual pH dependence of sorption process. *Journal of Radioanalytical and Nuclear Chemistry* 288: 447–454.

Kónya, J., N.M. Nagy, and M. Földvári. 2005. The formation and production of nano and micro particles on clays under environmental-like conditions. *Journal of Thermal Analysis and Calorimetry* 79: 537–543.

Kónya, J., N.M. Nagy, R. Király, and J. Gelencsér. 1998. The effect of complex-forming agent (EDTA) on the exchange of manganese ions on calcium-montmorillonite, II. Ca-montmorillonite-Mn(ClO$_4$)$_2$-Na$_2$EDTA system. *Colloids and Surfaces* 136: 309–317.

Kónya, J., N.M. Nagy, and K. Szabó. 1988. The study of the ion exchange processes in system of zinc ions and calcium-bentonite, clay soil, humate and sand by radioisotopic labelling method. *Reactive Polymers* 7: 203–209.

Kovács, E.M., E. Erdélyiné Baradács, P. Kónya, P. Kovács-Pálffy, S. Harangi, E. Kuzmann, J. Kónya, and N.M. Nagy. 2017. Preparation and structure's analyses of lanthanide (Ln)-exchanged bentonites, *Colloids and Surfaces A* 522:287–294.

Kovács, E.M., J. Kónya, and N.M. Nagy. 2019. Structural curiosities of lanthanide (Ln)-modified bentonites analyzed by radiochemical methods. *Journal of Radioanalytical and Nuclear Chemistry* 322: 1747–1754. doi: 10.1007/s10967-01906765-6.

Kozai, N., Y. Adachi, S. Kawamura, K. Inada, T. Kozaki, S. Sato, H. Ohashi, T. Ohnuki, and T. Banba. 2001. Characterization of Fe-montmorillonite: A simulant of buffer materials accommodating overpack corrosion product. *Journal of Nuclear Science and Technology* 38: 1141–1143.

Krupskaya, V.V., S.V. Zakusin, E.A. Tyupina, et al. 2017. Experimental study of montmorillonite structure and transformation of its properties under treatment with inorganic acid solutions. *Minerals* 7: 49.

Kukovskij, E.G. 1966. *Osobennostyi stroenyia i fiziko-khimichheskiye svoistva glinnistikh mineralov.* Kiev: Akademija Nauk SSSR.

Kuzmann, E., V.K. Garg, H. Singh, A.C. de Oliveira, S.S. Pati, Z. Homonnay, M. Rudolf, Á.M. Molnár, E.M. Kovács, E. Baranyai, S. Kubuki, N.M. Nagy, and J. Kónya. 2016b. Mössbauer study of pH dependence of iron-intercalation in montmorillonite. *Hyperfine Interaction* 237(1): 106.

Kuzmann, E., E.M. Kovács, Z. Homonnay, Sz. Csákvári, Z. Klencsár, P. Kónya, N.M. Nagy, and J. Kónya. 2019. Fe microenvironments in heat treated rare-earth exchanged montmorillonites, *Hyperfine Interaction* 240: 87–93. doi: 10.1007/s10751-019-1622-7.

Kuzmann, E., S. Nagy, A. Vértes, T. Weiszburg, and V.K. Garg. 1998. Applications of Mössbauer spectroscopy in mineralogy and geology. In *Nuclear methods in mineralogy and geology*, eds. A. Vértes, S. Nagy, and K. Süvegh, 285–377. New York: Plenum.

Kuzmann, E., L.H. Singh, V.K. Garg, A.C. de Oliveira, E.M. Kovács, Á.M. Molnár, Z. Homonnay, P. Kónya, M.N. Nagy, and J. Kónya. 2016a. Mössbauer study of the effect of rare earth substitution into montmorillonite, *Hyperfine Interaction* 237(1): 1–8.

Lacher, M., N. Lahav, and S. Yariv. 1993. Infrared study of the effects of thermal treatment on montmorillonite-benzidine complexes. II. Lithium-, sodium-, potassium-, rubidium- and cesium-montmorillonite. *Journal of Thermal Analysis and Calorimetry* 40: 41–57.

Lahav, N., M. Lacher, and S.Yariv. 1993. Infrared study of the effects of thermal treatment on montmorillonite-benzidine complexes. III. Magnesium-, calcium- and aluminum montmorillonite. *Journal of Thermal Analysis and Calorimetry* 39: 1233–1254.

Laudelout, H., R. van Bladel, G.H. Bolt, and A.L. Page. 1968. Thermodynamics of heterovalent cation exchange reactios in a montmorillonite clay. *Transactions of the Faraday Society* 64: 1477–1488.

Li, T.S., Z.H. Zhang, and Y.J. Gao. 1998. A rapid preparation of acylals of aldehydes catalysed by Fe^{3+}–montmorillonite. *Synthetic Communications* 28: 4665–4671.

Lück, R., R. Stösser, C. Gyepesová, H. Slosiariková, and L. Kolditz. 1993. Study of montmorillonite depostis. *Chemical Papers* 47: 79–84.

Maes, A., and A. Cremers. 1977. Charge density effects in ion exchange, part I-heterovalent exchange equilibria. *Journal of the Chemical Society Faraday Transactions* 73: 1807–1814.

Manjanna, J. 2008. Preparation of Fe(II)-montmorillonite by reduction of Fe(III)-montmorillonite with ascorbic acid. *Applied Clay Science* 42: 32–38.

Manjanna, J., T. Kozaki, and S. Sato. 2009. Fe(III)-montmorillonite: Basic properties and diffusion of tracers relevant to alteration of bentonite in deep geological disposal. *Applied Clay Science* 43: 208–217.

Marcus, Y., and A.S. Kertes. 1969. *Ion exchange and solvent extraction of metal complexes.* London, New York,Wiley-Interscience, pp. 321–340.

Marcussen, H., P.E. Holm, B.W. Strobel, and H. Chr.B. Hansen. 2009. Nickel sorption to goethite and montmorillonite in presence of citrate. *Environmental Science & Technology* 43: 1122–1127.

Martell, A.E., and R.M. Smith. 1974. *Critical stability constants, Vol.1.* New York: Plenum Press.

Martell, E.A., and R.M Smith. 1976. *Critical stability constants.* New York: Plenum.

Matijevic, E., J.P. Couch, and M. Kerker. 1962. Detection of metal ion hydrolysis by coagulation. IV. Zinc. *The Journal of Physical Chemistry* 66: 111–114.

McKenzie, R.M. 1980. The adsorption of lead and other heavy metals on oxides of manganese and iron. *Australian Journal of Soil Research* 18: 61–73.

Michot, L.J., and F. Villiéras. 2008. Surface area and porosity. In *Handbook of clay science*, 3rd ed., eds. F. Bergaya, B.K.G. Theng, and G. Lagaly, 965–978. Amsterdam: Elsevier.

Mogyorósi, K., I. Dékány, and J.H. Fendler. 2003. Preparation and characterization of clay mineral intercalated titanium dioxide nanoparticles. *Langmuir* 19: 2938–2946.

Nagy, N.M., M.A. Jakab, J. Kónya, and S. Antus. 2002. A convenient preparation of 1,1-diacetates from aromaticaldehydes catalysed by zinc-montmorillonite. *Applied Clay Science* 21: 213–216.

Nagy, N.M., A. Komlósi, and J. Kónya. 2004. Study of the change of the properties of Mn-bentonite by ageing. *Journal of Colloid and Interface Science* 278: 166–172.

Nagy, N.M., and J. Kónya. 1988. The interfacial processes between calcium-bentonite and zinc ion. *Colloids and Surfaces* 32: 223–235.

Nagy, N.M., and J. Kónya. 1998. Ion exchange processes of lead and cobalt ions on the surface of calcium-montmorillonite in the presence of complex forming agents I. The effect of EDTA on the sorption of lead and cobalt ion on calcium-montmorillonite. *Colloids and Surfaces* 137: 231–242.

Nagy, N.M., and J. Kónya. 2004. The sorption of valine on cation-exchanged montmorillonites. *Applied Clay Science* 25: 57–69.

Nagy, N.M., and J. Kónya. 2006. Acid-base properties of bentonite rocks with different origin. *Journal of Colloid and Interface Science* 295: 173–180.

Nagy, N.M., and J. Kónya. 2007. Study of interfacial reaction of palladium(II) ion and bentonite in the presence of complex forming agents. *Acta Geographica ac Geologica et Meterologica Debrecina* 2: 47–51.

Nagy, N.M., and J. Kónya. 2008. *Palladium-bentonites as catalysts, 4th Mid-European Clay Conference*, Zakopáne, 22–28. Sept. 2008: 119.

Nagy, N.M., J. Kónya, M. Beszeda, I. Beszeda, E. Kálmán, Zs. Keresztes, and K. Papp. 2001. Lead accumulation on montmorillonite. *Colloid and Polymer Science* 117: 117–119.

Nagy, N.M., J. Kónya, M. Beszeda, I. Beszeda, E. Kálmán, Zs. Keresztes, K. Papp, and I. Cserny. 2003a. Physical and chemical formations of lead contaminants in clay and sediment. *Journal of Colloid and Interface Science* 263: 13–22.

Nagy, N.M., J. Kónya, and T. Budai. 1998a. $Mn^{2+}/^{54}Mn^{2+}$ heterogeneous isotope exchange reaction on montmorillonite in the presence of complex-forming agents. *Colloids and Surfaces* 138: 81–89.

Nagy, N.M., J. Kónya, M. Földvári, and P. Kovács–Pálffy. 2003b. The adsorption of caesium–137 ion bentonites from the carpathian basin. *Czechoslovak Journal of Physics* 53: A103–111.

Nagy, N.M., J. Kónya, and I. Kónya. 1998b. Ion exchange processes of lead and cobalt ions on the surface of calcium-montmorillonite in the presence of complex forming agents. II. The effect of DTPA, tartaric acid and citric acid on the sorption of lead ion on calcium-montmorillonite. *Colloids and Surfaces* 137: 243–252.

Nagy, N.M., J. Kónya, and Z. Urbin. 1997. The competitive exchange of hydrogen and cobalt ions on calcium-montmorillonite. *Colloids and Surfaces* 121: 117–124.

Nagy, N.M., J. Kónya, and Gy. Wazelischen-Kun. 1999. The sorption and desorption of carrier-free radioactive isotopes on clay minerals and Hungarian soils. *Colloids and Surfaces* 152: 245–250.

Németh, J., I. Dékány, K. Süvegh, T. Marek, Z. Klencsár, A.Vértes, and J.H. Fendler. 2003. Preparation and structural properties of tin oxide-montmorillonite nanocomposites. *Langmuir* 19: 3762–3769.

Oliveira, L.C.A., R.V.R.A. Rios, J.D. Fabris, K. Sapag, V.K. Garg, and R.M. Lago. 2003. Clay–iron oxide magnetic composites for the adsorption of contaminants in water. *Applied Clay Science* 22: 169–177.

Oueslati, W., H. Ben Rhaiem, B. Lanson, and A. Ben Haj Amara. 2009. Selectivity of Na–montmorillonite in relation with the concentration of bivalent cation (Cu^{2+}, Ca^{2+}, Ni^{2+}) by quantitative analysis of XRD patterns. *Applied Clay Science* 43: 224–227.

Papp, S., and I. Dékány. 2002. Growth of Pd nanoparticles on layer silicates hydrophobized with alkyl chains in ethanol-tetrahydrofuran mixtures. *Colloid and Polymer Science* 280: 956–962.

Parks, G.A. 1965. The isoelectric points of solid oxides, solid hydroxides, and aqueous hydroxo complex systems. *Chemical Reviews* 65: 177–198.

Parks, G.A., and P.L. de Bruyn. 1962. The zero point of charge of oxides. *The Journal of Physical Chemistry* 66: 967–973.

Patel, M. 1982. Reduction of interlayer Ni^{2+} and Cu^{2+} in montmorillonite with hydrogen. *Clays Clay Minerals* 30: 397–399.

Peker, S., S. Yapar, and N. Besün. 1995. Adsorption behavior of surfactant onmontmorillonite. *Colloids and Surfaces* 104: 249–257.

Perez Zurita, M.J., G. Vitale, M.R. de Goldwasser, D. Royas, and J.J. Garcia. 1996. Fe-pillared clays: A combination of zeolite shape selectivity and iron activity in the CO hydrogenation reaction. *Journal of Molecular Catalysis* 107: 175–183.

Pillai, U.R., and E. Sahle-Demessie. 2003. Oxidation of alcohols over Fe^{3+}/montmorillonite-K10 using hydrogen peroxide. *Applied Catalysis A* 245: 103–109.

Pourbaix, M. 1966. *Atlas of electrochemical equilibria in aqueous solutions*. Oxford: Pergamon Press.

Rak, V.S., and Y.I. Tarasevich. 1982. Ion-exchange sorption of α-amino acids by montmorillonite with a divalent cation in the exchange complex. *Teoreticheskaya i Eksperimentalnaya Khimiya* 18: 166–174 (in *Chemical Abstracts*, Vol. 96, 223829k).

Ramchandani, R.K., B.S. Uphade, M.P. Vinad, R.D. Wakhankar, V.R. Choudhary, and A. Sudalai. 1997. Pd–Cu–exchanged montmorillonite K10 clay: An efficient and reusable heterogeneous catalyst for vinylation of aryl halides. *Chemical Communications* 2071–2072.

Rangel-Rivera, P., M. Belen Bachiller-Baeza, I. Galindo-Esquivel et al. 2018. Inclusion of Ti and Zr species on clay surfaces and their effect on the interaction with organic molecules. *Applied Surface Science* 445: 229–241.

Richards, L.A. 1957. *Diagnosis and improvement of saline and alkaline soils*. U.S. Dept. Agr. Handbook, N. 60. Washington, DC: United State Department of Agriculture.

Robinson, M., J.A. Pask, and D.W. Fuerstenau. 2006. Surface charge of alumina and magnesia in aqueous media. *Journal of the American Ceramic Society* 47: 516–520.

Rubiyanto, D., I. Prakoso Nurcahyo, et al. 2020. Microwave-assisted synthesized porous clay heterostructure-Zn/Si from montmorillonite for citronellal conversion into isopulegol. *Materials Research Express* 7: 105006.

Sawhney, B.L. 1972. Selective sorption and fixation of cations by clay minerals: A review. *Clays and Clay Minerals* 20: 93–100.

Schindler, P.W., B. Fürst, B. Dick, and P.U. Wolf. 1976. Ligand properties of surface silanol groups. I. Surface complex formation with Fe^{3+}, Cu^{2+}, Cd^{2+}, and Pb^{2+}. *Journal of Colloid and Interface Science* 55: 469–475.

Schindler, P.W., and H. Gamsjager. 1972. Acide-base reactions of the TiO_2 (anatase)-water interface and the point of zero charge of TiO_2 suspensions. *Kolloid–Zeitschrift und Zeitschrift für Polymere* 250: 759–765.

Schwarz, S., K. Lunkwitz, B. Keßler, U. Spiegler, E. Killmann, and W. Jaeger. 2000. Adsorption and stability of colloidal silica. *Colloids and Surfaces A* 163: 17–27.

Shangguan, W., and A. Yoshida. 2002. Photocatalytic hydrogen evolution from water on nanocomposites incorporating cadmium sulfide into the interlayer. *The Journal of Physical Chemistry, B* 106: 12227–12230.

Shrigadi, N.B., A.B. Shinde, and S.D. Samant. 2003. Study of catalytic activity of free and K10-supported iron oxyhydroxides and oxides in the Friedel-Crafts benzylation reaction using benzyl chloride/alcohol to understand their role in the catalysis by the Fe-exchanged/impregnated K10 catalysts. *Applied Catalysis A* 252: 23–25.

Shukla, U.C., and S.B. Mittal. 1980. Characterization of the zinc retained in some soils of India using pH gradient elution. *Soil Science Society of America Journal* 44: 29–33.

Silva, A.S., M.S. Kalmakhanova, B.K. Massalimova, et al. 2019. Wet peroxide oxidation of paracetamol using acid activated and Fe/Co-pillared clay catalysts prepared from natural clays. *Catalysts* 9: 705.

Singhal, J.P., and R.P. Singh. 1973. Thermodynamics of cobalt(II)-sodium exchange on montmorillonite clay. *Journal of Soil Science* 24: 271–276.

Siskens, C.A.M., H.N. Stein, and J.M. Stevels. 1975. Surfaces of silicates in aqueous alkaline solutions II. *Journal of Colloid and Interface Science* 52: 251–259.

Sondi, I., V. Tomasic, and N. Filipovic-Vincekovic. 2008. Release of silicon and aluminum from montmorillonite surfaces in aqueous systems. *Croatica Chemica Acta* 81: 623–629.

Sposito, G. 1981. *Thermodynamics of soil solutions*. London: Oxford Clarendon Press.

Stadler, M., and P.W. Schindler. 1993a. The effect of dissolved ligands upon the sorption of Cu(II) by Ca-montmorillonite. *Clays and Clay Minerals* 41: 680–692.

Stadler, M., and P.W. Schindler. 1993b. Modeling of H⁺ and Cu²⁺ adsorption on calcium-montmorillonite. *Clays Clay Minerals* 41: 288–296.

Stefani, V.F., R.V. Conceicao, L.C. Camiel, and M.L. Balzaretti. 2014. Stability of lanthanum-saturated montmorillonite under high pressure and high temperature conditions. *Applied Clay Science* 102: 51–59.

Stevens, J.G. (ed.) 1958–2002. *Mössbauer effect reference and data index (MERDI)*. New York: Plenum.

Stevens, J.G., H. Pollak, Zh. Li, V.E. Stevens, R.M. White, and J.L. Gibson. 1983. *The Mössbauer handbook of minerals*. North Carolina: Mössbauer Effect Data Center.

Strawn, D.G., and D.L. Sparks. 1999. The use of XAFS to distinguish between inner- and outer-sphere lead adsorption complexes on montmorillonite. *Journal of Colloid and Interface Science* 216: 257–269.

Stuckey, J.W., A. Neaman, R. Ravella, S. Komarneni, and C. Enid Martinez. 2008. Highly charged swelling mica reduces free and extractable Cu levels in Cu-contaminated soils. *Environmental Science and Technology* 42: 9197–9202.

Stucki, J.W. 2008. Properties and behavior of iron in clay minerals. In *Handbook of clay science*, 3rd ed., eds. F.B.K.G. Theng, and G. Lagaly, 423–475. Amsterdam: Elsevier.

Stucki, J.W., K. Lee, L.Zhang, and R.A. Larson. 2002. Effects of iron oxidation state on the surface and structural properties of smectites. *Pure and Applied Chemistry* 74: 2145–2158.

Szántó, Zs., L. Papp, J. Kónya, N.M. Nagy, and Zs. Lengyel. 1999. Iontophoretic delivery of calcium ions through guinea pig *in vivo* using different current systems. *Journal of Radioanalytical and Nuclear Chemistry* 241: 45–49.

Tateiwa, J.I., H. Horiuchi, and S. Vemura. 1994c. Mn²⁺-exchanged clay-catalysed oxidation of alkanes with tert-butyl hydroperoxide. *Chem. Commun.* 2567–2568.

Tateiwa, J.I., T. Nishimura, H. Hiriuchi, and S. Ulmura. 1994b. Rearrangement of alkyl phenyl ethers to alkylphenols in the presence of cation-exchanged montmorillonite (Mnⁿ⁺-mont). *Journal of the Chemical Society Perkin* 3367–3371.

Tateiwa, J.I., H. Horiuchi, K. Hashimoto, and T. Kamauchi. 1994a. Cation-exchanged montmorillonite-catalyzed facile friedel-crafts alkylation of hydroxy and methoxy aromatics with 4-hydroxybutan-2-one to produce raspberry ketone and some pharmaceutically active compounds. *The Journal of Organic Chemistry* 59: 5901–5904.

Theng, B.K.G. 1974. *The chemistry of clay-organic reactions*. Bristol: Hilger.

van Diemen, A.J.G., and H.N. Stein. 1978. Adsorption and electrokinetic potentials at solid/aqueous solution interfaces characterized by mutually stimulated adsorption of cations and anions. *Journal of Colloid and Interface Science* 67: 213–218.

Vlasova, M., G. Dominguez-Patiño, N. Kakazey, M. Dominguez-Patiño, D. Juarez-Romero, and Y. Enríquez Méndez. 2003. Structural-phase transformations in bentonite after acid treatment. *Science of Sintering* 35: 155–166.

Wang, X., and L. Andrews. 2005 Infrared spectrum of Hg(OH)2 in solid neon and argon. *Inorganic Chemistry* 44: 108–113.

Wanner, H., Y. Albinsson, O. Karnland, E. Wieland, P. Wersin, and L. Charlet. 1994. The acid/base chemistry of montmorillonite. *Radiochimica Acta* 66/67: 157–162.

Waterlot, C., D. Couturier, and B. Rigo. 2000. Montmorillonite-palladium-copper catalyzed cross-coupling of methyl acrylate with aryl amines. *Tetrahedron Letters* 41: 317–319.

www.princetonschools.net/site/handlers/filedownload.ashx?moduleinstanceid=714&dataid=1472&FileName=Solubility_Product_Constants.pdf (accessed 23.05.2021).

Yang, S., X. Duan, J. Liu, et al. 2020. Efficient peroxymonosulfate activation and bisphenol A degradation derived from mineral-carbon materials: Key role of double mineral-templates. *Applied Catalysis B-Environmental* 267: 118701.

Yariv, S., N. Lahav, and M. Lacher. 1994. Infrared study of the effects of thermal treatment of montmorillonite-benzidine complexes. IV. Mn-, Co-, Ni-, Zn-, Cd- and Hg-montmorillonite. *Journal of Thermal Analysis* 42: 13–30.

Yermiyahu, Z., A. Landau, A. Zaban, I. Lapides, and S. Yariv. 2003. Monoionic montmorillonites treated with Congo-Red. *Journal of Thermal Analysis Calorimetry* 72: 413–441.

Zhang, W.-H., J.-L. Shi, H.-R. Chen, Z.-L. Hua, and D.-S. Yan. 2001. Synthesis and characterization of nanosized ZnS confined in ordered mesoporous silica. *Chemistry of Materials* 13: 648–654.

Zhao, J., X. Zheng, Q. Liu, et al. 2020. Chitosan supported Pd-0 nanoparticles encaged in Al or Al-Fe pillared montmorillonite and their catalytic activities in Sonogashira coupling reactions. *Applied Clay Science* 195: 105721.

3 Interfacial Reactions at Rock and Soil Interfaces

This chapter presents some interfacial studies in natural geological systems that illustrate attempts to elucidate the interfacial reactions of rocks and soils. Rocks and soils are solids formed in nature, consisting of various mineral and organic components. As a result, the reactions that take place are complex, and the results obtained cannot be easily extrapolated to other environments. For this reason, investigations with original rock and soil samples are most appropriate to answer specific questions that arise for a given place to solve environmental, geological, agricultural, or raw material processing problems. However, when it is possible to define the samples from chemical, mineralogical, or soil science points of view at an appropriate level, general conclusions may also be drawn.

In this chapter, the relationship between geological origins and interfacial properties of bentonite clay will be demonstrated first. Then we will discuss the migration of water-soluble substances in rocks and soil, and the effect of sorption on the migration. A linear model will be derived by which the quantity of ion sorbed on rocks can be estimated when the mineral composition and the sorption parameters of the mineral components are known. The surface acid–base properties of soils will be discussed, and the sorption of anions (cyanide ion and phosphate) will be shown on different soils and sediments.

3.1 RELATIONSHIP BETWEEN THE INTERFACIAL PROPERTIES AND THE GEOLOGICAL ORIGIN OF BENTONITE CLAY

Bentonite rocks have many uses in the chemical and oil industries and also in agriculture and environmental protection. The usefulness of bentonite for each of these applications is based on its interfacial properties. These properties are determined by geological origin, chemical and mineral composition (especially montmorillonite content), and particle size distribution, and they include specific surface area (internal and external), cation-exchange capacity (CEC), acid–base properties of the edge sites, viscosity, swelling, water permeability, adsorption of different substances, and migration rate of soluble substances in bentonite clay.

Users and researchers are frequently interested in only some of the proceeding properties. For example, geologists deal with mineral composition and the details of mineral formation, and chemists study surface properties and catalytic effects. Swelling and viscosity are important for the oil industry, for environmental protection, etc. However, linking geological, chemical, and mechanical properties gives

DOI: 10.1201/9781003020080-3

very interesting insights and is important from a scientific and industrial point of view.

Because montmorillonite is the main component of bentonite rocks, montmorillonite content is the determining factor for most applications. As mentioned, however, in Chapter 2, Section 2.1.1, the chemical composition of rocks and the degree of isomorphic substitutions are determined by the geological site from which montmorillonite originates. For example, the accompanying minerals present in bentonite rocks may differ from one geological site to another. Moreover, different layers of bentonite from the same site can have different interfacial properties, even when montmorillonite and the other mineral compositions are similar. This will have a significant influence on the possible applications of bentonite. In this chapter we will also discuss what kind of differences may be seen in different bentonite layers found in the same place. The relations between the geological origin, mineral composition, and chemical, physical, and interfacial properties will be discussed as well.

3.1.1 GEOLOGICAL AND MINERAL CHARACTERISTICS OF SAJÓBÁBONY BENTONITES

To demonstrate the relationship between the interfacial properties of bentonites and their geological origin, some properties of bentonite samples from Sajóbábony (northern Hungary) will be discussed. The stratigraphy lithology of the Sarmatian sediment series with bentonite sites in the Kő valley around Sajóbábony (Hungary) are shown in Figures 3.1 and 3.2 (Püspöki et al. 2003a,b).

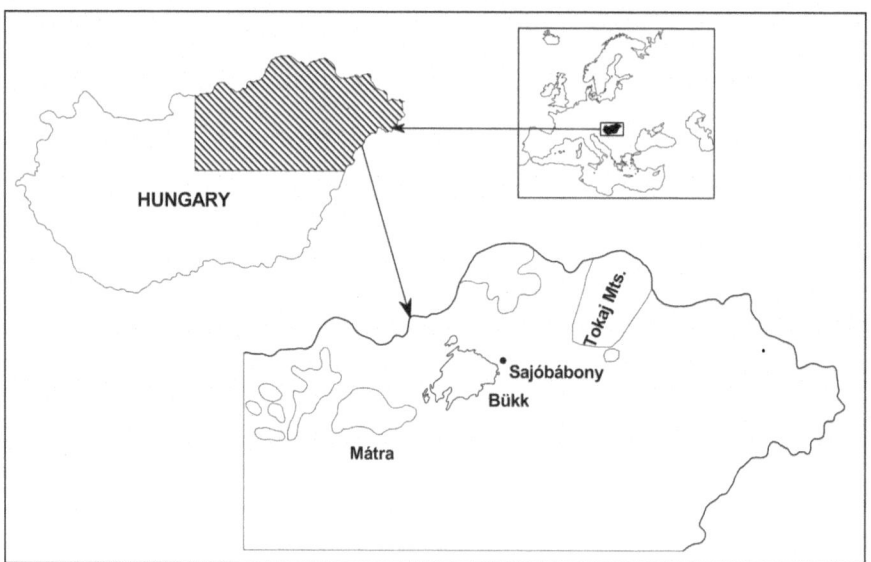

FIGURE 3.1 Sarmatian sediment series with bentonite sites in the Kő valley around Sajóbábony (Hungary) (Püspöki et al. 2003a, b). (Reprinted from Nagy and Kónya 2005, with permission from Elsevier.)

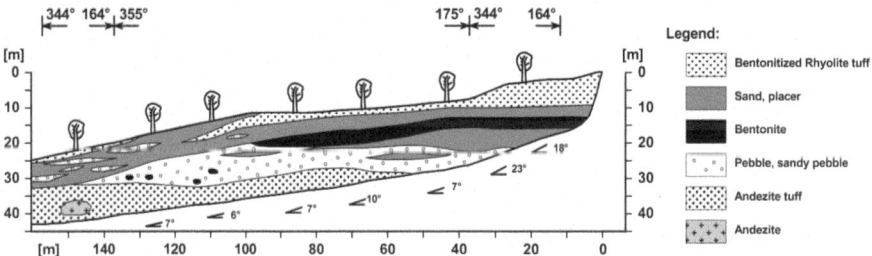

FIGURE 3.2 Stratigraphy litology and analytical results of the bentonitic Sarmatian series (Püspöki et al. 2003a, b). (Reprinted from Nagy and Kónya 2005, with permission from Elsevier.)

Lithological characteristics of the bentonite layers (Figures 3.1 and 3.2) show three main facies types of bentonites:

1. Bentonitic tuff represented by samples from B–II.b. layer.
2. Sedimentary bentonite with strong tuffogenic character (enclosing numerous small pumice fragments) represented by B–II.a. layer.
3. Sedimentary bentonite represented by B–I. bentonite layer.

As seen in Figure 3.2, the bentonite layers are placed one over the other. The analysis of the main trace element shows that there are genetic connections between the bentonitic tuff and sedimentary bentonites (Püspöki et al. 2003c).

The chemical and mineral composition of bentonite samples are shown in Table 3.1. X-ray diffractograms, used for the determination of mineral composition, are shown for a sedimentary bentonite and a bentonitic tuff sample in Figure 3.3.

The basal spacing d_{001} of the montmorillonite can be well seen at about 6° 2Θ, both for the sedimentary bentonite and the bentonitic tuff. It shows that bentonite samples contain calcium ion as exchangeable cation. The montmorillonite contents of bentonite samples are similar; the values are in a relatively narrow concentration range (35–48%; Table 3.1). However, as will be shown later, the properties of the sedimentary bentonite and the bentonitic tuff are rather different, depending on the geological origin and accompanying mineral components. These minerals are illite, plagioclase, quartz, and kalifeldspar. The concentration of the X-ray amorphous phase is 7–29%. The bentonitic tuff (B–II.b. upper and lower) contains a great amount of amorphous phase (22–29%). The amorphous phase is probably volcanic glass (Püspöki et al. 2005). Because of the presence of the amorphous phase, the X-ray diffractograms of the sedimentary bentonite and the bentonitic tuff are very different. Sedimentary bentonite (Figure 3.3 upper) has sharp peaks, while bentonitic tuff (Figure 3.3 lower) has diffuse peaks. So, the composition of sedimentary bentonite can be determined fairly well; the composition of the bentonitic tuff, however, has an uncertainty of ±10% (Granthoff and Moore 1996; Ryan and Reynolds 1997; Srodon et al. 2001).

TABLE 3.1
The Mineral and Chemical Composition of Bentonite Samples (%)

Sample	Montmorillonite	Illite/montmorillonite	Illite	Chlorite	Quartz	Kalifeldspar	Plagioclase	Gypsum	Crystobalite	Amorphous
B-II.b. upper	35		14		4	6	7		5	29
B-II.b. lower	48		11		2	4	9		4	22
B-II.a. upper	40		10	4	11	6	15		4	10
B-II.a. lower	47		10	2	15	3	13	tr	3	7
B-I.b.	42	3	9	2	1	8	21		4	10

	B-II.b.	B-II.a.	B-I.b.
SiO_2	67.4	63.8	58.1
TiO_2	0.27	0.60	0.65
Al_2O_3	15.1	16.0	17.4
Fe_2O_3	2.9	5.0	6.5
FeO	0.3	0.2	0.2
MnO	0.03	0.02	0.02
CaO	1.9	2.4	3.5
MgO	0.8	1.0	1.6

(Continued)

TABLE 3.1 (CONTINUED)
The Mineral and Chemical Composition of Bentonite Samples (%)

	B-II.b.	B-II.a.	B-I.b.
Na_2O	1.3	1.1	1.0
K_2O	2.6	0.8	1.0
P_2O_5	0.05	0.15	0.15
SO_3	<0.15	0.15	0.15
BaO	0.03	0.01	0.03
SrO	<0.01	0.01	0.01
Loss on igniting	7.4	8.9	10.4

Source: Reprinted from Nagy et al. 2006, with Permission from Springer.

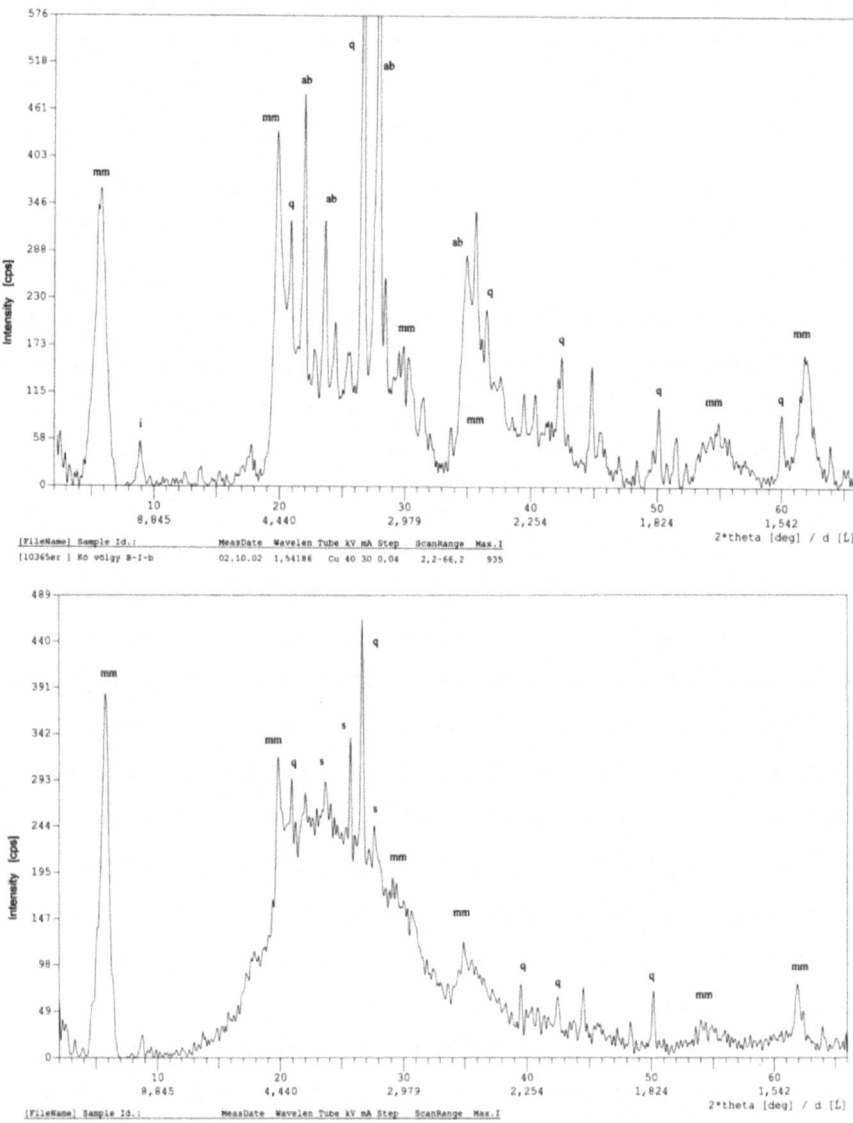

FIGURE 3.3 X-ray diffractograms of the B–I.b bentonite sample (sedimentary) (upper) and B–II.b upper bentonite sample (bentonitized tuff) (lower). mm = montmorillonite, ab = albite (plagioclase), q = quartz, i= illite, s = sanidine (feldspar). (Reprinted from Nagy and Kónya 2005, with permission from Elsevier.)

3.1.2 Interfacial Properties of Bentonite Samples from Sajóbábony (HU)

The various bentonite samples have different physical, chemical, and colloid chemical properties. The particle size distribution, specific surface area, CEC, and swelling in water are shown in Table 3.2. For an easier comparison, some specific surface area and CEC data measured on other bentonites from Chapter 2, Table 2.1 are repeated here.

The particle size distribution was measured by dynamic light scattering, a method that determines the particle size for spherical particles, which is a rather weak approach for clay particles. In addition, the clay particles are aggregated during the measurement, and thus the determined particle size distribution is only approximate and suitable for comparison between samples. The main part of the clay fraction is expected in the range <10 μm even when aggregates are present. The ratio of this fraction is in agreement with the clay mineral content (Table 3.1) in the case of B–I.b.; B–II.a. upper and lower; and B–II.b. upper samples. Only the B–II.b. lower sample shows a ratio in the range <10 μm, which is lower than expected compared to the montmorillonite content. This discrepancy is also seen with Ca-bentonite (from Istenmezeje, Hungary; Table 3.2).

Similarly, the specific surface area and CEC of B–II.b. lower sample (Table 3.2) is less than expected from the montmorillonite content even if a greater uncertainty in montmorillonite content determination (±20%) is taken into account. Similarly, low specific surface areas are observed for some other bentonite rocks (e.g., SWy-1, SWy-2, and MX-80; Table 3.2). In the case of SWy-2, the CEC is low as well. This bentonite consists of large particles with small layer charge.

On the basis of the specific surface area and CEC's (Table 3.2), the bentonite samples at the Sajóbábony site can be divided into two groups:

1. Sedimentary (B–I.b.), and sedimentary–volcano–sedimentary, bentonite samples (B–II.a.): the specific surface area (90–100 m²/g) is in the same range as usual for the bentonite samples (e.g., SWy-1, MX-80, FEBEX, Serrata de Nijar). The CEC is in agreement with the montmorillonite content.
2. Bentonitic tuff (B–II.b.): the specific surface area as well as the CEC is significantly smaller than expected from the montmorillonite content.

Similar classification can be made on the basis of the surface acid–base properties of bentonite samples (Chapter 1, Section 1.3.2.1.1 and Chapter 2, Section 2.4). The number and ratio of the edge silanol and aluminol sites, as well as the intrinsic stability constants of the protonation and deprotonation constants (Chapter 1, Equations 1.54–1.56; Chapter 2, Equations 2.3–2.5) are very different for sedimentary bentonites (layers B–I.b. and B–II.a.) and for the bentonitic tuff (B–II.b. layer; Table 3.3).

As seen in Table 3.3, the intrinsic stability constants of the protolytic processes of aluminol sites are approximately the same for all bentonite types. Only the intrinsic stability constants of deprotonation of aluminol sites show some differences. The error in the deprotonation constants of aluminol sites, however, is quite large because

TABLE 3.2

Particle Size Distribution, Specific Surface Area (S), Cation Exchange Capacity (CEC), Swelling in Water, and the Relative Amount of the Adsorbed $^{137}Cs^+$ Ion (x)

Sample	Montmorillonite content, %	<10 μm, v/v %	10–100 μm, v/v %	>100 μm, v/v %	A, m²/g	CEC, meq/100 g	Swelling, cm³	x
B–II.b. upper	35	25	66	9	17.94	17.7	2.1	0.7
B–II.b. lower	48	36	47	17	55.80	25.9	3.3	0.74
B–II.a. upper	40	50	40	10	93.01	42.1	3.6	0.89
B–II.a. lower	47	37	41	22	98.25	38.3	2.5	0.80
B–I.b.	42	50	46	4	94.07	47.8	3.6	0.86
Ca-bentonite[a] (Istenmezeje, HU)	91	56	33	11	93.5	104		
SWy-1[b]					21.4	93		
SWy-2[c]					29.4	26		
BSAB[b]	80				87	48		
MX-80[d]					31.53	108		
FEBEX[e]					62	100		
Serrata de Nijar (E)[f]					57	106		

Source: Reprinted from Nagy and Kónya 2005, with Permission from Elsevier.

[a] Nagy and Kónya 2004, [b] Stadler and Schindler 1993, [c] Barbier et al. 2000, [d] Wanner et al. 1994, [e] Fernandez et al. 2004, [f] Huertas et al. 2001.

TABLE 3.3

The Concentration of Edge Sites and Intrinsic Stability Constants of Protonation and Deprotonation of Silanol and Aluminol Sites of Bentonite Samples (Sajóbábony, Hungary) Calculated by the Surface Complexation Model

Sample	AlOH mol/g	SiOH mol/g	lgK $AlOH_2^+$	lgKAlO–	lgK SiO^-
B–II.b. upper	2.15E-04	1.29E-05	8.86	−5.78	−5.84
B–II.b. lower	6.75E-05	2.05E-05	8.33	−5.78	−5.84
B–II.a. upper	1.21E-04	1.55E-04	8.39	−7.01	−6.44
B–II.a. lower	1.00E-04	1.00E-04	8.28	−7.00	−6.48
B–I.b.	1.77E-04	2.12E-04	8.33	−6.95	−6.48

the sites practically do not deprotonate at pH<7 (in the pH range of the determination). The intrinsic stability constants of the deprotonation of the silanol site are different for sedimentary bentonites (B–I.b., B–II.a.) and the bentonitic tuff (B–II.b.).

Furthermore, the number and the ratio of silanol and aluminol sites are also very different. The ratio of silanol to aluminol is 1.3–1.5 for sedimentary bentonites; that is, there are more silanol sites. In the case of bentonitic tuff, the ratio of silanol/aluminol is reversed. For example, in the case of B–II.b. upper sample, the ratio of silanol to aluminol is 0.06. It is interesting that the bentonitic tuff containing volcanic glass in high concentration has less silanol sites than those in sedimentary bentonites. It can probably be explained by the higher particle size (smaller specific surface area) of the bentonitic tuff (Table 3.2).

It is worth noting here that using the value of specific surface area is problematic when the surface complexation model is used. In the case of the bentonitic tuff (B–II.b. upper), the parameter estimation (number of edge sites, intrinsic stability constants) fails when using the specific surface area as determined by the Brunauer–Emmett–Teller (BET) method (17.94 m^2/g; Table 3.2). The model is convergent only when at least 42 m^2/g is applied as an input parameter. This means that the other parameters estimated are only approximate. The same observation has been made for the titration of sandy soil and silica (Section 3.2.1) and discussed in the literature (e.g., Stadler and Schindler 1993). As seen in Table 3.1, the sample B–II.b. upper contains a high-amount amorphous phase. The difference can be explained by recognizing that the specific surface area is determined by nitrogen gas (Chapters 1, Sections 1.1.3 and 1.3.4.1.5), while the protolytic processes of the edge sites are studied by a titration process with acid, that is, hydrogen ions. Hydrogen ions are much smaller than nitrogen molecules; consequently, they can occupy interfacial places that nitrogen molecules cannot. This difference plays an even more significant role in the amorphous phase. Furthermore, the mechanisms of the interfacial process are also different: nitrogen molecules are adsorbed as a result of physical interactions with the external surface, whereas hydrogen ions participate in chemical reactions with the edge sites.

When comparing the number of edge sites (aluminol+silanol sites; Table 3.3) with the CEC (Table 3.2) of the different bentonite samples, we can see that the number of surface sites is in the range of 5–8% of the CEC values for the sedimentary bentonites and the bentonitic tuff B–II.b. lower. However, the number of edge sites is about 13% of the CEC for the bentonitic tuff B–II.b. upper. This is in contradiction with the small value measured for the specific surface area (Table 3.2) in this sample. It means that there may be some differences in the crystal structure (see the X-ray diffractograms in Figure 3.3), or the X-ray amorphous phase is present in high concentration (Table 3.1), or the inner surface can be reached by hydrogen ions. A large number of edge sites originate from aluminol sites (Table 3.3).

The surface acid–base parameters of bentonites from Sajóbábony can be compared to the similar parameters of other bentonites samples (Chapter 2, Section 2.4.3, Table 2.4). As mentioned in Chapter 2, Section 2.4.3, the parameters of SWy-2, BSAB, and MX-80, from the literature, cannot be compared to the data of bentonite samples from Sajóbábony because the deprotonation of silanol and aluminol sites are treated together. Calcium-bentonite (Istenmezeje, HU) shows similar characteristics as sedimentary bentonites, while the ratio of the amount of the edge site and the deprotonation intrinsic stability constant of SWy-1 bentonite is similar to the data bentonitic tuff.

The swelling of the bentonite samples (Table 3.2) is usually linked to the montmorillonite content (Table 3.1), except for the B–II.a lower sample in which the fraction of particles of <10 μm is low (Table 3.2). The presence of the large aggregates prohibits the swelling.

The relatively adsorbed amount of $^{137}Cs^+$ ion (x) (Chapter 1, Section 1.3.4.2.2) on bentonite samples is listed in Table 3.2. The two types of the bentonite samples can also be differentiated by the amount of the sorbed caesium. The x values vary between 0.8–0.9 on the sedimentary bentonite, while this value is about 0.7 for the bentonitic tuff. It means that the cation sorption of the bentonitic tuff is less than expected from the CEC.

3.1.3 RELATIONS OF GEOLOGICAL ORIGIN AND INTERFACIAL PROPERTIES

As seen previously, bentonite samples from the same bentonite deposit (Sajóbábony) show no significant differences in the montmorillonite content, but the circumstances of their geological formation are different. The sedimentary bentonites (B–I.b. and B–II.a.) show similar interfacial properties, while the bentonitic tuff (B–II.b.) behaves differently. This is probably caused by the large amount of X-ray amorphous phase.

Some characteristic properties of bentonites (CEC, sorption properties) are mainly governed by the montmorillonite content and the layer charge of montmorillonite. Other properties, however, depend on the circumstances under which the rock is formed. These are the particle size distribution, the external specific surface area, and surface acid–base properties. The quantity of the edge sites mainly depends on the specific surface area. The protonation and deprotonation reactions take place on the edge sites of other silicates and aluminosilicates present beside montmorillonite, so their effects manifest via surface reactions. Consequently, the origin of bentonite determines all properties that are related to external surfaces. The differences among these properties indicate the various stages of the formation processes for sedimentary bentonites and the bentonitic tuff.

The swelling of bentonites in water and, as will be discussed in Section 3.2.2, the migration of a nonadsorbing ion shows no direct relationship to the montmorillonite content or other geological characteristics in this narrow range (35–48% montmorillonite content; Table 3.1). However, they are influenced by several other factors, for example, the quality of the exchangeable cation (Chapter 2, Section 2.1), particle size distribution, aggregation, density, free pore size, other minerals, etc.

As a consequence of this, the properties and possible application of bentonites are determined by both the montmorillonite content and the geological origin. The origin appears in the particle size and external surface properties.

Although what has been discussed is specific to a particular bentonite deposit, namely the Kő Valley around Sajóbábony, the conclusions are general and valid for other bentonites, too.

3.1.4 Applications of Bentonites of Different Interfacial Properties

Before discussing the practical applications of bentonite, its different properties have to be studied in order to determine how the various bentonites rocks can be used. In the scope of this book, some agricultural and environmental applications are mentioned. In these applications, natural or chemically modified bentonite rocks are in direct contact with the soil or the geological environment. As an example, we can mention the bentonite barriers used for waste disposals.

The agricultural and environmental applications are mostly based on the water permeability (swelling) and sorption properties. Therefore, the montmorillonite content (CEC) and the quality of interlayer cations are the most important characteristic that influence the usability of montmorillonite in these cases. In addition, the role of the cation exchange in the sorption of anionic species is shown in Chapter 2, Section 2.12.

The addition of bentonite to soils can increase agricultural yields via improving the water permeability of soils and increasing the concentration of macro- (calcium, magnesium, sodium, potassium ions) and micronutrients (such as copper, zinc, manganese, cobalt ions). With regard to macronutrients, the presence of calcium ions in the interlayer space is especially desirable. When the soil does not contain micronutrients in suitable quantity, cation-exchanged bentonites (Chapter 2, Section 2.3), as trace element fertilizers can be added. In these cases, the application of bentonites with high CEC (high montmorillonite content) is desirable to ensure a higher concentration of the trace elements. Different cation- exchanged (copper, zinc) bentonites can be applied as seed disinfectants.

As seen earlier, cation exchange of montmorillonite has an important role in agricultural applications. In these cases, cations sorbed previously on bentonite are added to the soil. The opposite process, however, is also important in environmental applications, namely, when polluting cations have to be sorbed on bentonite. Such applications are the clay barriers of waste disposals. The high CEC is desirable. The sorption of cationic species decreases the migration rate of wastes, too (Section 3.2). The migration of anions can be reduced by cation exchange bentonite, as discussed in Chapter 2, Section 2.12.

Bentonite can also be added to addle because it can sorb ammonia or ammonium ions. Moreover, clay minerals in bentonite can catalyze the transformation of organic compounds to humic substances, or the oxidation of ammonia to nitrite and nitrate. In this case, the surface properties, namely, the catalytic activity plays an important role.

3.2   MIGRATION OF WATER-SOLUBLE SUBSTANCES IN ROCKS

Migration of water-soluble substances is a common process occurring in rocks and soils (Chapter 1, Section 1.3.6.2). This process is the net result of different hydrological and sorption processes. Hydrological processes (Chapter 1, Equation 1.143), that is, the movement of water, are governed by the water levels. Sorption processes can decrease the rate of water movement via interfacial processes (Chapter 1, Equations 1.144–1.147). In this chapter, the effect of sorption will be shown on the migration rate of water-soluble substances.

3.2.1   SORPTION AND MIGRATION OF CARRIER-
FREE RADIOACTIVE ISOTOPES IN ROCKS

The most problematic aspects of dealing with radioactive isotopes are ensuring their safe treatment and the storage of radioactive wastes. Most radioactive waste is produced in nuclear power plants (Chapter 1, Section 1.3.6.4). These wastes are classified on the basis of the radioactivity and the half-life of the isotopes. The classification is different in the different countries. Nuclear waste has to be stored safely for a very long time, for hundreds, even thousands of years. Therefore, the question is how good any extrapolation can be for such a long time. Natural geological analogies may be able to answer this question. For example, there are places where the radioactive ores (e.g., thorium, uranium, radium, etc.) has been isolated for millions of years. It was made possible by suitable geological environments. The natural and artificial barriers for nuclear (or other dangerous) wastes should be chosen on the basis of what we can learn from these examples. For example, in the Cigar Lake uranium mine in Canada (Cramer and Smellie 1994), salt mines provide suitable conditions.

3.2.1.1   A Model Predicting Migration Rate on the
Basis of Mineral Composition

To find a suitable place for the safe storage of nuclear wastes, the interactions of radioactive isotopes with the structural materials of the repositories and surrounding geological formations, as well as their transport phenomena, have to be studied. Geological formations, rocks, and soils, adsorb the radionuclides. The sorption processes significantly influence the migration of the radioactive isotopes. The sorption of radioactive isotopes on geological formations strongly depends on the mineral composition. The composition of the geological formations, however, is extremely variable even in the same place. Therefore, it is important to know how sorption depends on the mineral composition of rocks.

This problem may be treated by a linear model (Kónya et al. 2005; Nemes et al. 2006) by which the quantity of radioactive isotopes sorbed on a rock can be evaluated if we know the mineral composition of the rock and the distribution coefficients (Chapter 1, Equations 1.94 and 1.95) of the different minerals.

As discussed in Chapter 1, Section 1.3.3, cations can usually be sorbed onto geological formations (rocks and soils) by adsorption (Chapter 1, Section 1.3.3.1) and cation exchange (Chapter 1, Section 1.3.3.2).

Both adsorption and ion exchange can be treated by isotherms equations; Equations 1.73, 1.74, 1.94, and 1.95 can be used for the competitive adsorption or cation exchange, respectively. Radioactive ions are usually present in extremely low concentration in the geological environment. In Chapter 1, Equations 1.73 and 1.94, $c_1 = 0$: So, Equations 1.73 and 1.94 can be simplified as follows:

$$\frac{c_1}{a_1} = \frac{m}{V}\frac{y_1}{x_1} = \frac{m}{V}\frac{1}{k_1} = \frac{1}{z}\left(K_1 + \frac{K_1}{K_2}c_2\right) \tag{3.1}$$

$$\frac{c_1}{a_1} = \frac{m}{V}\frac{y_1}{x_1} = \frac{m}{V}\frac{1}{k_1} = \frac{1}{\zeta}\left(\frac{K_1}{K_2}c_2\right) \tag{3.2}$$

In addition, the concentration of the competing ion (c_2) is constant under a certain geological environment. So, Equations 1.73 and 1.94 can be expressed as follows:

$$\frac{c_1}{a_1} = \frac{m}{V}\frac{y_1}{x_1} = \frac{m}{V}\frac{1}{k_1} = \frac{K_{ads}}{z} \tag{3.3}$$

$$\frac{c_1}{a_1} = \frac{m}{V}\frac{y_1}{x_1} = \frac{m}{V}\frac{1}{k_1} = \frac{K_{ie}}{\zeta} \tag{3.4}$$

where c_1 and a_1 are the equilibrium concentrations of the radioactive ion in solution and in the solid, respectively; m is the mass of solid, V is the volume of the solution, x and y are the relative radioactivity in solid and solution $(x_1 + y_1 = 1)$, respectively; z and ζ are the number of adsorption and ion-exchange sites, respectively, in mol/g; c_2 is the concentration of the competing ion in solution; and K_{ads} and K_{ie} are the isotherm parameters for adsorption and ion exchange. Equation 3.4 is valid for heterovalent ion exchange because of the very low concentration of the exchanging ion.

As seen from Equations 3.3 and 3.4, the distribution coefficient of the ion (k_1) can be expressed both for adsorption and ion exchange as:

$$k_1 = \frac{x_1}{y_1} \tag{3.5}$$

and from here,

$$x_1 = \frac{k_1}{1+k_1} \tag{3.6}$$

The advantage of the application of distribution ratios is that results at different solution concentration can be compared directly, and transport equations take adsorption into account via the distribution ratio.

Equations 3.5 and 3.6 are valid for pure minerals. Rocks obviously consist of more than one mineral, and the relative sorbed quantity of the radioactive ion on any i-th mineral component is proportional to the mass ratio ($n(i)$) of the given mineral in the rock:

$$x_i(i) = \frac{k_i(i)\dfrac{m(i)}{m}}{1+k_i(i)\dfrac{m(i)}{m}} = \frac{k_i(i)n(i)}{1+k_i(i)n(i)} \tag{3.7}$$

where m and $m(i)$ are the mass of the rock and the i-th mineral component, respectively; and $k_i(i)$ is the distribution coefficient of the sorbing radioactive ion for the i-th mineral component. When the rock consists of two minerals, the net value of x_1 is the sum of the $x_1(i)$ values for the two components ($x_1(1)$ and $x_1(2)$), assuming no synergetic or inhibiting effects between the sorption properties of the minerals. It means that the sorption sites can linearly be summarized. This treatment can be used in the case of minerals because ions are sorbed on geological formations mostly by electrostatic forces, and the surface charge of the minerals is negative.

Mathematically, this is expressed as follows:

$$x_i = x_1(1) + x_1(2) = \frac{k_1(1)n(1)}{1+k_1(1)n(1)} + \frac{k_1(2)n(2)}{1+k_1(2)n(2)} \tag{3.8}$$

When the rock consists of n components,

$$x_i = x_1(1) + x_1(2) + x_1(3) + \ldots + x_1(n)$$

$$= \frac{k_1(1)n(1)}{1+k_1(1)n(1)} + \frac{k_1(2)n(2)}{1+k_1(2)n(2)} + \frac{k_1(3)n(3)}{1+k_1(3)n(3)} + \ldots \tag{3.9}$$

$$+ \frac{k_1(n)n(n)}{1+k_1(n)n(n)} = \sum_{i=1}^{n} \frac{k_i(i)n(i)}{1+k_1(i)n(i)}$$

When a radioactive ion is sorbed both by adsorption and ion exchange on a given mineral, it has two members for the different sorption mechanisms in Equation 3.9.

Equation 3.9 shows when the total sorbed quantity of the radioactive ion (x_1) and the mineral composition ($n(i)$'s) are determined, the distribution coefficients for the mineral components $(k_I(i))$ can be calculated. The reverse procedures are also possible; that is $k_I(i)$ values for pure minerals can be determined directly if simple rocks containing more than 90% of the components in question are available and these components are responsible for the sorption in a high degree. In this way, a relatively simple equation can be used for the calculation.

3.2.1.2 The Application of the Linear Model for the Sorption of Cs-137 and Sr-85 Ions

The application of the linear model will be shown for cesium-137 and strontium-85 ions. Cs-137 and the different strontium isotopes, especially Sr-90, are important components of nuclear wastes. As seen previously in Table 3.2, the cesium ion has a different sorption property on bentonite samples from the Sajóbábony deposit, depending on the geological origin and composition. Similarly, different bentonite rocks from the Carpathian Basin (Table 3.4) show different sorption properties, including kinetics and equilibrium (Figure 3.4; Table 3.5; Nagy et al. 2003b; Kónya et al. 2005).

Since cation exchange kinetics is usually very fast, occurring in millisecond time scales (Tang and Sparks 1993), the kinetics of cesium-137 sorption on bentonite is determined by diffusion steps: diffusion through the adhesion layer and in the pores, including the interlayer space of montmorillonite. Therefore, the kinetics can be treated by using Equation 1.142 (Chapter 1). Two members of the equation are taken into account, according to the number of diffusion steps. As seen in Table 3.5, the first step, diffusion through the adhesion layer is fast, and k_I values are high. The diffusion in the pores (k_2 values) is much slower. In the case of Egyházaskesző bentonite, only one diffusion step can be shown. The results of strontium sorption of bentonite have been discussed in Chapter 1, Section 1.3.6.1.

The sorption properties of the radioactive ions can be different when the composition of the rocks is more complex than that of bentonites. The mineral composition of the bentonites from the Carpathian Basin and some other Hungarian rocks are shown in Table 3.6.). Table 3.7 illustrates the relative sorbed quantity of ^{137}Cs ion in equilibrium (x_{Csexp}) for these bentonite and rocks.

From the data of Table 3.7 and the mineral composition of the rocks (Tables 3.4 and 3.6), the distribution coefficients of the minerals $(k_{Cs}(i))$ can be determined by using Equation 3.9. The distribution coefficients for the different groups of minerals and the calculated relative sorbed quantity of ^{137}Cs ion (x_{Cscalc}) and the differences between the experimental and calculated values (Δx_{Cs}) are shown in Tables 3.7 and 3.8.

As seen in Table 3.7, the differences between the experimental and calculated values (< 0.07) are within the deviation of the determination of the mineral composition and relative sorbed quantities of ^{137}Cs ion (5–10%). So, the linear model fairly well describes the relationship between the mineral composition and sorbed quantity of cesium ions on different rocks.

The distribution coefficients determined by Equation 3.9 (Table 3.8) show that the different minerals can be classified from the point of view of cesium ion sorption

TABLE 3.4
Mineral Composition, External Specific Surface Area, and Cation Exchange Capacity of Bentonite Rocks from the Carpatian Basin

	Montmorillonite	Illite	Rectorite	Paligorscite	Kaolinite	Biotite	Quartz	Plagioclase (Albite)	Kalifeldspar	Amorphous	Cristobalite+opal CT	Muscovite	Zeolite	Specific surface area (BET) m²/g	Cation exchange capacity, meq/100g
Mád Újhegy (Hungary)	21				3		58			3		5		24	44
Kuzmice (Slovakian)	41		10				1			3	54		1	18	70
Muideni (Romanian)	42									4	54			18	106
Valea Chrioarului (Romanian)	48	8							10	5	25		4	15	40
Sajóbábony (Hungary)	50	10			4		6	5		20	5				
Istenmezeje black (Hungary)	54				tr		4	1		8	33	tr		23	44
Istenmezeje yellow (Hungary)	74	2					1	1		5	17			24	66
Egyházaskesző (Hungary)	89			4		1				6				27	78

Source: Reprinted from Nagy et al. 2003b, with Permission from Springer.

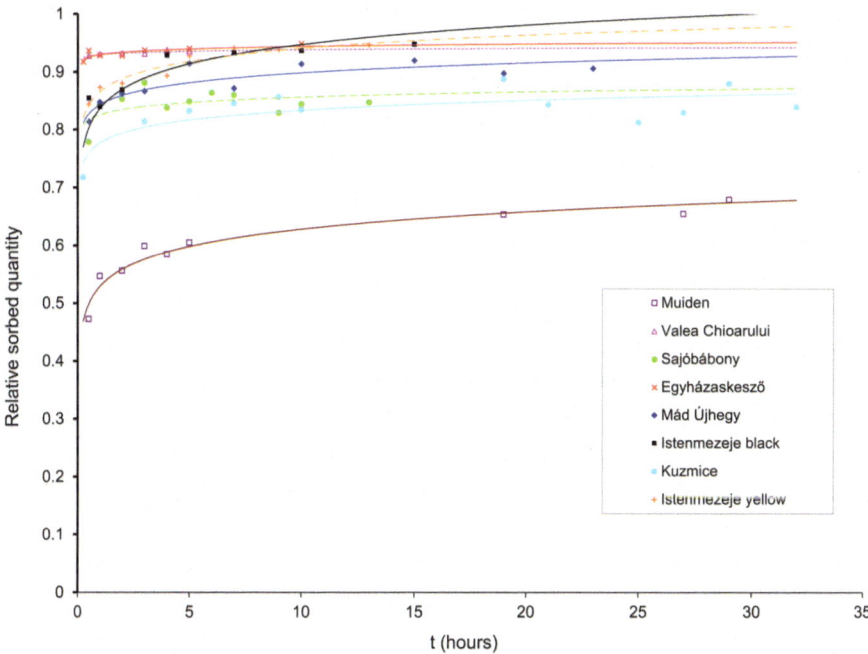

FIGURE 3.4 The relative sorbed quantity of cesium-137 on bentonite from the Carpathian Basin as a function of time. (Reprinted from Nagy et al. 2003b, with permission from Springer.)

(Table 3.8). As expected from the high CEC, montmorillonite sorbs cesium ion in the highest degree. Besides cation exchange in the interlayer space of montmorillonite, cations, including cesium ion, can be adsorbed on the deprotonated silanol and aluminol sites of silicates (Chapter 2, Section 2.5). The other clay minerals, illite and rectorite, also has some cesium sorption, but their distribution coefficients are about two orders of magnitude lower than that of montmorillonite because cation sorption is possible only on the deprotonated silanol and aluminol sites of illite and rectorite. Tectosilicates (quartz, cristobalite, kalifeldspar, plagioclase) and some phyllosilicates (chlorite) show similar cesium ion sorption. The mechanism of sorption is also the same: cation sorption on the deprotonated edge charges.

Another phyllosilicate mineral, namely, biotite, has a larger sorption capability as expected. It is likely related to the iron content of biotite. Similarly, the carbonates containing iron and magnesium (ankerite and dolomite; Table 3.8) show more significant cesium sorption as calcite (calcium carbonate), which practically does not adsorb cesium. The low sorption ability of calcite can be explained by the Hahn adsorption rule (Chapter 1, Section 1.2.4); that is, the sorption is low when the sorbate (cesium carbonate) has great solubility.

The sorption of $^{85}Sr^{2+}$ ion was studied on bentonites from the Carpathian Basin (Table 3.4; Nemes et al. 2006). Similar to the sorption of Cs-137, the distribution coefficients of the mineral components (Table 3.10) were calculated from the net

TABLE 3.5
Kinetic Parameters of Cesium-137 Sorption on Bentonite from Carpathian Basin Calculated by Eq. 1.135

	Mád Újhegy (HU)	Kuzmice (SK)	Muideni (RO)	Valea Chioarului (RO)	Sajóbábony (HU)	Istenmezeje black (HU)	Istenmezeje yellow (HU)	Egyházaskeszõ (HU)
x_∞	0.90	0.85	0.67	0.94	0.85	0.92	0.96	0.93
A_1	0.76	0.70	0.54	0.93	0.46	0.61	0.84	0.93
A_2	0.14	0.15	0.13	0.02	0.39	0.31	0.12	
$k_1\ (h^{-1})$	>10	>10	4	>10	>10	>10	>10	8.9
$k_2\ (h^{-1})$	1.00	0.48	0.13	0.17	3.35	1.00	0.24	

TABLE 3.6
Mineral Composition, External Specific Surface Area, and Cation Exchange Capacity of Some Hungarian Rock Samples

	Montmorillonite	Illite/smectite	Illite	Paligorscite	Kaolinite	Chlorite (Fe)	Biotite	Quartz	Plagioclase (Albite)	Kálifeldspare	Calcite	Dolomite	Ankerite	Amphibolite	Goethite	Hematite	Amorphous	Cristobalite+opal CT	Specific surface area (BET) m²/g	Cation exchange capacity meq/100g
1	3		18			16		7	7	14				25	6		4		9	8
2	3	3				13			31	18	7	4	18						8	8
3	4	3				3		1	1	1	87		1						8	9
4	4					31		5	8	4	39	2							16	9
5	5	2				16		7	7	14				25	6		4		18	10
6	14			37		7		7	5	4	16	8			3		5		21	8
8		5				6	30	16	24	12	2			3		2			12	<5
9						10	37	1	20	18									10	<5
10						41		1	6	23				26					3	<5
11		4						6	2	1	1	86							11	

Source: Reprinted from Nagy et al. 2003b, with Permission from Springer.

TABLE 3.7
The Experimental (x_{Csexp}) and Calculated (x_{Cscalc}) Values (Equation 3.9) of the Relative Sorbed Quantities of Cesium Ions and Their Differences
$(\Delta x_{Cs} = x_{Csexp} - x_{Cscalc})$

Rock	x_{Csexp}	x_{Cscalc}	$\Delta x_{Cs} = x_{Csexp} - x_{Cscalc}$
Mád Újhegy (H)	0.950	0.933	0.017
Kuzmice (SK)	0.923	0.963	−0.040
Mujdeni (RO)	0.923	0.964	−0.041
Valea Chioarului (RO)	0.936	1.000	−0.064
Istenmezeje black (H)	0.971	0.969	0.002
Istenmezeje yellow (H)	0.973	0.976	−0.003
Egyházaskesző (H)	0.973	1.014	−0.041
1	0.812	0.815	0.003
2	0.708	0.716	−0.008
3	0.371	0.396	−0.025
4	0.666	0.636	0.030
5	0.851	0.811	0.040
6	0.848	0.846	0.002
8	0.866	0.825	0.042
9	0.797	0.774	0.024
10	0.153	0.096	0.056
11	0.533	0.464	0.069

Source: Reprinted from Kónya et al. 2005, with Permission from Elsevier.

TABLE 3.8
Distribution Coefficients (k_{Cs}) of ^{137}Cs Ion on Different Minerals

Group of minerals	Mineral	k_{Cs}, dm³/g
Tectosilicates	Quartz, cristobalite	0.1
	Kalifeldspar, plagioclase	0.2
Phyllosilicates	Biotite	6
	Chlorite	0.1
Clay minerals	Illite, rectorite	0.4
	Montmorillonite	25
Carbonates	Calcite	0.01
	Dolomite, ankerite	0.9

Source: Reprinted from Kónya et al. 2005, with Permission from Elsevier.

TABLE 3.9
The Experimental (x_{Srexp}) and Calculated (x_{Srcalc}) Values (Eq. 3.9) of the Relative Sorbed Quantities of Sr Ions and Their Differences ($\Delta x_{Sr} = x_{Srexp} - x_{Srcalc}$)

Samples	x_{Srcalc}	x_{Srcalc}	$\Delta x_{Sr} = x_{Srexp} - x_{Srcalc}$
Istenmezeje black (H)	0.666	0.699	−0.033
Istenmezeje yellow (H)	0.778	0.754	0.024
Mád Újhegy (H)	0.737	0.74	−0.003
Egyházaskesző (H)	0.778	0.779	−0.001
Sajóbábony (H)	0.7	0.698	0.002
Kuzmice gel (SK)	0.837	0.645	0.192
Mujdeni gel (RO)	0.872	0.65	0.222

Source: Reprinted from Nemes et al. 2006, with Permission from Elsevier.

relative sorbed amount of Sr ion (x_{Srexp}; Table 3.9) and the mineral composition (Table 3.4) by using Equation 3.9. The calculated values of the relative sorbed quantity of strontium ion (x_{Srcalc}) and the differences of the experimental and calculated values (Δx_{Sr}) are listed in Table 3.9.

As seen in Table 3.9, the difference between the experimental and calculated values is below 0.033 in the case of Egyházaskesző, Istenmezeje (black and yellow), Mád Újhegy, and Sajóbábony samples. This value is in the range of the deviation of the determination of the mineral composition and the sorption experiments (5–10%). So, the model, similarly to cesium ion, fairly well describes the relationship between the mineral composition and the adsorbed quantity of strontium ion on different rocks. The deviation is higher in the case of the two samples (Kuzmice, Muiden), the

TABLE 3.10

Distribution Coefficients ($k_{Sr}(i)$) of Sr Ion on Different Minerals

Mineral	k_{Sr} (dm3/g)
Quartz	0.03 + 5%
Cristobalite	0.03 ± 5%
Montmorillonite	1.98 ± 5%
Rectorite	1.75 ± 5%
Illite	0.15 ± 5%

Source: Reprinted from Nemes et al. 2006, with permission from Elsevier.

main component of which is cristobalite, and it can be explained with the presence of a gel phase (Chapter 1, Section 1.3.6.1).

Similar to Cs-137, minerals can be classified on the basis of $k_{Sr}(i)$ values determined by the linear model (Table 3.10). From clay minerals, montmorillonite and rectorite adsorb the largest quantity of strontium ions. The greater strontium sorption comes from the cation-exchange properties of montmorillonite and rectorite (van Olphen 1977): they both can uptake cations in the interlayer space by cation exchange. In addition, cation sorption is also possible on the deprotonated edge sites (van Olphen 1977; Missana et al. 2008, Chapter 2, Section 2.5).

The other clay mineral, illite also has some strontium sorption, but the distribution coefficient is about one order of magnitude lower than that of montmorillonite and rectorite. Since, in illite, the layer charge is compensated by nonexchangeable cations (Chapter 1, Table 1.2), cation sorption can take place only on the deprotonated edge sites. This is the case for tectosilicates (quartz, cristobalite).

The construction and application work of the linear model of sorption has important results. First, the distribution coefficients of the different minerals for radioactive ions in very low concentration can be determined. The minerals with similar structure obviously have similar distribution coefficients because of the similar sorption mechanism. The differences are the result of a specific property of the minerals; for example, the presence of iron or magnesium in carbonates has a significant effect on the sorption of cesium ion.

Another result is that the equilibrium-sorbed quantities of radioactive ions in extremely low concentration can be predicted on any rock if we know the mineral composition and the distribution coefficients of the minerals or mineral groups. This has huge importance, given the great variety of geological formations (Söderlund et al. 2011).

Obviously, the linear model of sorption on rocks can be used not only for radioactive isotopes but also for other wastes (e.g., heavy-metal ions) when the concentration is extremely low. Their sorption can also be predicted in any geological environment. With the aid of the model, we construct the lines of a given area (e.g., in the area of waste disposals) where the sorption is equal ("isosorption lines").

The sorption parameters allow the prediction of the migration properties, too, because the sorption properties are simply added to the hydrological properties (Equations 1.128–1.131, Chapter 3.2.2).

3.2.2 EFFECT OF SORPTION ON THE MIGRATION OF IONS IN BENTONITE

As mentioned in Chapter 1, Section 1.3.6.2, the migration of water-soluble substances in geological formations is determined by hydrological and sorption properties. Under natural conditions, the movement of water (hydrological part), in the first approximation, can be considered as a stationary state (constant temperature, humidity, free pore volume) established as the result of exposure to the same environmental factors during a long time period. Of course, this approximation is only right when the rock is not in direct connection with the atmosphere and biosphere, so the daily climatic changes can be neglected (e.g., underground conditions). The migration rate of water represents the maximum rate of migration of water-soluble substances. Sorption can only decrease the rate of migration.

In laboratory experiments, the migration of the water (Sánches et al. 2008) and water-soluble substances in geological formations can be studied under different conditions (temperature, pressure, humidity, bulk density, free pore volume, experimental technique, etc.), but only those results can be compared in which all conditions were the same. A very illustrative example is when the migration rate of water is studied in a dry rock or soil and compared to the migration rate of a nonsorbing substance (e.g., chloride ion) in a wet rock or soil. In these cases, the migration rate can be very different, especially when strong physical and chemical interactions occur between the water and the rock or soil. The wetting and swelling of dry bentonite in water is a much slower process than the migration of chloride ion in a wet bentonite sample. So, it is not correct to say that the migration rate of chloride is higher than that of water.

Besides the migration rate of the nonsorbing substance, the effect of sorption on migration has to be studied on wet rocks and soils. Through these studies, the migration rate of the nonsorbing and sorbing substances can be compared, and the difference will provide information on the effect of sorption affinity (distribution coefficients). Migration can be described by Equation 1.147, in which all effects can be combined by the effective migration coefficient (D_m).

As mentioned in Chapter 1, Section 1.3.6.1, Fick's second law (including Equation 1.146) cannot be solved generally, only partial solutions exist under well-determined boundary and initial conditions. In Figure 3.5, the so-called migration cell type, a frequently used experimental setup, is shown.

The migration cell (Figure 3.5) contains three compartments. The rock sample is placed in the middle part between two membranes. Because of the previously discussed reasons, the rock has to be saturated with the solution in which the migration will be studied. To achieve this, both the donor and acceptor cells are filled with the solution. After saturation, the solution in the donor cell is exchanged for the solution containing the substance – the migration that we intend to study. If necessary, the

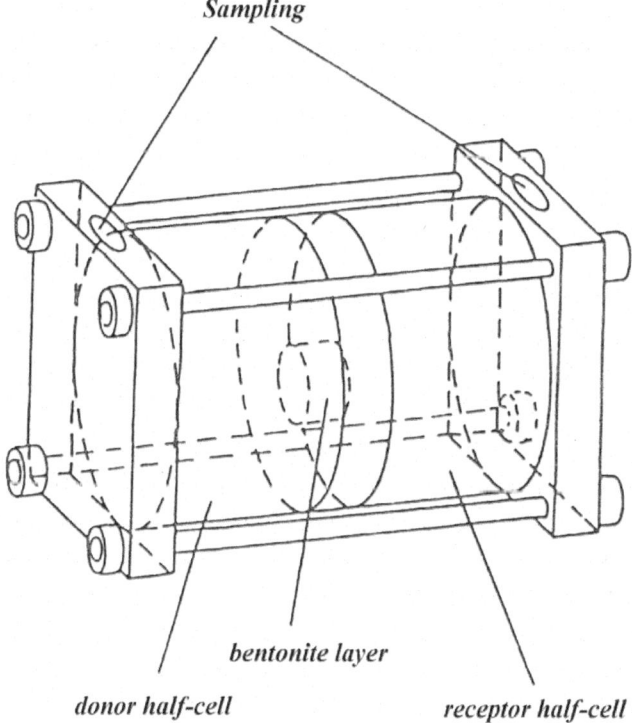

Sampling

bentonite layer

donor half-cell

receptor half-cell

FIGURE 3.5 Migration cell. (Reprinted from Nagy and Kónya 2005, with permission from Elsevier.)

acceptor cell is also filled with a fresh solution (without the substance to be studied). Migration can be studied in different sampling techniques:

1. Solution samples are taken at different times from the receptor cell, and the concentration of the migrating substances is determined as a function of time.
2. After a given time, the rock sample is cut into thin layers, and the concentration of the migrating substances is determined as a function of distance.

The correct solution to Fick's second law for the cell (Figure 3.5) can be given (Crank 1956), assuming the boundary condition

$$C = C_0 \quad x = 0 \quad t > 0 \tag{3.10}$$

and the initial conditions

$$C = 0 \quad x > 0 \quad t = 0 \tag{3.11}$$

$$C_{(x,t)} = C_0 erfc \frac{x}{2\sqrt{D_m t}} \tag{3.12}$$

where $erfc\ z = 1 - erfz$.

When the migration rate in the solution of the donor cell is much faster than that in the solid rock (the solution is always homogeneous), the approximate solution to Fick's second law can be applied as follows:

$$C_{(x,t)} = \frac{M}{(\pi D_m t)^{1/2}} \exp\left(-\frac{x^2}{4D_m t}\right) \tag{3.13}$$

Both Equation 3.12 and 3.13 can be used for the data obtained with techniques 1 and 2; in technique 1, x is constant, while in technique 2, t is constant.

The effective migration coefficients of bentonite samples from Sajóbábony deposit (Table 3.1) are listed in Table 3.11.

In Table 3.11, the data of a nonsorbing (chloride ion) and two sorbing ions (cesium and strontium ions) are summarized, all ions being in extremely low concentration. As seen from the data in Table 3.11, the highest values of effective migration coefficients are obtained when samples are taken from the acceptor cell. This method is the simplest and, for this reason, most frequently used. This method can have significant errors because the solution can transfer from the donor to the acceptor cell along the periphery of samples without migrating through the bentonite. The degree of accidental solution transfer differs from sample to sample. The other disadvantage of this method is that the well-sorbing substances (e.g., cesium and strontium ions) may require a very long time to pass through the rock sample and, during this time, the structure of the rock and the environmental conditions may change. Therefore, more exact migration coefficients can be obtained by technique 2, by cutting the rock samples into thin layers.

The results obtained with Equations 3.12 and 3.13 are within the same order of magnitude; the correct mathematical equation (Equation 3.12), however, always gives higher values of the effective migration coefficients. For the good estimation of migration rate, Equation 3.12 is needed.

It is also seen from Table 3.11 that the sorption significantly decreases the effective migration coefficients. In the case of cesium ions, for which the distribution coefficients are high (Table 3.2), it diminishes by about an order of magnitude. The effective migration coefficients of strontium ion are between the migration coefficients of chloride ion (no sorption) and cesium ions (strong sorption). This can be explained by the sorption affinities of the cesium and strontium ion; however, quantitative comparison is difficult since there is a homovalent (Sr^{2+}-Ca^{2+}-bentonite) and a heterovalent (Cs^+-Ca^{2+}-bentonite) exchange.

3.2.3   Effect of Precipitation in Migration

As shown in Chapter 2, Section 2.10.3, the system of lead ion/calcium-montmorillonite lead enrichments may form when, based on thermodynamic conditions, their formation is not expected. The process was considered as surface precipitation or coprecipitation, depending on the chemical composition of the lead enrichments.

TABLE 3.11

Effective Migration Coefficients of Chloride, Cesium, and Strontium Ions in Bentonites from Sajóbábony Deposit (Table 3.1)

	Chloride ion			Cesium ion		Strontium ion		Bulk density (g/cm³)
	Samples taken from the acceptor cell, D (m²/hour)	Samples from the cutting bentonite samples into layers						
	Eq. 3.13	Eq. 3.13	Eq.3.12	Eq. 3.13	Eq.3.12	Eq. 3.13	Eq.3.12	
BIB	3.6E-08	2.0E-09	3.0E-09	4.4E-10	5.8E-10	9.7E-10	4.8E-09	1.17
BIIA upper	8.2E-08			3.5E-10	9.2E-10	1.1E-08		1.35
BIIA lower	6.3E-08			5.0E-10	4.8E-09	5.7E-09		1.36
BIIB upper	4.5E-08			1.1E-10	4.6E-10	7.1E-09		1.24
BIIB lower	8.8E-08			5.5E-11	3.2E-10	2.1E-09	3.6E-08	1.4

FIGURE 3.6 Scanning electron microscopic picture of a mica plate immersed into the suspension of a natural clay sediment. Mean lead concentration of the clay is 36 mg/kg. (Reprinted from Nagy et al. 2003a, with permission from Elsevier.)

Since lead enrichments are fairly bulky (a few hundred nanometers to micrometers), their migration rate is practically zero. Therefore, if they form under natural conditions, they should be observed on the surface of natural clay. In fact, lead enrichments were found on clay mineral surfaces from lake sediments (Figure 3.6; Nagy et al.2003a).

Under natural conditions, the thermodynamics conditions are obviously unknown, so the lead enrichment formation mechanism is also unknown. However, the formation of nano- and microparticles containing lead ions under environmental conditions is especially intriguing. At least a portion of lead ions in nature will form similar enrichments. The process, all types of precipitation, can decrease the migration rate of lead ions. If precipitation of any substance occurs, its migration rate can decrease.

3.3   INTERFACIAL ACID–BASE PROPERTIES OF SOILS

As mentioned several times in this book (e.g., Chapter 1, Sections 1.2.2 and 1.2.3; Chapter 2, Sections 2.4 and 2.7), the hydrogen and hydroxide ions formed through the dissociation of water play a determining role in all interfacial processes of rocks and soils. In soils, not only hydrogen ion but aluminum ion, producing hydrogen ions by hydrolysis (Chapter 1, Equations 1.26–1.36, Figure 1.6) is taken as an important source of acidity. Accordingly, different forms of acidity in soils are classified as follows (Ulrich and Sumner 1991):

1. Active acidity: the H^+ ion concentration (pH) of the soil solution.
2. Exchangeable acidity includes exchangeable Al^{3+} as well as exchangeable H^+ on the permanent charges of mineral components. Usually, there is a small amount in acid mineral soils, but it is more abundant in organic soils. It can be extracted with neutral salt solutions such as KCl, $CaCl_2$, or NaCl (Sevink et al. 1998).
3. Nonexchangeable or residual acidity consists of weak acids that cannot be replaced by neutral salt solutions. The source of this acidity is the acidic (mainly, carboxyl and OH groups) of soil organic matter (Chapter 1, Table 1.5) and the neutral and protonated edge sites of mineral components (e.g., $-SiOH$, $-AlOH$, $-FeOH$, $-AlOH_2^+$, $-FeOH_2^+$).

Soil acidity is usually characterized by the pH of the soil solution or by exchanging the hydrogen ions by a cationic (e.g., sodium, potassium, or calcium) salt (Sevink et al. 1998) and the titration of the dissolved hydrogen ions by a basic solution. In this way, exchangeable acidity can be determined.

Nonexchangeable or residual acidity can be characterized by the surface acid–base properties, that is, with the number of surface sites and their intrinsic stability constants (Chapter 1, Section 1.3.2.1.1; Chapter 2, Section 2.4.3) (Nagy and Kónya 2007). The acid–base properties of the edge sites, silanol and aluminol, and soil organic matter can be studied by potentiometric titrations of soil suspensions and the results can be interpreted based on the surface complexation model considering the possible interfacial reactions. Usually, soil components extracted from the soil are studied in this way; however, to draw practically applicable conclusions, the soil in its original form should be studied. The main problem with studying the extracted soil components is that a number of surface-related parameters are considered constant: for example, intrinsic stability constants of deprotonation and protonation processes ($lgK_{protonation} = 7.35$, $lgK_{deprotonation} = -8.95$), the number of edge sites (21 μmol/ dm^3 at 1 g soil/25 cm^3 solution ratio which is equal to ca. 5.10^{-7} mol/g [Goldberg and Sposito 1984], and the total number of reactive surface hydroxyl groups), which cannot be valid for all types of soils because the external specific surface area varies within an order of magnitude (Goldberg et al. 2005). Some examples of such treatments are the sorption of arsenate, boron, molybdenum, phosphate, etc. (Goldberg and Criscenti 2008).

To study the nonexchangeable acidity, active and exchangeable acidity has to be blocked (Hargrove and Thomas 1982; Thomas and Hargrove 1984). In potentiometric titration, the main source of exchangeable acidity is permanent charge of aluminosilicates, which can be neutralized by a support electrolyte in high concentration (e.g., 0.1 mol/dm^3 sodium salt) (Chapter 2, Section 2.4.3). The application of the support electrolyte makes it possible to use the constant capacitance model, too.

The characteristic properties of some soils, studied by potentiometric titration and the surface complexation model, are shown in Table 3.12. The mineral composition of some soils is also provided (Table 3.13).

Interpretation of results of the potentiometric titration of soil samples present the same problems as in the case of bentonitic tuff from the Sajóbábony deposit (Section

TABLE 3.12
Characteristic Parameters of Soil Samples (Hungary)

Sample	Soil type	pH (water)	pH (KCl)	Humus (%)	Sand (%) >0.05 mm	Silt (%) 0.05–0.002 mm	Clay (%) <0.002 mm	Carbonate (%)
Agyagosszergény	Calcarous meadow soil	7.5	7.1	6.6	50.4	26.6	23.0	1.3
Ebes	Brown forest soil with clay illuvation	6.6	6.2	3.3	9.2	46.1	44.7	
Eger	Humid soil	6.3	5.8	2.9	7.2	45.9	46.9	
Gagyvendégi	Brown forest soil with clay illuvation	6.0	5.3	2.0	3.6	68.9	27.5	
Hajdúböszörmény	Meadow soil	6.5	5.8	6.4	3.6	60	36.4	
Homokszentgyörgy	Brown forest soil with clay illuvation	4.8	4.0	1.1	54.0	36.6	9.4	
Jármi	Humous sand	4.5	4.2	0.7	85.2	10	4.8	
Jászladány	Clayey meadow soil	7.5	7.0	2.6	11.3	8.2	80.5	2.2
Karcag	Meadow solonetz	6.7	4.8	4.0	2.4	52.8	44.8	
Örbottyán	Calcarous sandy soil	7.7	7.6	1	81.4	13.2	5.4	3.3
Ragály	Acidic brown forest soil with clay illuvation	4.3	3.4	3.4	17.1	51.1	31.8	
Aranyosapáti	Loamy soil	8.4	7.8	2.5	55	30	15	5
Záhony	Sandy soil	8.0	7.9	0.6	64	27	3	
Dombrád	Loamy soil	8.4	8.0	2.9	47	44	9	
Tiszabercel	Loamy soil	7.7	7.7	4.2	61	34	5	1
Tiszaeszlár	Loamy soil	8.7	7.9	1.8	64	30	6	1
Tiszalök	Sandy soil	9.0	8.0	1.0	56	39	5	2

Source: Reprinted from Nagy and Kónya. 2007, with Permission from Elsevier.

TABLE 3.13

Mineral Composition of Soil Samples Collected along River Tisza (Hungary)

	Montmorillonite	Illite/montorillonite	Illite	Muscovite	Chlorite	Quartz	Kalifeldspar	Plagioclase	Calcite	Dolomite	Amfibol	Hematite	Goethite	Amorphous
Aranyosapáti	13	2		8	6	52	2	5	2	3			2	5
Záhony	3		8		6	64	2	11			1		2	3
Dombrád	9			11	9	47	2	16	1			1	2	3
Tiszabercel	5	3		11	6	60	2	6	1				2	4
Tiszaeszlár	6		6		5	63	2	13	2				1	3
Tiszalök	5			7	5	55	6	15	9				2	3
Tiszavasvári	15			6	6	44	2	7		2			4	5

Source: Reprinted from Nagy and Kónya. 2007, with Permission from Elsevier.

Note: For Other Soil Properties see Table 3.12.

3.1.2), That is, the parameter adjustment is not convergent when the specific surface areas determined by the BET method are applied. At least 40 m²/g for sandy soils (similar value to bentonitic tuff, Section 3.1.2) and 80–100 m²/g for clayey soils are needed as an input parameter for reaching convergence during the adjustment. The source of this problem is the same as in the case of bentonitic tuff: the difference between the size of nitrogen molecules used to determine external specific surface area and that of hydrogen ion as main reagent in potentiometric titrations. The size of hydrogen ion is much less than the size of nitrogen molecules, so hydrogen ions can introduce themselves into the pores where nitrogen molecules cannot. As already mentioned in Section 3.1.2, the mechanisms of the interfacial process are also different: pure physical adsorption takes place for nitrogen gas, and chemical reaction with the surface sites for hydrogen ion.

As seen in Table 3.12, the humus content of soils varies within a rather wide concentration range (0.6–6.6%). However, parameter adjustment is only successful when the protolytic processes of humus are neglected. Consideration of the protonation and deprotonation of aluminol and silanol sites (Chapter 1, Equations 1.54–1.56; Chapter 2, Sections 2.3–2.5) is sufficient. It is likely caused by the cations of the support electrolyte and the divalent and trivalent (aluminum and ferric) cations dissolved from the soil that react with the acidic functional groups of soil organic matter, limiting the protonation of functional groups (Hargrove and Thomas 1982, Sparks 2003).

The concentration of the aluminol and silanol sites and intrinsic stability constants of protonation and deprotonation are listed in Table 3.14.

The data in Table 3.14 show that the number of surface silanol and aluminol sites is different for each soil, confirming that it is important to take the actual surface sites into consideration.

The intrinsic stability constants of protonation and deprotonation (Table 3.14) for most soils are the same within the experimental errors. Therefore, it can be concluded that these are thermodynamic parameters characterizing the surface aluminol and silanol sites. The values, however, are different to those used by Goldberg et al. (2005). It can be explained by the modifying effect of silanol site, neglected by Goldberg et al. (2005).

The results of Table 3.14 show that there is a relationship between the concentration of surface sites and composition (Table 3.12). The number of silanol sites is proportional to the sand content, except for freshly deposited alluvial soils with high primary silicate content (e.g., Tiszalök, Záhony). The sandy soils from wetland areas (soils near River Tisza) do not fit the usual tendencies; that is, the concentrations of silanol and aluminol sites are significantly lower, as expected from similar data of other sandy soils.

There is also a relationship between the number of aluminol sites and the pH of natural soil solution (Table 3.12). The acidity of natural soil solution is the compound of all of the processes influencing soil acidity mentioned previously in this chapter.

The parameters for characterizing the edge sites of soils fall into two distinct groups: (1) those depending on the quality and composition of soils (i.e., the

TABLE 3.14

Concentration of Silanol– (SiOH) and Aluminol Sites (AlOH), Intrinsic Stability Constants of the Deprotonation (lg K (SiO−), lg K (AlO−)) and protonation (lgK (AlOH₂−), the Specific Surface Area Used for the Parameter Adjustment

Soil	−AlOH mol/g	lg K AlOH$_2^+$	lg K AlO⁻	−SiOH mol/g	lg K SiO⁻	S m²/g	WSOS/DF*
Aranyosapáti	9.00E-05	8.4	−6.9	7.50E-04	−7.5	100	87
Tiszabercel	1.75E-04	8.8	−6.8	8.50E-04	−7.2	100	57
Tiszaeszlár	1.10E-04	8.2	−6.9	1.40E-04	−6.6	100	18
Tiszalök	1.40E-04	6.9	−4.7	1.10E-03	−7.2	60	22
Dombrád	2.75E-04	8.2	−6.9	6.00E-04	−7.3	100	220
Záhony	2.75E-05	7.1	−6.5	2.85E-05	−6.1	100	77
Agyagosszergény	1.45E-04	8.3	−6.9	6.50E-05	−7.3	100	9
Ebes	1.50E-04	8.2	−6.9	2.25E-04	−7.3	100	23
Eger	9.50E-05	8.3	−6.9	1.30E-04	−7	100	13
Gagyvendég	8.00E-05	8.1	−6.9	8.50E-05	−7.4	60	27
Hajdúböszörmény	1.10E-04	8.2	−6.9	3.15E-04	−7.5	100	13
Homokszentgyörgy	1.35E-04	8.1	−6.8	4.75E-04	−7.3	40	57
Jármi	8.00E-05	8.3	−6.9	6.00E-05	−7.3	60	36
Jászladány	1.95E-04	9.3	−6.8	9.00E-05	−6.9	80	30
Karcag	1.60E-04	6.7	−7.3	1.75E-04	−7.3	100	13
Örbottyán	4.50E-04	9.4	−7.2	2.95E-04	−7.6	100	252
Ragály	4.10E-05	8.3	−6.9	3.30E-04	−7.4	100	4

Source: Reprinted from Nagy and Kónya. 2007, with Permission from Elsevier.

* Weighted sum of squares divided by degrees of freedom (Herbelin and Westall 1996).

concentration of surface sites) and (2) those depending on the thermodynamically well-defined acid–base processes, independent of soil composition.

It is important to understand the acid–base reactions that involve surface sites because an important part of inorganic and organic pollutants can be adsorbed on these sites. For example, lead ions are adsorbed on these sites by inner-sphere complexation, initiating other transformation processes on the external surfaces (Section 3.2.3; Chapter 2, Sections 2.51.4 and 2.10.3). In addition, organic molecules can also bond to the protonated or deprotonated sites, depending on their charges (Chapter 2, Section 2.9). This process can initiate the formation of mineral-organic or metal ion-mineral-organic complexes.

3.4   SORPTION OF CYANIDE ANION ON SOIL AND SEDIMENT

As mentioned in Chapter 1, Sections 1.3.2.1 and 1.3.3.2 and Chapter 2, Section 2.9, anions can also be sorbed on rocks and soils. The sorption of anions can take place on the protonated sites of minerals (e.g., on the $AlOH_2^+$ and $FeOH_2^+$) at acidic pH values. Precipitation reactions of anions can also occur. In addition, chemical reactions can take place with the functional groups of soil organic matter.

The sorption of anions has much importance because some essential (e.g., boron, phosphorous, Section 3.5) and polluting (e.g., arsenic) elements are present in anionic forms (borate, phosphate, arsenate, etc.).

A highly toxic anion is cyanide, which is fortunately not present in nature; it can occur only as anthropogenic pollution. Some years ago, a large amount of cyanide ion from a rare-metal mine entered into rivers. The cyanide caused significant environmental damages during and right after the pollution took place. In addition to the direct damage at the time of pollution, the question of whether cyanide can cause long-term effects arises. These effects depend on the adsorption and desorption of cyanide on the sediment and the soils next to the river. The question is if the sorbed cyanide can reappear in the river water or in the soil solution, and then from there if it could contaminate drinking water sources; that is, if it can have long-lasting effects. For this reason, the sorption of cyanide ion on sediments and soils has to be studied. Since cyanide ion can also form complexes with metal ions, the sorption of metal-cyanide complexes is also interesting.

When dissolved cyanide ions, for example, in river water, interact with the geological environment (e.g., sediment, soil, air), the cyanide concentration of the solution may decrease if

1. The quantity of the solvent (river water) increases (e.g., rain, other incoming rivers), and therefore the solution becomes diluted.
2. The pH of river water and soil solutions decreases (i.e., becomes more acidic). CO_2 in the air or acids stronger than hydrogen cyanide form HCN with the cyanide ion, which evaporates into the air like a gas. Since HCN is a rather weak acid (pK = 9.14), it forms even in weakly alkaline solution. The pH of river water and soil solutions is usually neutral or weakly

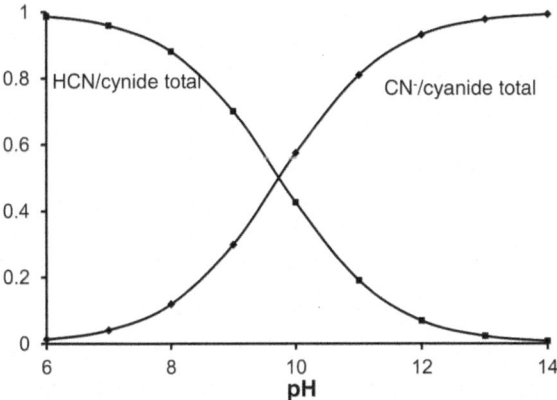

FIGURE 3.7 Relative quantity of cyanide ion and hydrogen cyanide versus pH (total cyanide concentration 0.001 mol/dm³).

alkaline, and the evaporation of cyanide can occur. The ratio of relative quantity of cyanide ion and hydrogen cyanide versus pH is shown in Figure 3.7.

As seen in Figure 3.7, the major part of cyanide is present as hydrogen cyanide at the pH range of the rivers and soil solutions. (It is usually neutral or weakly alkaline; pH = 6–8). The rate of evaporation decreases as pH increases. When studying these systems, it is important to decrease the rate of evaporation as much as possible, or else the evaporated quantity must be taken into consideration. The evaporation of cyanide, however, is slow; the rate-determining step is diffusion, and it can be described with an exponential relation similar to a formally first-order kinetic equation. The evaporation of cyanide slightly increases as temperature increases.

3. Since cyanide pollution usually originates from metal mines or galvanic plants, heavy-metal-ion pollution (e.g., copper, zinc ion) occurs simultaneously with that of cyanide. So, the complex formation and redox reactions of these metal ions and cyanide ion have to be considered, too (Sharpe 1976; Yoshimura et al. 1980). With zinc and copper ions, the reactions take place as follows:

$$Zn^{2+} + 2CN^- \rightleftarrows Zn(CN)_2, \quad \log K = 11.7 \qquad (3.14)$$

$$Zn(CN)_2 + CN^- \rightleftarrows \left[Zn(CN)_3\right]^-, \quad \log K = 5 \qquad (3.15)$$

$$\left[Zn(CN)_3\right]^- + CN^- \rightleftarrows \left[Zn(CN)_4\right]^{2-}, \quad \log K = 4.9 \qquad (3.16)$$

$$Cu^{2+} + 2\,CN^- \rightarrow Cu(CN)_2 \tag{3.17}$$

$$Cu(CN)_2 \rightarrow 2CuCN + (CN)_2 \tag{3.18}$$

$$Cu^+ + 2CN^- \rightleftarrows \left[Cu(CN)_2\right]^- \quad \log K = 21.7 \tag{3.19}$$

$$\left[Cu(CN)_2\right]^- + CN^- \rightleftarrows \left[Cu(CN)_3\right]^{2-} \quad \log K = 5.1 \tag{3.20}$$

$$\left[Cu(CN)_3\right]^{2-} + CN^- \rightleftarrows \left[Cu(CN)_4\right]^{3-} \quad \log K = 1.1 \tag{3.21}$$

The relative quantity of the different chemical forms of cyanide in the presence of zinc and copper ions are shown in Figure 3.8.

As seen in Figure 3.8, complex formation decreases but does not inhibit the formation of hydrogen cyanide, and thus its evaporation into the air. It is important to note, however, that Figures 3.7 and 3.8 refer to thermodynamic equilibrium, which does not exist in the natural systems, so only the tendencies can be seen. Experiments show that the evaporation of cyanide is practically not effected by Zn– and Cu(I) ions. As seen in Figure 3.9, in the presence of Zn^{2+} and Cu(I) ions, there is HCN in a high amount at pH = 7–9 because the dominant complex form is the 1:3 complex, in the case of both Zn^{2+} and Cu(I). The evaporation of free HCN may occur at a similar rate as in the KCN solution.

4. Cyanide ion and cyanide complexes may sorb on the sediment of the river or the surrounding soil or may react with soil organic matter (chemisorption). Later, the sorbed cyanide may reappear as a polluting agent.

The sorption of cyanide ions and copper and zinc cyanide complexes was studied on soils (Table 3.13) and sediments (Table 3.15). Distilled water and natural river water were applied as the solution phase. Sorption was studied by radioisotopic labeling with $^{14}CN^-$ ions.

The results show that the sorption of cyanide on soils and sediments is fast; it reaches equilibrium within 10 minutes. The sorbed quantity, however, is low. From 10^{-4}–10^{-3} mol/dm³ cyanide solutions, it is about 10^{-7} mol/g. This means an approximately 10^{-3} dm³/g distribution ratio for cyanide ion. This value is typical for anion sorption of soils, and it is explained by the interfacial properties of soil components. The main mineral components of soils (primary silicates, clay minerals, oxides) have negative surface charges at pH applied (about 8.5), inhibiting the sorption of negative cyanide ion. The metal complexes of cyanide are also negative. Consequently,

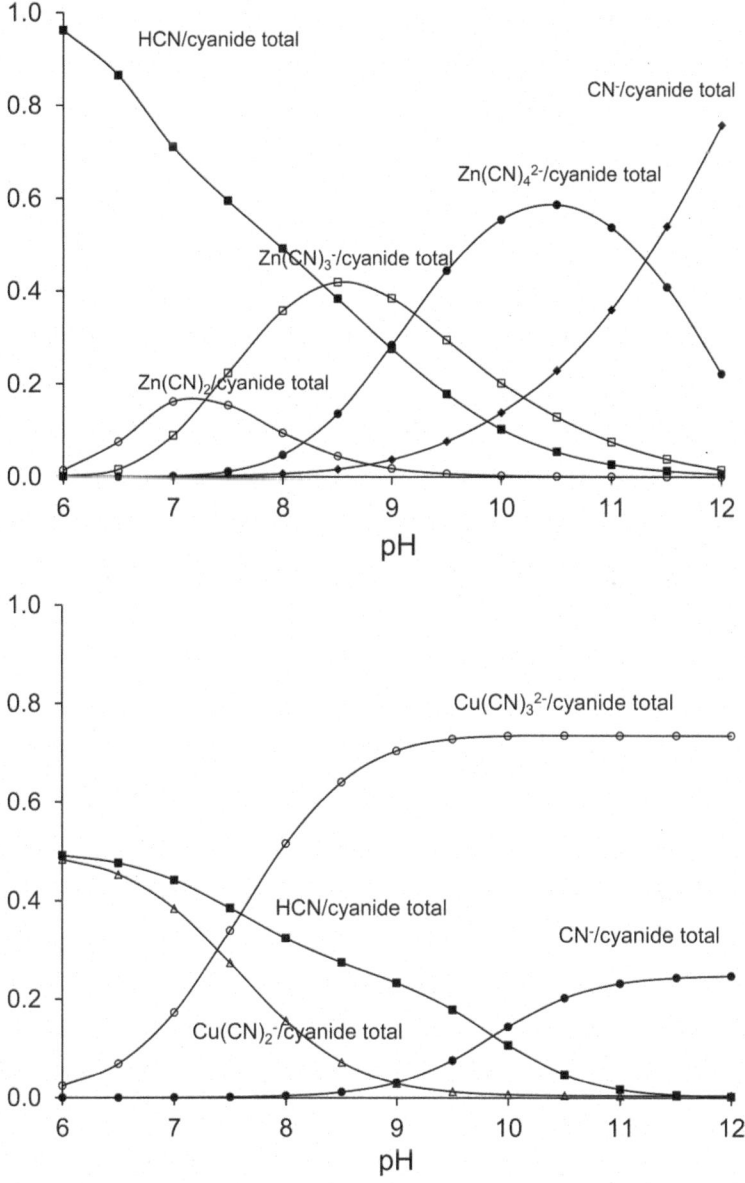

FIGURE 3.8 The relative quantities of the chemical species of cyanide ion in the presence of zinc (upper) and copper (I) (lower) ion (Zn: CN = 1:4, Cu: CN = 1:4).

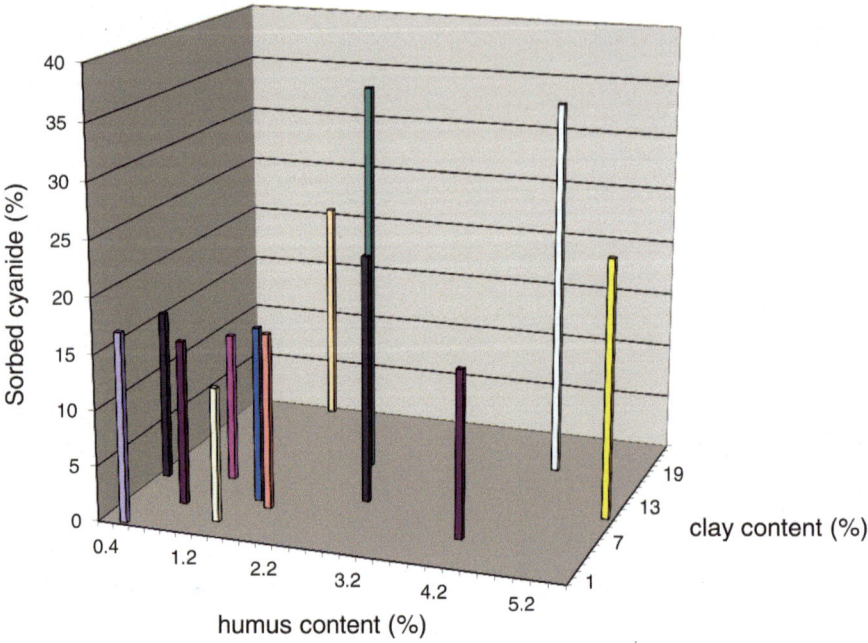

FIGURE 3.9   Sorption of cyanide from Cu(I)-containing solution vs. clay and humus content.

copper(I) and zinc ions have no effects on cyanide adsorption. Only amorphous iron oxide hydroxides can have small positive charges.

As it was mentioned, instead of distilled water, sorption experiments were also performed using river water originated from the same place as the soil and sediment samples. It can be stated, in general, that the degree of adsorption is similar to or a bit higher in the presence of river water than in distilled water. This is explained by the presence of colloid particles floating in natural water to which cyanide can be sorbed, and it is measured together with the solid phase after filtration.

In the presence of cations, it can be hypothesized that cyanide can be adsorbed to the mineral components via a cation-bridge, but its effect could not be measured.

The functional groups of organic material (humus) can react with cyanide as described in the following equations. Under natural conditions (pH near neutral, 0–40°C), cyanide ion can react with aldehyde and keto groups, and carboxylic salt is formed in the presence of metal ions:

$$R-\overset{H}{\underset{O}{\big\langle}} \;+\; HCN \rightleftharpoons R-\overset{H}{\underset{CN}{\Big\langle}}OH \tag{3.22}$$

Aldehydealdehyde-cyanohydrine

TABLE 3.15
Mineral Composition of Sediment Samples Collected along River Tisza (Hungary)

Sediment	Montmorillonite	Illite/montorillonite	Illite	Muscovite	Kaolinite	Chlorite	Quartz	Kalifeldspare	Plagioclase	Calcite	Dolomite	Amfibol	Hematite	Goethite	Amorphous
Aranyosapáti	1			5		6	67	2	14			1		2	2
Záhony	3		7			5	61	3	11	1	2	2	1	1	4
Dombrád	7			8	1	4	58		16	1				1	4
Tiszabercel	1	2		8		3	76	2	4					1	3
Tiszaeszlár	13		11			7	41	5	15					2	6
Tiszalök	2	2	5			5	76		6	tr				1	3
Tiszavasvári	6		5	15	5	5	42	2	14						5

Sediment	Humus (%)	pH (KCl)	pH (water)	Sand (%) >0.05 mm	Silt (%) 0.05–0.002 mm	Clay (%) <0.002 mm	Carbonate (%)
Aranyosapáti	0.55	7.6	8.01	67	22	1	
Záhony	1.08	7.8	8.12	58	18	10	3
Dombrád	1.58	7.82	8.42	59	33	7	1
Tiszabercel	1.28	7.75	8.29	76	20	3	
Tiszaeszlár	1.23	7.5	8.22	41	34	24	
Tiszalök	0.44	7.84	8.97	76	16	9	
Tiszavasvári	5.68	6.9	7.96	42	41	11	

$$R{\Large\diagdown}\!\!=\!\!O + HCN \rightleftharpoons R{\Large\diagup}{\overset{OH}{\underset{CN}{\diagdown}}}R \qquad (3.23)$$

Ketoneketone-cyanohydrine

$$R{\overset{\diagdown}{\underset{OH}{}}}=O + M^+ \rightleftharpoons R{\overset{\diagdown}{\underset{OM}{}}}=O + H^+ \qquad (3.24)$$

Carboxylic acidCarboxylic salt

The adsorbed amount of cyanide is proportional to the clay, humic acid, and iron content.

It is difficult to tell which soil component is mainly responsible for cyanide adsorption because, in the case of examined samples, clay, humus, and Fe content increase simultaneously. More informative is the three-dimensional diagram, in which the amount of adsorbed cyanide versus clay number and humus content is plotted (Figure 3.9).

The long-lasting damaging effect of cyanide may be mainly determined whether its sorption on soil or sediment is reversible or not. If the adsorption is reversible and quick, no pollution will remain after the polluted stream passes. If the sorption is irreversible, the sorbed cyanide becomes the source of pollution later, with the decomposition of soil components. The desorption of cyanide was examined in two soil and two sediment samples (Tiszavasvári and Tiszalök). The samples were chosen in such a way that one of them had high concentration of clay and humus (Tiszavasvári), and the other was very sandy (Tiszalök). The results (Table 3.16) show that the sorption is reversible in 97–100% of cases. We can conclude that cyanide is not adsorbed in a significant degree on the sediment and soil, and long-lasting effects are unlikely.

TABLE 3.16

Relative Sorbed and Desorbed Quantity of Cyanide on Tiszavasvári and Tiszalök Soil and Sediment Samples

	Sorbed quantity, %	Desorbed quantity, %	Not desorbed/total sorbed cyanid, %
Tiszavasvári soil	10.4	86.0	1.3
Tiszavasvári sediment	16.3	84.6	1.0
Tiszalök soil	6.6	91.5	1.3
Tiszalök sediment	5.6	92.8	1.4

Note: Cyanide Concentration $2*10^{-4}$ mol/dm^3, solvent is distilled water.

3.5   SORPTION OF PHOSPHATE ANION ON SOILS

As mentioned in Chapter 1, Sections 1.3.4.4 and 1.3.6.2, heterogeneous isotope exchange is a useful tool to obtain information on the equilibrium state of interfacial reactions and the transport rate of any substances between the phases without disturbing the chemical equilibrium or steady state. An example is shown here: how the phosphate ion uptake by soils is studied using heterogeneous isotope exchange.

It is known that the nutrient cycle of soils is driven mainly by the quantity of macro and micronutrients in the soil solution. The concentration of the nutrients is affected by the interfacial reactions between the solid phase of soil and the soil solution. These interfacial reactions are especially significant in the case of those nutrients that can bind on soil components in several different ways. An example is phosphate anion. At the typical pH range of soil solution (pH~6–7), the dominant phosphate species is $H_2PO_4^{2-}$.

The sorption reactions of $H_2PO_4^{2-}$ on the soil components are expressed in the equations below. A portion of phosphate is present as weakly or fast sorbed phosphate on the soil (Barrow and Shaw, 1975). The weakly sorbed and dissolved phosphate is in equilibrium or steady state. The weak sorption of phosphate ions occurs on the surface sites of soil (S-OH), and, as suggested by Goldberg and Sposito (1984), can be described as:

$$S-OH + H_2PO_4^- \rightleftarrows S-PO_4H^- + H_2O \tag{3.25}$$

where S means the surface of soil. The weakly sorbed phosphate can be transformed to tightly sorbed phosphate (Mansell et al. 1977; Sparks 1989) via a relatively slow process (Shuai et al. 2014), affecting the steady state between the dissolved and weakly sorbed phosphate. The significant portion of the tightly sorbed phosphate is produced by the esterification of the alcoholic groups of organic matter resulting in phosphate esters:

$$S-OH + H_2PO_4^- \rightleftarrows S-O-PO_3H^- + H_2O \tag{3.26}$$

Moreover, phosphate ions can precipitate with some metal cations (aluminum, iron, and calcium ions) in the soil. For example:

$$3Ca^{2+} + 2H_2PO_{4=}^- = Ca_3(PO_4)_2 + 4H^+ \tag{3.27}$$

In Figure 3.10, the sorption processes of phosphate on soil are summarized.

The processes in Figure 3.10 are affected by many factors, such as soil type and composition, phosphate quantities of the original soil and phosphate fertilizer doses, the time elapsed since the fertilizer addition, and the agricultural use (plant culture), etc. Since the quantity of phosphate in the soil solution significantly controls the nutrient uptake by plants, the information gained on the phosphate species can be utilized to plan phosphate fertilization. The optimal fertilization rate is especially important because of decreasing reserves of raw phosphate.

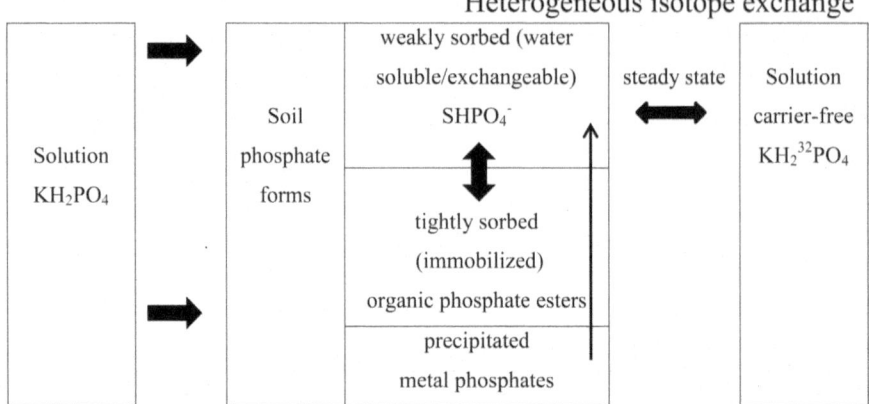

Total radioactivity:
in the weakly sorbed phosphate and in the solution

FIGURE 3.10 Sorption processes of phosphate on soil. (Adapted from Kónya and Nagy 2015, with permission from Elsevier.)

Many cultivated soils contain a high amount of phosphate but mainly as tightly sorbed species. Thus, a large part of soil phosphate is not available for uptake by plants. For this reason, studies on phosphorous mobilization processes become more and more important, including the kinetics and equilibrium of the mobilization process from soil as well as the pathways of phosphorous in the complex systems of the soil, soil solution, and plants. An important method to study this latter process is heterogeneous isotope exchange.

During these laboratory studies, a carrier-free radioactive $H_2{}^{32}PO_4{}^-$ ions (its concentration is $<10^{-8}$ μg/g soil) are added to the soil solution in equilibrium with the soil. In a short period of time, the radioactive phosphate ions from the soil solution exchange with the weakly sorbed phosphate of soil:

$$S-PO_4H^- + H_2{}^{32}PO_4^- \rightleftarrows S-{}^{32}PO_4H^- + H_2PO_4^- \qquad (3.28)$$

This process is the heterogeneous isotopic exchange, described in Chapter 1, Sections 1.3.4.4 and 1.3.6.2 and shown in Figure 3.10 (right side). As seen in Equation 3.28, there is no chemical reaction, i.e., so the process is directed by the mixing entropy (Chapter 1, Equation 1.131). The total radioactivity is distributed between the soil solution and the weakly sorbed phosphate. The kinetics of the exchange (Kónya and Nagy, 2012, 2015, 2018) can be described as:

$$x_{HIE} = \frac{m_2}{m_1+m_2}\left(1-exp\left(-\frac{C}{m_1}\frac{m_1+m_2}{m_2}t\right)\right) \qquad (3.29)$$

$$y_{HIE} = \frac{m_1}{m_1+m_2} - \left(\frac{m_1}{m_1+m_2} - 1\right)\left(exp\left(-\frac{C}{m_1}\frac{m_1+m_2}{m_2}t\right)\right) \qquad (3.30)$$

where t is time, x_{HIE} and y_{HIF} are the relative portions of radioactive phosphate in the soil and the soil solution, respectively, related to the total added radioactivity ($x_{HIE} + y_{HIE} = 1$); m_1 is the mass of the phosphorus in the solution, m_2 is the mass of weakly sorbed phosphorus in the soil under a steady state; C is the rate of the phosphate exchange in an equilibrium or steady state. In equilibrium or steady state, the transfer rate of phosphate from solution to the soil and vice versa are equal.

From the relative radioactivities of P–32 in the soil (x_{HIE}) or the soil solution (y_{HIE}) vs time plot, $m_2/(m_1 + m_2)$ and C/m_1 can be calculated; and when m_1 is measured by an independent analytical method, m_2 (weakly sorbed or water soluble/exchangeable phosphorus in the soil) and C (the exchange rate of phosphorus between the soil and solution under equilibrium or steady state) can be calculated.

The $m_1 + m_2$ give the total quantity of dissolved + weakly sorbed phosphorus in the whole system (soil + soil solution). This means that the tightly sorbed phosphate is not involved in the heterogeneous isotope exchange process. Thus:

$$x_{HIE} = \frac{m_2}{m_1+m_2} \qquad (3.31)$$

and

$$y_{HIE} = \frac{m_1}{m_1+m_2} \qquad (3.32)$$

The quantity of the tightly sorbed (immobilized/organic) phosphorus ($m_{tightly}$) can be obtained by subtracting the sum ($m_1 + m_2$) from the total phosphorus quantity of the system (P_{total}, the sum of the phosphorus content of the original soil and the added phosphorus):

$$m_{tightly} = P_{total} - (m_1 + m_2) \qquad (3.33)$$

If a plant is grown in the soil before heterogeneous isotope exchange studies, the phosphorous taken up by the plant (m_{uptake}) has also to be considered:

$$m_{tightly} = P_{total} - (m_1 + m_2 + m_{uptake}) \qquad (3.34)$$

As seen, the heterogeneous isotope exchange studies provide information on the quantities of phosphate species and they allow to study the effect of different factors in time. Using short-lived radioactive isotopes (e.g., P–32, half-life about 14 days) even long-term processes can be investigated since the changes in soil happen before the addition of the radionuclide (Nagy and Kónya, 2018). The studied momentary state, i.e., the quantity of phosphate species, shows the effect of soil composition and preliminary treatments such as phosphate dosage, time elapsed from phosphate addition, plant growing, microbiological activity, etc. The application of short-lived radioactive isotope is advantageous from radiation protection points of view, as well.

The effects of soil type (Arenosol [calcareous and non-calcarous], Chernozem, Meadow, Marshy meadow, Rendzina; classified according to the World Reference Base (WRB) for Soil Resources [IUSS Working Group, 2014; European Soil Bureau Network European Commission, 2005]), phosphorous doses (0–320 mg P/kg soil), time (up to 22 weeks), and plant culture (ryegrass, *Lolium perenne L.*) on the different phosphate species were studied by heterogeneous isotope exchange (Kónya and Nagy 2015; Nagy and Kónya 2018; Nagy et al. 2019; Balla Kovács et al. 2021). The main results are summarized as follows.

The ratio of the phosphorus species is mainly determined by the humus content of the soil. The sum of the water-soluble and weakly sorbed/isotopically exchangeable phosphate is inversely proportional to humus content. The effect of clay content is less pronounced. Clay plays a role in the formation of soil types because of swelling and nutrient content.

There is a good correlation between Al-P (phosphate extracted by ammonium lactate, Egner et al. 1960) and the sum of the dissolved and weakly sorbed/isotopically exchangeable phosphate ($m_1 + m_2$). Less tight correlation was found between isotopically exchangeable phosphate (m_2) and phosphorous uptake of plant (m_{uptake}) because the P uptake of plant is determined by both the dry matter production of plant and P concentration of plant tissue. Closer relationships were found between the $m_1 + m_2$ and m_{uptake}. This result suggests that the sum of the dissolved and weakly sorbed/isotopically exchangeable phosphate ($m_1 + m_2$) is a good indicator of plant available P (Balla Kovács et al. 2021).

The transport rate of phosphate between soil and soil solution, that is the exchange of dissolved and weakly sorbed phosphate, at a steady state is determined by the diffusion of phosphate ions, slightly influenced by the soil type, that is, the chemical reaction.

The portion of phosphate added to the soil is converted for tightly sorbed species. The rate of the transformation of weakly sorbed phosphate to tightly tsorbed species (the half-lives are about 22, 20–30, 11–15, 14, 10, and 7–8 weeks for Marshy meadow, Meadow, Chernozem, Rendzina, Arenosoil (non-calcarous), and Arenosoil (calcareous), respectively) is inversely proportional to the humus content, except for Rendzina.

In these studies, the type of soil determined the quantity of crop. The effect of phosphate fertilization was very small. Instead, we have to transform the tightly sorbed phosphate of soil to weakly sorbed phosphate before or during the vegetation period, e.g., by microbiological treatments.

The weakly sorbed phosphorus added as fertilizer transformed to tightly sorbed phosphate or leached from soils with low humus content to groundwater – leading to environmental pollution. It was determined that all the studied soils contained enough phosphorus, and it was proportional to the humus content of the soil. Phosphorus addition did not have a beneficial effect on the yield. Instead, the transformation of the tightly sorbed phosphate of soil to weakly sorbed phosphate before or during the vegetation period, e.g., by microbiological treatments is preferable.

REFERENCES

Balla Kovács A., R. Kremper, J. Kátai, I. Vágó, D. Buzetzky, E. M. Kovács, J. Kónya, N. M. Nagy. 2021. Characterisation of soil phosphorus forms in the soil-plant system using radioisotopic tracer method. *Plant, Soil and Environment*, 67: 367–375. doi.org/10.17221/458/2020-PSE.

Barbier, F., G. Duc, and M. Petit-Ramel. 2000. Adsorption of lead and cadmium ions from aqueous solution to the montmorillonite/water interface. *Coll. Surfaces* 166: 153–159.

Barrow, N.J., and T.C. Shaw. 1975. The slow reactions bretween soil and anions. II. Effects of time and temperature ont he decrease in phosphate concentration int he sol solution. *Soil Science* 119: 167–177.

Cramer, J.J., and J.A.T. Smellie. 1994. *Final report of the AECL/sKB Cigar Lake analog study*. Pinawa, MB: Whiteshell Laboratories.

Crank, J. 1956. *The mathematics of diffusion*. Oxford: Clarendon Press.

Egnér, H., H. Riehm, and W. Domingo. 1960. Untersuchungen über die chemische Bodenanalyse als Grundlage für die Beurteiling des Nährstoffzustandes der Böden: II. Chemische Extraktionsmethoden zur Phosphor und Kaliumbestimmung. *Kungliga Lantbrukshögskolans Annaler* 26: 199–215.

European Soil Bureau Network European Commission. 2005. *Soil atlas of Europe* Luxembourg: Office for Official Publications of the European Communities, 128 pages.

Fernandez, A.M., B. Baeyens, M. Bradbury, and P. Rivas. 2004. Analysis of the porewater chemical composition of a Spanish compacted bentonite used in an engineered barrier. *Physics and Chemistry of the Earth* 29: 105–118.

Goldberg, S., and G. Sposito. 1984. A chemical model of phosphate adsorption by soils. I. Reference oxide minerals. *Soil Science Society of America Journal* 48: 772–778.

Goldberg, S., and L.J. Criscenti. 2008. Modeling adsorption of metals and metalloids by soil components. In *Biophysico-chemical processes of heavy metals and metalloids in soil environments*, eds. A. Violante, P.M. Huang, G.M. Gadd, 215–264. Hoboken, NJ: Wiley-Interscience.

Goldberg, S., S.M. Lesch, D.L. Suarez, and N.T. Basta. 2005. Predicting arsenate adsorption by soils using soil chemical parameters in the constant capacitance model. *Soil Science Society of America Journal* 69: 1389–1398.

Granthoff, G.H., and D.M. Moore. 1996. Illite polytype quantification using WILDFIRE(c) calculated X-ray diffraction patterns. *Clays and Clay Minerals* 44: 835–842.

Hargrove, W.L., and G.W. Thomas. 1982. Titration properties of Al-organic matter. *Soil Science* 134: 216–225.

Herbelin, A.L., and J.C. Westall. 1996. *FITEQL3.2: A program for the determination of chemical equilibrium constants from experimental data*. Corvallis, OR: Oregon State University.

Huertas, F.J., E. Caballero, C. JimeÂnez de Cisneros, F. Huertas, and J. Linares. 2001. Kinetics of montmorillonite dissolution in granitic solutions. *Applied Geochemistry* 16: 397–407.

IUSS Working Group WRB. 2014. World reference base for soil resources 2014. International soil classification system for naming soils and creating legends for soil maps. *World Soil Resources Reports* No. 106. FAO, Rome.

Kónya, J., N.M. Nagy, and Z. Nemes. 2005. The effect of mineral composition on the adsorption of caesium ion on geological formations. *Journal of Colloid and Interface Science* 190: 350–356.

Kónya, J., and N.M. Nagy. 2012. *Nuclear and radiochemistry*. Oxford: Elsevier. Hardcover ISBN: 9780123914309; Paperback ISBN: 9780323282451; eBook ISBN: 9780123914873.

Kónya, J., and N.M. Nagy 2015. Determination of water-soluble phosphate content of soil using heterogeneous exchange reaction with ^{32}P radioactive tracer. *Soil and Tillage Research* 150: 171–179.

Kónya, J., and N.M. Nagy. 2018. *Nuclear and radiochemistry*, 2nd ed. Oxford: Elsevier. Paperback ISBN: 9780128136430; eBook ISBN: 9780128136447.

Mansell, R.S., H.M. Selim, and J.G. Fiskell. 1977. Simulated transformation and transport of phosphorous in soil. *Soil Science* 124: 102–109.

Missana, T., M. Garcia-Gutierrez, and U. Alonso. 2008. Sorption of strontium onto illite/smectite mixed clays. *Physics and Chemistry of the Earth, Parts A/B/C* 33: S156–S162.

Nagy, N.M., J. Kónya, M. Beszeda, I. Beszeda, E. Kálmán, Zs. Keresztes, K. Papp, and I. Cserny. 2003a. Physical and chemical formations of lead contaminants in clay and sediment. *Journal of Colloid and Interface Science* 263: 13–22.

Nagy, N.M., J. Kónya, M. Földvári, and P. Kovács–Pálffy. 2003b. The adsorption of caesium–137 ion bentonites from the Carpathian Basin. *Czechoslovak Journal of Physics* 53: A103–111.

Nagy, N.M., and J. Kónya. 2004. The sorption of valine on cation-exchanged montmorillonites. *Applied Clay Science* 25: 57–69.

Nagy, N.M., and J. Kónya. 2005. The relations between the origin and some basic physical and chemical properties of bentonite rocks illustrating on the example of Sarmatian bentonite site at Sajóbábony (HU). *Applied Clay Science* 28: 257–267.

Nagy, N.M., and J. Kónya. 2006. Acid-base properties of bentonite rocks with different origin. *Journal of Colloid and Interface Science* 295: 173–180.

Nagy, N.M., and J. Kónya. 2007. Study of pH-dependent charges of soils by surface acid-base properties. *Journal of Colloid and Interface Science* 305: 94–100.

Nagy, N.M., and J. Kónya. 2018. Study of fast and slow consecutive processes by heterogeneous isotope exchange using P-32 radiotracer. *Journal of Radioanalytical and Nuclear Chemistry* 318: 2349–2353.

Nagy, N.M., D. Buzetzky, E.M. Kovács, A.B. Kovács, J. Kátai, I. Vágó, and J. Kónya, 2019. Study of phosphate species of chernozem and sand soils by heterogeneous isotope echange with ^{32}P radioactive tracer. *Applied Radiation and Isotopes* 152: 64–71.

Nemes, Z., N.M. Nagy, A. Komlósi, and J. Kónya. 2006. The effect of mineral composition on the sorption of strontium ion on geological formations. *Applied Clay Science* 32: 172–178.

Püspöki, Z., P. Kovács-Pálffy, M. Kozák, G. Szöőr, and J. Deák. 2003a. Eustatic and tectonic control on bentonite formation (Sarmatian, NE Hungary). *EUROCLAY 2003*, pp. 232–233. University of Modena Reggio Emilia, Modena, Italy.

Püspöki, Z., M. Kozák, Á. Csámer, R. McIntosh, and L. Vincze. 2003b. Paleogeographic conditions and sequence stratigraphy of the Sarmatian sediment series in the Tardona hills. *Földtani Közlöny* 133: 169–190.

Püspöki, Z., M. Kozák, B. Kiss, P. Kovács-Pálffy, and A. Bartha. 2003c. Descriptive characteristics of bentonite layers in a sedimentary bentonite site with tuffogenic origin (Sajóbábony-NE Hungary). *Anuarul Institutului Geologic Al Romaniei* 73: 36.

Püspöki, Z., M. Kozák, P. Kovács-Pálffy, M. Födvári, R.W. McIntosh, and L. Vincze. 2005. Eustatic and tectonic/volcanic control in sedimentary bentonite formation-a case study of miocene bentonite deposits from the Pannonian basin. *Clays and Clay Minerals* 53: 71–91.

Ryan, P.C., and R.C. Reynolds. 1997. The chemical composition of serpentine/chlorite in the tuscaloosa formation, Unites States Gulf Coast: EDX vs. XRD determinations, implications for mineralogic reactions and the origin of anatase. *Clays and Clay Minerals* 45: 339–352.

Sánchez, F.G., L.R. Van Loon, T. Gimmi, A. Jacob, M.A. Glaus, and L.W. Diamond. 2008. Self-diffusion of water and its dependence on temperature and ionic strength in highly compacted montmorillonite, illite and kaolinite. *Applied Geochemistry* 23: 3840–3851.

Sevink, J., J.M. Verstraten, and J. Jongejans. 1998. The relevance of humus forms for land degradation in Mediterranean mountainous areas. *Geomorphology* 23: 285–292.

Sharpe, A.G. 1976. *The chemistry of cyano complexes of the transition metals.* Cambridge: University Chemical Laboratory.

Shuai, X., R.S. Yost, and O. Wendroth. 2014. State-space estimation of the intrinsic soil phosphorus pools fromsoil phosphorus tests. *Geoderma* 214–215: 239–245. doi: 10.1016/j.geoderma.2013.09.006

Söderlund, M.J., M. Lusa, J. Lehto, M. Hakanen, K. Vaaranmaa, and A.-M. Lahdenperä. 2011. Sorption of iodine, chlorine, technetium and cesium in soil. Working Report; No. 2011-04. *Posiva.* www.posiva.fi/files/1524/WR_2011-04_web.pdf (accessed: 19.04.2021).

Sparks, D.L. 1989. *Kinetics of chemical processes.* San Diego: Academic Press Inc. p. 180.

Sparks, D.L. 2003. *Environmental soil chemistry.* Amsterdam: Academic Press.

Srodon, J., V.A. Drits, D.K. McCarty, J.C.C. Hsieh, and D.D. Eberl. 2001. Quantitative X-ray diffraction analysis of clay-bearing rocks from random preparations. *Clays and Clay Minerals* 49: 514–528.

Stadler, M., and P.W. Schindler. 1993. Modeling of H^+ and Cu^{2+} adsorption on calcium-montmorillonite. *Clays and Clay Minerals* 41, 288–296.

Tang, L., and D.L. Sparks. 1993. Cation exchange kinetics on montmorillonite using pressure-jump relaxation. *Soil Science Society of America Journal* 57: 42–46.

Thomas, G.W., and W.L. Hargrove. 1984. The chemistry of soil acidity. In *Soil acidity and liming*, ed. F. Adams, 2nd ed. *Agronomy Monographs* 12: 3–56. Madison, WI: Am. Soc. Agron.

Ulrich, B., and M.E. Sumner. 1991. *Soil acidity.* New York: Springer Verlag.

van Olphen, H. 1977. *An introduction to clay colloid chemistry*, 2nd ed. New York: Wiley and Sons.

Wanner, H., Y. Albinsson, O. Karnland, E. Wieland, P. Wersin, and L. Charlet. 1994. The acid/base chemistry of montmorillonite. *Radiochimica Acta* 66/67: 157–162.

Yoshimura, S., A. Katagiri, Y. Degughi, and S. Yoshizawa. 1980. Studies of the anodic oxidation of the cyanide ion in the presence of the copper ion. IV. The kinetics and mechanism of the decomposition of the intermediatetetracyanocuprate(II) ion. *Bulletin of the Chemical Society of Japan* 53: 2437–2442.

4 Experimental Methods in Studying Interfacial Processes of Rocks and Soils

Studies of interfacial reactions include the analysis of both the bulk phases and their interface. Analysis means using not only qualitative and quantitative analytical methods, but also structural studies of the phases and an investigation of chemical species present by means of thermodynamic calculations and/or experimental techniques. For the interfacial studies on rocks and soils, many different classical and novel methods can be used. In this chapter, the most important methods used for the analysis of solid, liquid, and interface will be presented.

4.1 ANALYSIS OF THE SOLID PHASE

In order to describe a rock, its chemical and mineral composition, the structure, and the characteristic physical properties of its particles have to be studied. In the case of soils, the soil organic matter and its properties also have to be investigated.

4.1.1 METHODS OF CHEMICAL ANALYSES

The chemical composition can be determined by wet and dry procedures. With wet procedures, the rock or soil has to be dissolved. The dissolution can be achieved by basic and acidic digestion. Digestion with basic solutions or solids results in a significantly diluted sample, and it also introduces into the mixture a matrix in quite a large quantity. Acidic digestion is more advantageous, and silicates can also be dissolved if the digesting agent contains hydrogen fluoride. It is frequently performed at elevated pressure (e.g., in Teflon bombs). Digestion can be accelerated by ultrasound or microwave activation. Commercial instruments are available for this purpose.

Another method to dissolve rocks and soil involves using different sequential extraction methods (Hlavay et al. 2004; Bacon and Davidson 2008), where the chemical species of the elements are identified as:

1. Water soluble
2. Exchangeable
3. Specifically adsorbed

DOI: 10.1201/9781003020080-4

    4. Bound to carbonates
    5. Organically bound
    6. Mn-oxides
    7. Amorphous Fe-oxides
    8. Crystalline Fe-oxides
    9. Sulfides
  10. Residuals (silicate bound)

The listed species can be dissolved in different solvents. Depending on the composition and aim of the studies (e.g., the study of the speciation of a toxic element), different solvents and procedures can be used for the sequential extraction (e.g., see Lakanen and Erviö 1971; McGrath 1996).

In both total and sequential dissolutions, the result is a solution containing the components of rocks and soils. This solution is then analyzed by different methods. Mostly, spectroscopic methods are used: atomic absorption and emission spectroscopic methods, ultraviolet, atom fluorescence, and X-ray fluorescence spectrometry. Multielement methods (e.g., inductively coupled plasma and microwave plasma optical emission spectroscopy) have obviously some advantages. Moreover, electroanalytical methods, ion-selective electrodes, and neutron activation analysis can also be applied. Spectroscopic methods can be combined with mass spectrometry.

Some of these methods (X-ray fluorescence spectrometry, neutron activation analysis) can be applied directly to the analysis of rocks and soils. No dissolution is needed, and therefore they are called *dry methods*. The chemical composition of rock and soil samples can be analyzed directly by microprobe techniques (e.g., scanning electron microscopy). The resolution of these methods is about 1 μm, so the composition of individual particles can be investigated separately.

The results of total chemical analysis are usually expressed in oxides. For example, the % of silicon dioxide, aluminum oxide, ferrous and ferric oxides, etc. is given (Chapter 1, Section 1.3.2.2). A special use of chemical analysis is the determination of cation- or anion-exchange capacity (Chapter 1, Section 1.3.3.2).

The methods mentioned thus far are well-known tools for chemical analysis, so they are not discussed here in detail. Only a special type of neutron activation analysis, the so-called prompt gamma activation analysis (PGAA), is discussed briefly.

In traditional neutron activation analysis, radioactive isotopes are produced by neutron irradiation of the sample, and the gamma radiation (i.e., delayed gamma radiation) of the produced radioactive isotopes is measured. During neutron activation, however, the emitted gamma radiation of (n,γ) nuclear reaction (i.e., prompt gamma radiation) can also be measured at the same time as the irradiation. This phenomenon is used in prompt gamma activation analysis, which measures the radiation emitted by the excited nucleus formed by neutron capture. The excitation energy is about the same as the binding energy of the neutron (7–8 MeV). During transmission to the lowest energy state, the nucleus emits one gamma photon, with high energy, or cascades consisting of gamma photons with lower energies. The energy of gamma photons is characteristic of the nuclei.

The potential of PGAA can best be explored when guided beams of cold neutrons are used because the cross section of (n,γ) can increase significantly.

The advantages of prompt gamma activation analysis are similar to those of neutron activation analysis, that is, it is a sensitive multielement method for many elements. The samples can be examined without digestion. In addition, prompt gamma activation analysis is suitable for all elements. The method is especially important in the measurement of light elements (H, B, N, etc.), which are difficult to measure with other methods. For example, the typical sensitivity of the methods is 1–10 ng for boron, or 0.04–1 μg for hydrogen (Molnar and Lindstrom 1998). This sensitivity renders it to be particularly applicable to the analysis of trace elements (Cd, Hg, etc.). The method has some disadvantages, however. It is expensive and requires the release of neutron beam from the nuclear reactor, decreasing the neutron flux in order to observe prompt gamma photons. Furthermore, gamma photons have to be measured directly in the neutron beam; that is, the background radiation is high and, as it was mentioned previously in this chapter, cascades of gamma photons with low energies can also be produced, and so the spectrum is rather complicated.

4.1.2　STUDY OF MINERAL COMPOSITION

The method for the study of crystal structure, including mineral composition, is X-ray diffraction. It is a semiquantitative method (Klug and Alexander 1954; Warren 1999) for the study of mineral composition. Powder samples of rocks are studied, and the place of the peaks (in nanometer or diffraction angle) and the relative intensity are observed. The diffractogram is usually compared to that of known substances that can be found in databases (the International Centre for Diffraction Data – ICDD 2009). In order to accurately quantify the rock samples, X-ray diffraction must be complemented with other quantitative techniques, including chemical analysis (Section 4.1.1), thermal, infrared analyses (Chapter 2, Section 2.1.2), and surface area (Chapters 1, Section 1.3.4.1.5; Section 4.1.3) measurements.

Thermal analysis is suitable for the study of the changes under heat, including the changes in mass and enthalpy (Smykatz-Kloss and Warne 1991; Liptay, 1973). In addition, when the minerals are qualitatively or semiquantitatively known from X-ray diffraction studies, their quantity can be determined more precisely on the basis of characteristic thermal changes of the minerals. Thermal analytical methods give very useful information on the types of water in clay minerals, and on the valency of the interlayer cation (Chapter 2, Section 2.1.2; Földvári 1991; Földvári et al. 1998).

Thermal analysis, together with other methods (X-ray diffraction, X-ray photoelectron spectroscopy), provides more accurate information on metal speciation than extraction methods (Section 4.3).

Infrared spectra, including near-infrared spectra, also give useful information on the crystal structures of minerals and functional groups of soil organic matter (Farmer 1974; Petit 2008). Infrared spectra are usually interpreted empirically because the infrared absorption bands depend on several factors: the atomic mass, length and forces of chemical bonds, and the symmetry of the overall unit cell and each atom in the unit cell.

Near-infrared spectroscopy (NIR) works in the 800 nm–2.5 μm (12,500–4,000 cm^{-1}) range. The advantage of NIR is that it can typically penetrate much farther into a sample than the mid-infrared radiation (30–1.4 μm, 4,000–400 cm^{-1}). It can be used for the quantitative measurement of organic functional groups of soil organic matter, especially O–H, N–H, and C=O (Siesler et al. 2002). In addition, the structural modifications under the effect of chemical treatments (e.g., acidic treatments) can also be studied by NIR (Madejova et al. 2009).

The water-rock/soil and organic matter-minerals (including soil organic matter-clay), interactions can also be studied by infrared spectroscopy (Chapter 2, Section 2.1.2).

4.1.3 STUDY OF MORPHOLOGY

The morphology of the particles in rocks and soils means the size or size distribution, shape of the particles and pores, and the specific surface area of the external and internal surfaces.

Size and size distribution can be studied by classical sedimentation techniques. The classification of soils is based on the particle sizes (Chapter 1, Section 1.1.3, Table 1.6). The size and shape of the particles can be observed using different microscopes, from the traditional light microscopes to scanning and transmission electron microscopes. These microscopes, equipped with a microprobe (scanning and transmission microscope) are suitable for chemical analysis of the sample. The nanometer-sized particles can be observed by atomic force microscope.

The size distribution of solid samples can be determined by static and dynamic light scattering techniques; the size ranges from a few nanometers to a few microns. The concept uses the idea that small particles in a suspension move in a random pattern called *Brownian motion*. When the moving particles are irradiated by a coherent beam of light (e.g., laser) with a known frequency, the light is scattered with a different frequency. The change of the frequency is directed by the Doppler theory of moving particles. The size of the particles is related to the shift in light frequency, and thus can be determined.

Traditionally, the porosity of solid samples can be studied by mercury porosimetry. The total volume, specific surface area of the pores, bulk density, and particle size can be determined in 1.8 nm–300 μm pore size and 15 nm–3 mm particle size. The principle of the method is that there is a relationship between the pressure of mercury and the size of the pores filled with mercury. The pressure of mercury (p) required for its introduction into pores of a given radius (r) can be expressed by the Washburn's equation:

$$p = \frac{-2\gamma \cos \Theta}{r} \tag{4.1}$$

where γ is the surface tension of mercury, and Θ is the contact angle of mercury and solid.

Another traditional method for porosity studies is the water vapor adsorption method based on Kelvin's capillary condensation equation:

$$\ln \frac{p}{p_0} = -\frac{2\gamma V}{RTr}$$ (4.2)

where p is the vapor pressure of water vapor condensed in a capillary with r radius, p_0 is the saturation vapor pressure of water, γ is the surface tension of water, V is the molar volume of water, and R and T are the same as defined previously. The radius is determined by water adsorption isotherms. From the water saturation, the total pore volume is measured (Kate and Gokhale, 2006).

Both mercury intrusion and the water vapor adsorption method are applied for dry samples. However, the pore size of many rocks frequently depends on the moisture content. In this case, the liquid state nuclear magnetic resonance (NMR) method is useful to study the pore structure (Gallegos et al. 1987). ^1H isotope as water or other compounds containing hydrogen is used as an NMR active probe nucleus.

The NMR relaxation rate depends on the tumbling motion of the molecules. When the molecules adsorb onto a solid surface, their tumbling motion, characterized by the rotation correlation time, slows down. The slower the motion of the adsorbed molecule is, the faster the relaxation rate of the proton in it. The application of NMR for pore size distribution is based on relaxation. There are two methods, namely, NMR relaxometry and NMR cryoporometry, which are used for this purpose. NMR diffusiometry provides information related to the permeability of porous material, namely average surface/volume ratio (S/V).

Relaxometry utilizes the phenomenon that the protons of the probe liquid have extremely different rates of relaxation when adsorbed on a solid surface or imbibed in small pores of solids or located in the bulk phase (Hills et al. 1993, 1995). The measured transverse relaxation rate constant ($1/T_2$) can be expressed by the weighted average of the relaxation rate of liquid in the bulk-like ($1/T_{2bulk}$) and the surface region ($1/T_{2s}$):

$$\frac{1}{T_2} = \frac{V_s}{V}\frac{1}{T_{2s}} + \frac{V_{bulk}}{V}\frac{1}{T_{2bulk}}$$ (4.3)

where V is the total volume of the liquid, and V_s and V_{bulk} are the volumes of the liquid on the surface and in the bulk phase, respectively. T_2 is the measured average transverse relaxation time, while T_{2bulk} and T_{2s} are the characteristic relaxation time constants of liquid molecules in the bulk-like and surface region, respectively. By titrating the porous material with the probe liquid and measuring T_2, the process of wetting and saturation can be followed (Allen et al. 1997).

The surface relaxation strength, ξ, can also be calculated from the surface relaxation time (T_{2s}) and the liquid layer thickness on the surface (l):

$$\xi = l\left(\frac{1}{T_{2s}} - \frac{1}{T_{2bulk}}\right)$$ (4.4)

By substituting Equation 4.4 into Equation 4.3, and supposing spherical pore geometry, pore size (r) can be calculated (Barrie 2000):

$$\frac{1}{T_2} = \frac{1}{T_{2bulk}} + \left[l \left(\frac{1}{T_{2s}} - \frac{1}{T_{2bulk}} \right) \frac{3}{r} \right] \tag{4.5}$$

In this manner, T_2 relaxation distributions reflect the pore size distribution as well.

The relaxation signal is usually converted into continuous distribution of relaxation components (Fang et al. 2018).

The weakness of the method lies in that there is varying thickness of the adsorbed layer and that, in some cases, the T_s value is usually uncertain and strongly depends on the surface composition. An accepted procedure is that the NMR relaxometry is calibrated with another method where mercury intrusion porometry was used for this purpose.

The theory and method of NMR cryoporometry have been elaborated by Strange (1993) and developed by Petrov and Furó (2009). It is a modification of thermoporometry based on the Gibbs-Thomson theory postulating that the melting and freezing points of a liquid depend on the size of the pore filled that it fills. For the characterization of porous solid materials, the pores are filled with a test liquid (water, cyclohexane, etc.) and the change in the NMR peak intensity as a function of temperature is recorded. Due to the very fast relaxation, the frozen material is not detected, only the molten part of the test liquid is observed. From the intensity, the frozen to molten ratio can be calculated. The primary result is a melting-freezing curve of the liquid; and the melting and freezing point depressions ($T_{m/f}$) provide information on the size distribution of the pores (r_p) when they are completely filled:

$$\Delta T_{m/f} = T_{m/f} - T_0 = -\frac{nK_c}{r_p} \tag{4.6}$$

where T_0 is the transition temperature of the bulk liquid, n is a factor characteristic for the pore geometry, K_c is the cryoporometric constant of the liquid, and r_p is the radius of the pore. The hysteresis between the freezing and melting curves refers to the geometry of the pores, thus n equals to 3, 2, and 1 for spherical, cylindrical, and slit like geometry, respectively, for freezing, while 2, 1, and 0, respectively, for melting. An example of freezing/melting curves of test liquids is shown in Figure 4.1 (a). From the measured transition temperature, the pore size is calculated and the distribution can be plotted (b) (Kéri et al. 2021).

The results obtained from NMR pore size distribution measurements often compare favorably with gas adsorption results; nevertheless, gas adsorption data are occasionally used for the calibration of the NMR methods. When the gas sorption and cryoporometric experiments give different results, they hold different or complementary information. The critical value is that of K_c (Cryptometric constant), for which one can use a literature value or it can be determined for a test liquid and a solid adsorbent.

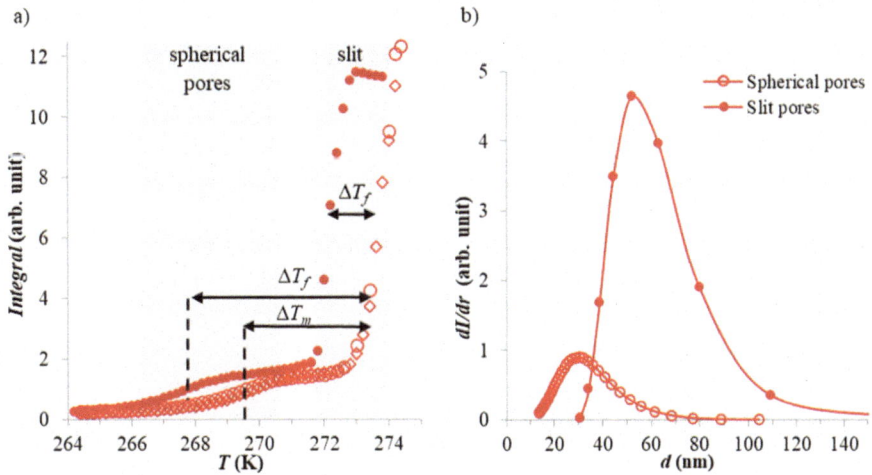

FIGURE 4.1   a) Freezing (●) and melting (○,◊) curves of water in the polymer aerogel (PA) sample. The different empty symbols show the reproducibility of the curves (second cycle). Double arrows show the melting (ΔT_m) and freezing point (ΔT_f) depressions. b) Pore size distributions of spherical (○) and slit-like (●) pores in PA. Solid lines are a guide to the eyes. (Reprinted from Kéri et al. 2021.)

Another method of porosity studies is positron annihilation technique.

The positron, emitted during the β^+–decay of several radioactive atomic nuclei, is an unstable particle, the antiparticle of the electron (Kónya and Nagy 2012, 2018). After the emission, the positron losses its energy, the half-time of thermalization (the time needed for the velocity of the positron decreases to the energy of thermal movement) is about 10^{-10} s. If the positron meets an electron in this interval, a short-life light element, positronium, is produced, whose nucleus is the positron. Positronium can be considered as an atom with an atomic number of zero.

Positronium has two forms: *para-* and *ortho-* positronium, depending on the spins of the positron and the electron. In *para*-positronium, the spins are antiparallel; the lifetime in vacuum is 1.25×10^{-10} s. In *ortho*-positronium, the spins are parallel; the lifetime in a vacuum is 1.4×10^{-7} s.

Then the two particles (electron and positron) transform into electromagnetic radiation, gamma photons. The process is called "annihilation". The rest mass of the positron (β^+–particle) is 0.51 MeV, equal to the rest mass of the electron, so 2×0.51 MeV energy is emitted in the annihilation process. Usually, two gamma photons with 0.51 MeV energies are emitted at an angle of 180°. The probability of the formation of two photons is about 90%. Gamma photons with 0.51 MeV energy originate from the *para*-positronium. In about 10% of the annihilation process, only one photon with 1.02 MeV is formed. In some cases, three photons are emitted, and the total energy of them is also 1.02 MeV. The odd photons are produced from the *ortho*-positronium.

In media other than vacuum, the electron density is higher, positron–electron encounters are more frequent. Thus, the lifetime of positrons becomes shorter. This can be utilized in pore analysis for the determination of characteristic pore sizes (Eldrup and Smith, 1997; Mukherjee et al. 2021).

The measurements of external and internal specific surface area have already been discussed in Chapter 1, Section 1.1.3. The principles and the isotherm equation of the Brunauer–Emmett–Teller theory (BET) method to measure external specific surface area, including macro- and mesopores, have been presented in Chapter 1, Section 1.3.4.1.5. The external specific surface area is usually determined by the nitrogen gas adsorption at the temperature of liquid nitrogen. Both static (one-point) and dynamic (five-point) methods are applied. The calculations are made using Equation 1.75 (Chapter 1), using one or five different pressure values. The external specific surface area is calculated from the maximum number of surface sites, that is, monolayer and the cross-sectional area of nitrogen molecules.

4.1.4   STUDY OF SOIL ORGANIC MATTER

As seen in Chapter 1.1.2, soil organic matter is an undefined system, consisting of many components. So, the determination of soil organic content and the study of humic substances are challenging. For practical purposes, the humus content, or more correctly, the soil organic matter content is determined by standard procedures; however, the results cannot be considered as absolutely real values, and can only be compared to each other.

Soil organic matter can be measured with or without the separation from the mineral phases. In case of there being no separation, the usual procedure is the oxidation of soil organic matter, for example, by potassium dichromate in acidic medium. In this way, the carbon content is determined, from which soil organic matter content is calculated using the average elementary composition of humic substances (Kononova 1972).

Elemental analysis and the detailed characterization require the extraction and purification of soil organic matter. The International Humic Substances Society (2008) recommends procedures for the isolation of humic substances both from soils and natural waters. The different fractions of humic substances (fulvic acid, humic acid, humic substances) are separated by acidic and basic extraction. Separation and purification are done by ion-exchange resins and dialysis. The fractions of soil organic matter can be characterized by elemental analysis, infrared spectroscopy, thermoanalytical studies, etc.

4.2   ANALYSIS OF THE LIQUID PHASE

The same methods can be used for determining the chemical composition of the liquid phase as those used for the composition of the solid after digestion or extraction (absorption and emission spectroscopic, electroanalytical methods, ion-selective electrodes, neutron activation analysis, mass spectrometry, etc.). The humic substances of natural waters can also be analyzed.

As mentioned several times in this book, pH is the determining property of the liquid phase. So, the measurement of pH is crucial. From the pH and the total concentration of different substances, the ratio of the thermodynamically stable species can be calculated. It is very useful both in the planning of the experiments and evaluation of the results of laboratory and field studies.

The pH can be measured by the usual potentiometric method. Furthermore, potentiometric measurements give information on the redox conditions of the solution. These can be done in the suspension of rocks/soils in the liquid phase. In both cases, oxidized or reduced species are studied, and the results are introduced into the thermodynamic calculation of stable species.

Potentiometric measurements are simple; the redox potential is measured compared to a reference electrode. For pH measurements, commercially available electrodes, comprising the working (glass electrode) and the reference electrodes, can be used. For redox potential measurements, the working electrode is usually an inert (e.g., platinum) electrode, and the reference electrode can be a hydrogen electrode, calomel, or other electrodes. Ion-selective electrodes are also based on potential measurements.

Natural waters always contain colloid particles. They originate from the rocks and soils and contain mineral particles (clay, sand, silt, etc.) and organic matter (humic substances). Since the size of the particles is small, they have great specific surface area and high adsorption capacity. The concentration of colloids is usually studied by the measurement of turbidity.

Colloid particles can be formed by the hydrolysis of cations. In addition, complex formation with other species (e.g., carbonate) can also result in colloid formation. The sorption properties of such hydroxide, carbonate, etc., colloid particles are different from that of hydrated cations because the size and charge are different. Colloid formation can play a very important role in interfacial processes and the migration of different substances in the geological environment. As a guiding principle, in all studies of interfacial processes of rocks and soils, chemical conditions have to be adjusted so that the chemical species are known and well defined. This is especially important in the case of extremely diluted solutions (Chapter 1, Section 1.2.4).

4.3   STUDY OF INTERFACES

The basic information in the study of sorption processes is the quantity of substances on the interfaces. In order to measure the sorbed quantity accurately, very sensitive analytical methods have to be applied because the typical number of particles (atoms, ions, and molecules) on the interfaces is about 10^{-5} mol/m^2. In the case of monolayer sorption, the sorbed quantity is within this range. As the sorbed quantity is defined as the difference between quantities of a given substance in the solution and/or in the solid before and after the sorption processes (surface excess concentration, Chapter 1, Section 1.3.1), all methods suitable for the analysis of solid and liquid phases can be applied here, too. These methods have been discussed in Sections 4.1 and 4.2. In addition, the radioisotopic tracer method can also be applied for the accurate measurements of the sorbed quantities. On the basis of the radiation of properties of the available isotopes, gamma and beta spectroscopy can be used as an analytical method. Alpha spectroscopy may also be used, if needed; however, it necessitates more complicated techniques and sample preparation due to significant absorption of the alpha radiation. The sensitivity of the radioisotopic labeling depends on the half-life of the isotopes. With isotopes having medium half-time (days–years), 10^{-14}–10^{-10} mol can be measured easily.

With irradiation methods, the chemical composition and the structural features of the interface are studied directly (Bisdom and Ducloux 1983; Walls 1990; Somorjai 1994; O'Day 1999; D'Amore 2005). The sample is irradiated with electromagnetic waves or particles, and the properties of the produced radiation or particles are examined. The most important of these in rock and soil studies are shown in Table 4.1.

The depth of introduction of radiation determines whether the properties of bulk phases or the interface may be studied, or, in other words, how thick the studied layer is. The energy of the radiation and the size of the particles influence the depth of the introduction and, consequently, the thickness of the studied layer. For example, X-ray radiation is introduced into the phases depending on the energy: at high X-ray energies the radiation is introduced deeply, so the properties of the bulk can be studied by X-ray diffraction or X-ray fluorescence analysis. At smaller X-ray energies (up to 1000 eV), the structure of the surface layer can be studied by X-ray absorption near edge structure (XANES) and extended X-ray absorption fine structure (EXAFS) methods. The same stands for electron and ion radiation.

The depth of introduction can be varied by technical tools as well. Transmission techniques can be applied for bulk analysis, while reflection techniques are appropriate to the study of the interfaces. Transmission and reflection techniques can be applied using a wide range of electromagnetic radiation, from infrared to X-ray ranges. Moreover, particle radiation (e.g., neutron scattering technique) can also be used for the study of the structure of interfaces.

As seen in Table 4.1, there are many techniques for the study of chemical composition of the bulk and interfaces. Practically, all these methods can be used to study rocks and soils. A part of these methods, however, demands special instrumentation, for example, nuclear reactors or cyclotrons for the studies with neutrons and charged particles. Techniques using electromagnetic or electron radiation can usually be carried out in most chemical, geological, and agricultural laboratories.

Some methods provide similar information, namely, the chemical composition of the bulk phases and the interface. The detectable elements and the sensitivity, however, can differ. When choosing an analytical method, the following factors have to be taken into account:

1. Elements to be studied
2. Sensitivity
3. Thickness of the studied layer (bulk or interface)
4. Vertical and horizontal resolution of the technique
5. Whether structural information can be gained or not
6. Whether chemical bonds and species can be determined or not

The study of chemical speciation requires that the irradiation energy be in the range of chemical bonds. There are some methods in which chemical structure can influence the properties of the nuclei. The most important method is the nuclear magnetic

TABLE 4.1
Interface Analytical Methods

Induced process	Irradiation			
	Photon	Electron	Ion	Neutron
Transmission, reflection, or absorption	Spectroscopic method, depending on the wavelength: NMR ESR IR, NIR visible UV Mössbauer spectroscopy			Neutron absorption (including nuclear reaction, NAA, PGAA)
Scattering	Dynamic light scattering XRD	EELS LEED RHEED	RIBS ISS	Neutron scattering SANS Inelastic neutron scattering
Photon emission	XANES or NEXAFS EXAFS XRF	EMP	IMXA IEX PIXE CPINRA CPAA	NAA PGAA
Electron emission	AES XPS (ESCA) UPS	AES SAM SEM TEM	INS	
Ion emission	LAMMA	EIID	SIMS IMMA CPINRA	Nuclear reactions (e.g., (n,p), (n,α))

α: alpha particle, AES: Auger electron spectroscopy, CPINRA, CPAA: charged particle activation analysis, EIID: electron induced ion desorption, ELS: electron-energy-loss spectroscopy, EMP: electron microprobe, ESCA: electron spectroscopy for chemical analysis, ESR: electron spin resonance, EXAFS: extended X-ray absorption fine structure, IEX: ion excited X-ray fluorescence spectroscopy, IMMA: ion microprobe mass analyzer, IMXA: ion microprobe X-ray analysis, INS: ion neutralization spectroscopy, IR: infrared spectroscopy, ISS: ion scattering spectrometry, LEED: low energy electron diffraction, LAMMA: laser microprobe mass analysis, n: neutron, NAA: neutron activation analysis, NEXAFS: near edge X-ray absorption fine structure, NIR: Near-infrared spectroscopy, NMR: nuclear magnetic resonance, p: proton, PGAA: prompt gamma activation analysis, PIXE: particle induced X-ray emission, RHEED: reflection high energy electron diffraction, RIBS: Rutherford backscattering spectroscopy, SAM: scanning Auger microanalysis, SANS: small-angle neutron scattering, SEM: scanning electron microscopy, SIMS: secunder ion mass spectroscopy, UPS: ultraviolet photoelectron spectroscopy, UV: ultraviolet spectroscopy, XPS: X-ray photoelectron spectroscopy, XRD: X-ray diffraction analysis, XRF: X-ray fluorescence analysis, XANES: X-ray absorption near edge structure

TABLE 4.2
Some Characteristic Properties of Analytical Methods

Irradiation	Method	Thickness of the studied layer	Typical sensitivity	Primary information	Detectable elements and species
Photon	NMR	Bulk	—	Chemical state of bulk and adsorbed molecules	Magnetically active nuclei (with 1/2 spin isotopes, cca. 80)
	ESR	Bulk	—	Electron structure	Paramagnetic species
	IR, NIR	0.5–2.5 μm	0.1–0.5%	Bonding geometry and strength of bulk and adsorbed molecules	Functional groups
	Visible, UV spectroscopy	0.1 μm	0.001–1000 ppm	Elementary and molecular analysis	Li-U
	UPS, XPS (ESCA)	3 nm	0.1%	Species, surface elemental composition, valency, chemical bond	Li-U
	LAMMA	—	0.5% inorganic, 0.1–10 ppm organic	Microelement and molecular analysis	Na-U
	AES	1 nm	0.1%	Elemental composition, adsorbate analysis	Li-U
	EXAFS, XANES, NEXAFS	50 nm	500 ppm	Oxidation state, species, coordination number, some structure	Li-U
	XRF	10^4 nm	1–10 ppm	Chemical composition of bulk and near surface region	Na-U

(Continued)

TABLE 4.2 (CONTINUED)
Some Characteristic Properties of Analytical Methods

Irradiation	Method	Thickness of the studied layer	Typical sensitivity	Primary information	Detectable elements and species
	XRD	10^4 nm	—	Structure of bulk and surface, mineral composition	
	Mössbauer	bulk	1–1000 ppm	Site locations, structure, bonding, chemical environment	Isotopes with Mössbauer transitions
Electron	Electrondiffraction, including LEED, RHEED	1 nm	—	Identification of microcrystalline phases	Li-U
	AES	1 nm	0.1%	Elemental composition, adsorbate analysis	Li-U
	SAM	1 nm	0.1%	Elemental composition, adsorbate analysis	Li-U
	EMP	10^3 nm	0.1%	Chemical composition of bulk and near surface region	Na-U
	EELS	100 nm	<0.1%	Elemental composition, species like IR, bonds, structure	Li-U
	SEM	5 nm		Surface and bulk morphology	

(Continued)

TABLE 4.2 (CONTINUED)
Some Characteristic Properties of Analytical Methods

Irradiation	Method	Thickness of the studied layer	Typical sensitivity	Primary information	Detectable elements and species
	TEM	1–100 μm	100 ppm	Surface and bulk morphology	
Ion	ISS			Elemental composition, location of adsorbed species	Li-U
	SIMS	3–10 nm	0.1–10 ppm	Elemental, isotopic, and molecular composition	H-U
	IMMA	3–10 nm	0.1–10 ppm	Elemental, isotopic, and molecular composition	H-U
	CPINRA	10^4 nm	0.1–10 ppm	Elementary composition	
	IEX, PIXE	10^4 nm	0.1–10 ppm	Elemental composition, location of adsorbed species	Na-U
	IMXA	10^4 nm	0.1–10 ppm	Elementary composition	
	RIBS	10^3 nm	0.01–1%	Elemental composition, location of adsorbed species	Li-U
Neutron	NAA	bulk	0.001–0.1 ppm	Elemental analysis of bulk	Li-U
	PGAA	bulk	0.001–0.1 ppm	Elemental analysis of bulk	H-U
	Neutron scattering	bulk		Structure and morphology	
	SANS	bulk		Structure and morphology	
	Nuclear reactions	bulk	0.001–0.1 ppm	Elemental analysis of bulk	Li-U

resonance spectroscopy (NMR). Another example is Mössbauer spectroscopy, in which the oxidation state and the chemical environment affect the energy levels of nucleons. This method is mostly used for the study of iron in geological species (Kuzmann et al. 1998).

In Table 4.2, some important characteristics of the methods listed in Table 4.1 are shown. These characteristics can assist in the selection of appropriate methods for a given purpose.

REFERENCES

Allen, S.G., P.C.L. Stephenson, and J.H. Strange. 1997. Morphology of porous media studied by nuclear magnetic resonance. *The Journal of Chemical Physics* 106: 7802–7809.

Bacon, J.R., and C.M. Davidson. 2008. Is there a future for sequential chemical extraction? *Analyst* 133: 25–46. doi: 10.1039/b711896a.

Barrie, P.J. 2000. Characterization of porous media using NMR methods. *Annual Reports on NMR Spectroscopy* 41: 265–316.

Bisdom, E.B.A., and J. Ducloux. 1983. *Submicroscopic studies of soils.* Amsterdam: Elsevier.

D'Amore, J.J., S.R. Al-Abed, K.G. Scheckel, and J.A. Ryan. 2005. Methods for speciation of metals in soils: A review. *Journal of Environmental Quality* 34: 1707–1745.

Eldrup, M., and B.N. Singh. 1997. Studies of defects and defect agglomerates by positron annihilation spectroscopy. International workshop on defect production, accumulation and materials performance in an irradiation environment location: DAVOS, OCT 02–08, 1996. *Journal of Nuclear Materials* 251:132–138.

Fang, T., L. Zhang, N. Liu, L. Zhang, W. Wang, L. Yu, C. Li, and Y. Lei. 2018. Quantitative characterization of pore structure of the Carboniferouse Permian tight sandstone gas reservoirs in eastern Linqing depression by using NMR technique. *Petroleum Research* 3:110–123.

Farmer, V.C. 1974. *The infrared of minerals.* London: Mineralogical Society.

Földvári, M. 1991. Measurement of different water species in minerals by means of thermal derivatography. In *Thermal analysis in the geosciences*, eds. W. Smykatz-Kloss, and S. St.J. Warne, 84–100. Berlin: Springer.

Földvári, M., P. Kovács-Pálffy, N.M. Nagy, and J. Kónya. 1998. Use of the second derivative of TG curves for investigation of the exchanged interlayer cation in montmorillonite. *Journal of Thermal Analysis* 53: 547–558.

Gallegos, D.P., K. Munn, D.M. Smith, and D.L. Stermer. 1987. A NMR technique for the analysis of pore structure: Application to materials with well-defined pore structure. *Journal of Colloid and Interface Science* 119: 127–140.

Hills, B.P., P.S. Belton, and V.M. Quantin. 1993. Water proton relaxation in heterogeneous systems 1. Saturated randomly packed suspensions of impenetrable particles. *Molecular Physics* 78: 893–908.

Hills, B.P., and J.E.M. Snaar. 1995. Water proton relaxation studies of pore microstructure in monodisperse glass bead beds. *Molecular Physics* 84:141–157.

Hlavay, J., T. Prohaska, M. Weisz, W.W. Wenzel, and G.J. Stingeder. 2004. Determination of trace elements bound to soils and sediment fractions. *Pure and Applied Chemistry* 76: 415–442.

ICDD. 2009. International Centre for Diffraction Data. www.icdd.com.

International Humic Substances Society. 2008. *Isolation of IHSS samples.* http://ihss.gatech.edu/ihss2/isolation.html.

Kate, J.M., and C.S. Gokhale. 2006. A simple method to estimate complete pore size distribution of rocks. *Engineering Geology* 84: 48–69.

Kéri, M., B. Nagy, K. László, and I. Bányai. 2021. Structural changes in resorcinol formaldehyde aerogel seen by NMR. *Microporous and Mesoporous Materials* 317: 110988.

Klug, H.P., and L.E. Alexander. 1954. *X-ray diffraction procedures*. New York: Wiley.

Kononova, M. 1972. *Organic matter of natural and agricultural soils. Experimental data and methods of investigation*. Moskva: Nauka.

Kónya, J., and N.M. Nagy. 2012. *Nuclear and radiochemistry*. Oxford: Elsevier. Hardcover ISBN: 9780123914309; Paperback ISBN: 9780323282451; eBook ISBN: 9780123914873.

Kónya, J., and N.M. Nagy. 2018. *Nuclear and radiochemistry*, 2nd ed. Oxford: Elsevier. Paperback ISBN: 9780128136430; eBook ISBN: 9780128136447.

Kuzmann, E., S. Nagy, A. Vértes, T. Weiszburg, and V.K. Garg. 1998. Applications of Mössbauer spectroscopy in mineralogy and geology. In *Nuclear methods in mineralogy and geology*, eds. A. Vértes, S. Nagy, and K. Süvegh, 285–377. New York: Plenum.

Lakanen, E., and R. Erviö. 1971. A comparison of eight extractants for the determination of plant available micronutrients in soils. *Acta Agr. Fehn* 123: 223–232.

Liptay, G. 1973. *Atlas of thermoanalytical curves*. Budapest: Akadémiai Kiadó.

Madejová, J., M. Pentrák, H. Pálková, and P. Komadel. 2009. Near-infrared spectroscopy: A powerful tool in studies of acid-treated clay minerals. *Vibrational Spectroscopy* 49: 211–218.

McGrath, D. 1996. Application of single and sequential extraction procedures to polluted and unpolluted soils. *Science of the Total Environment* 178: 37–44.

Molnár, G.L., and R.M. Lindstrom. 1998. Nuclear reaction prompt gamma-ray analysis. In *Nuclear methods in mineralogy and geology. Techniques and applications*, eds. A. Vértes, S. Nagy, K. Süvegh, 145–164. New York: Plenum Publishing Corporation.

Mukherjee, S., N. Pathak, D. Das, and D. Dutta. 2021. Engineering defect clusters in distorted NaMgF3 perovskite and their important roles in tuning the emission characteristics of Eu3+ dopant ion. *RSC Advances* 11: 5815–5831. doi: 10.1039/d0ra10008k.

O'Day, P.A. 1999. Molecular environmental geochemistry. *Reviews of Geophysics* 37: 249–274.

Petit, S. 2008. Fourier transform infrared spectroscopy. In *Handbook of clay science* 3rd ed., eds. F.B.K.G. Theng, and G. Lagaly, 909–918. Amsterdam: Elsevier.

Petrov, O.V., and I. Furó. 2009. NMR cryoporometry: Principles, applications and potential. *Progress in Nuclear Magnetic Resonance Spectroscopy* 54: 97–122.

Siesler, H.W., Y. Ozaki, S. Kawata, and H.M. Heise. 2002. *Near-infrared spectroscopy: Principles, instruments, applications*. New York: Wiley.

Smykatz-Kloss, W., and S.St.J. Warne. 1991. *Thermal analysis in the geosciences*. Berlin: Springer.

Somorjai, G.A. 1994. *Introduction to surface chemistry and catalysis*. New York: John Wiley and Sons, Inc.

Strange, J.H., M. Rahman, and E.G. Smith. 1993. Characterization of porous solids by NMR. *Physical Review Letters* 71: 3589–3591.

Walls, J.M. 1990. *Methods of surface analysis: Techniques and applications*. Cambridge: Cambridge University Press.

Warren, B.E. 1999. *X-ray diffraction*. New York: Dover Publications.

Index

Adsorption isotherms, 46–51
    BET isotherm, 50–51
    competitive Langmuir adsorption isotherm,
        49–50
    Gibbs isotherm, 46
    isotherms, heterogeneous surfaces, 49
    Langmuir adsorption isotherm, 46–50
    multilayer adsorption, 50–51
Albite, 6, 192, 202, 205
Aldehyde, 12, 222
Alkali metal ions, 23
Aluminol sites, 111
Aluminum oxides, stability diagram, 23
Amhbol, 215, 223
Amide, 12, 69
Amine, 12, 44, 104, 143
Amorphous aluminum hydroxide, 10
Amorphous ferric hydroxide, 10
Amorphous hydroxide, 10, 24–25
Amorphous manganese hydroxide, 10
Amphibolite, 205
Analytical method properties, 244–246
Anhydrite, 11, 128
Anhydrous sulfate, 11
Ankerite, 203, 205–206
Annihilation, 239
Aragonite, 11
Arsenite, 173

Bentonite, 174
    chemical modification of, 172–177
    migration of ions in, 208–210
Bentonite clay, geological origin, 187–198
    bentonites of interfacial properties, 197–198
    geological origin, 196–197
    interfacial properties, 193–196
    Sajóbábony bentonites, geological
        characteristics, 188–189
Bentonite rocks, Carpathian basin
    external specific surface, 202
    mineral composition, external specific
        surface area, 202
Bi- and trivalent ions, 62
Biotite, 5, 202, 203, 205–206
Boehmite, 10–11
Böhmite, 24
Brunauer–Emmett–Teller theory (BET)
    method, 240

Calcite, 11, 98, 203, 205–206, 215, 223
Calcium ions, 23

Calcium-lead cation exchange, selectivity
    coefficient, 163
Calcium-montmorillonite
    cation-exchanged montmorillonites prepared
        from, 106
    cation-exchange processes of, 101
Ca-montmorillonite/hydrogen-ion exchange, 124
Carbonates, 2–3, 11, 16, 45, 108, 206, 207, 234
Carbonate with compound formula, 11
Carboxyl, 12, 44, 213
Carpathian basin, bentonite rocks from
    external specific surface, 202
    mineral composition, external specific
        surface area, 202
Carrier-free radioactive isotopes, 198–208
    linear model application, 201–208
    model predicting migration rate, 198–201
Catalytic effects of clays, 67–69
Cation exchange, outer-sphere complexation,
    98–103
Cation-exchange capacity, Carpathian basin
    rocks, 202
Cation-exchanged montmorillonites
    characterization, 103–107
    synthesis, 103–107
Cesium, Sajóbábony deposit bentonite
    migration, 211
Cesium-137 sorption, kinetic parameters, 203
Characterization of external surfaces, 33–40
    application of surface complexation models,
        33–39
    deficiencies of surface complexation models,
        39–40
Characterization of interface of geological
    system/groundwater, 33–41
    external surfaces, 33–40
        deficiencies of surface complexation
            models, 39–40
    internal surfaces, 40–41
    surface complexation models for external
        surfaces, 33–39
Characterization of internal surfaces, 40–41
Chemical analysis methods, 233–235
Chemical composition of bentonite samples, 190
Chloride, bentonites from Sajóbábony deposit,
    migration, 211
Chlorite, 8, 40, 190, 203, 205–206, 215, 223
Clay minerals, 2, 4–8, 14, 33, 40–44, 67–69, 76,
        92, 95, 103, 116, 143–144, 149–150,
        198, 203, 206–207, 222, 235
Cobalt-calcium ion exchange, 127

Cobalt ions, equilibrium distribution of, 101
Colloids, 23
Competitive adsorption isotherm, model
        calculation, 65
Competitive Langmuir adsorption isotherm,
        49–50
Complexation agents, effect of, 128–142
    EDTA Ca-montmorillonite, 132–136
    EDTA/Ca-montmorillonite/lead ion system,
        140–142
    EDTA/Ca-montmorillonite/manganese ion
        system, 132–138
    interlayer space, cation composition, 138–140
Components, rocks, soil, 1–79
    adsorption, 41
    adsorption isotherms, 46–51
    alkali earth metal ions, 23
    alkali metal ions, 23
    aluminum, 23
    aluminum oxides, stability diagram, 23
    aqueous solution interfaces, 29–79
    BET isotherm, 50–51
    calcium ions, 23
    catalytic effects of clays, 67–69
    colloids, 23
    comparison of law of mass action, 58–61
    competitive adsorption isotherm, model
        calculation, 65
    competitive Langmuir adsorption isotherm,
        49–50
    deficiencies of surface complexation models,
        39–40
    disposal of nuclear waste in geological
        formations, 77–79
    electric double layer model, 32
    evaluation of interfacial processes, 45–46
    external, internal surfaces, 41–46
    external surfaces, 33–40
    geological system/groundwater, 33–41
    Gibbs isotherm, 46
    heterogeneous isotope exchange, 66–67
    hydrocarbonate, as function of pH, chemical
        species, 15
    hydroxide, 23
    internal surfaces, 40–41
    ion exchange, 42–45
        model calculation, 64
    ion-exchange isotherms, 61–62
    ion-exchange processes, 51–65
    iron, 23
        potential-pH equilibrium, 18
    isotherms, heterogeneous surfaces, 49
    kinetics of heterogeneous isotope
        exchange, 75
    kinetics of interfacial processes, 69–79
    Langmuir adsorption isotherm, 46–50

law of mass action, 52–58
    magnesium ions, 23
    manganese, potential-pH equilibrium, 18
    migration of water-soluble substances, 75–77
    mineral/composition, 2–9
    multilayer adsorption, 50–51
    organic matter in soil, 9–12
    oxides, 9–10, 23
    particle size
        external surface, 12–14
        internal surface, 12–14
    pitfall of applications of adsorption, 61–62
    pore size
        external surface, 12–14
        internal surface, 12–14
    potassium ions, 23
    precipitation, 45
    quantitative treatment of interfacial
        processes, 46–67
    separation of interfacial processes, 45–46
    silicates, 3–9
    silicon, 23
    silicon oxides, stability diagram, 24
    simultaneous ion-exchange, adsorption
        processes, 65–66
    sodium ions, 23
    solids, soil, rock, 1–14
    sorption isotherms, 58–61
    steps in interfacial reactions, 70–75
    structure of interface, 30–32
    surface complexation model characteristics,
        35–38
    surface complexation models for external
        surfaces, 33–39
    water, potential-pH equilibrium, 16
Composition of minerals, 2–9
Constant capacitance, 35, 37, 108, 110, 213
Copper, chemical species of cyanide ion, 220
Corundum, 10, 24–25
Course gravel, 13
Cristobalite, 25–26, 72, 202, 203, 205–206
Crystalline hydroxides, 10
Crystalline oxide hydroxide, 10
Crystalline oxides, 10
Cyanide anion, soil, sediment, 218–224
Cyanide ion *vs.* pH, 219

Deficiencies of surface complexation models,
        39–40
Deprotonation
    aluminol, 111, 195
    silanol, 111
Disposal of nuclear waste in geological
        formations, 77–79
Distribution coefficients, ^{137}Cs ion on
        minerals, 206

Dolomite, 11, 203, 205–206, 215, 223
  ankerite, 206

Electric double layer model, 32
Electron paramagnetic resonance (EPR) spectra,
    171, 174
Enol, 12
Ester, 12, 69
Ether, 12, 14, 44, 69
Ethylene diamin tetraacetic acid (EDTA), 105
Experimental methods, interfacial processes,
    rocks, soils, 233–247
  analytical method properties, 244–246
  chemical analysis methods, 233–235
  interface analytical methods, 243
  interface study, 241–247
  liquid phase analysis, 240–241
  mineral composition study, 235–236
  morphology study, 236–240
  soil organic matter study, 240
  solid phase analysis, 233–240

Ferrihydrites, 10
Fresh Mn-bentonite, 150

Geological origin, 196–197
  bentonite clay, 187–198
    bentonites of interfacial properties,
      197–198
    geological characteristics, 188–189
    geological origin, 196–197
    interfacial properties, 193–197
    mineral characteristics, 188–189
Geological systems, montmorillonite model
    substance, 91–177
  aluminol sites, 111
  atomic percent, 163
  calcium-lead cation exchange, selectivity
    coefficient, 163
  calcium-montmorillonite
    cation-exchanged montmorillonites
      prepared from, 106
    cation-exchange processes of, 101
  Ca-montmorillonite/hydrogen-ion exchange,
    124
  cation exchange, outer-sphere complexation,
    98–103
  cation-exchanged montmorillonites
    characterization, 103–107
    synthesis, 103–107
  cobalt-calcium ion exchange, 127
  cobalt ions, equilibrium distribution of, 101
  complexation agents, effect of, 128–142
    EDTA/Ca-montmorillonite, 132–136
    EDTA/Ca-montmorillonite/lead ion
      system, 140–142

EDTA/Ca-montmorillonite/manganese
    ion system, 132–138
  stability constants on cation composition
    of interlayer space, 138–140
deprotonation
  aluminol sites, 111
  silanol sites, 111
effect of chemical modification of bentonite
    on sorption properties, 172–177
elements of lead-montmorillonite,
    concentration profiles, 166
formation of iron oxyhydroxide nanolayer,
    interlayer space of montmorillonite,
    153–158
hydrogen ion-exchange isotherm, 124
hydrogen ions in dissolution processes of
    montmorillonite, acidic destruction of
    montmorillonite, 127–128
hydrogen ions in interfacial processes of
    montmorillonite, 120–128
  hydrogen ion on cation-exchange
    processes, 122–127
hydrogen ions in interfacial processes of
    montmorillonite, acidic destruction of
    montmorillonite, 127–128
ion adsorption, external surfaces, 113–118
  dissolved quantity of sodium ion as
    function of sorbed palladium ion, 118
  ion sorption processes, 114–118
  lead ions on montmorillonite, 118
  manganese ion, montmorillonite, 116
  palladium ion, montmorillonite, 116–118
  zinc ion, montmorillonite, 114–116
lead enrichments on montmorillonites, 165
montmorillonite as model substance, 92–98
  clay-water interactions, 95–96
  crystal structure, 92–94
  edge charges of montmorillonite, 96–97
  layer charge of montmorillonite, 92–94
  montmorillonite as model substance,
    97–98
Mössbauer parameters, bentonites, 156
old Mn-bentonite, 150
organic matter on minerals, 142–149
  EDTA on montmorillonite, 143–144
  valine on montmorillonite, 143–149
protonation
  aluminol sites, 111
  silanol sites, 111
separation of interfacial processes of
    montmorillonite, 118–120
silver-montmorillonite
  scanning electron microscopic, 159
  thermoanalytical curves, 159
structure of montmorillonite, 94

surface acid-base properties of
  montmorillonite, 107–113
  acid–base properties, cation-exchanged
    montmorillonites, 109–113
  formation of edge sites, 107–108
  permanent charge, edge charges study,
    108–109
transformations initiated by interfacial
    processes, 149–171
  Fe ion reduction, 160
  heterogeneous nucleation, 160–169
  ion reduction, 158–160
  lanthanide ion exchange, 169–171
  lead oxide fine particle, 160–169
  light lanthanide and ittrium ions, 169–171
  nanolayer, 150–153
  oxidation of Mn ion, 150–153
  palladium ion reduction, 158–160
  silver reduction, 158–160
  structural change upon heating
    lanthanide-bentonites, 171
  valine species, cation-exchanged
    montmorillonites, 148
  zinc ion/calcium-montmorillonite interfacial
    reactions, 118–119
Gibbs isotherm, 46
Gibbs-Thomson theory, 238
Goethite, 10, 22, 205, 215, 223
Gouy–Chapman model, 31
Groundwater solutions, 15–29
  compositions of natural water, 14–15
  dissolution, 22–27
  parameters affecting chemical speciation,
    15–22
  precipitation, 22–27
  properties of very dilute solutions, 27–29
Gypsite, 10
Gypsum, 11, 128, 190

Halite, 11
Hematite, 10, 205, 215, 223
Heterogeneous isotope exchange, 66–67
Hungarian rock samples, cation-exchange
    capacity, 205
Hydrated carbonate, 11
Hydrated phosphates, 11
Hydrated sulfates, 11
Hydrocarbonate, as function of pH, chemical
    species, 15
Hydrogen cyanide *vs.* pH, 219
Hydrogen ion-exchange isotherm, 124
Hydrogen ions
  dissolution processes of montmorillonite, acidic
    destruction of montmorillonite, 127–128
  interfacial processes of montmorillonite,
    120–128

hydrogen ion in cation-exchange
    processes, 122–127
  interfacial processes of montmorillonite,
    acidic destruction of montmorillonite,
    127–128
Hydroxides, 10, 23

Illite, 7, 40, 76, 128, 189–190, 192, 202, 203,
    205–206, 215, 223
  rectorite, 206
Inosilicates, 5
Interface analytical methods, 243
Interface of rock/soil-aqueous solutions, 29–79
  catalytic effects of clays, 67–69
  interface of geological system/groundwater,
    33–41
    characterization of external surfaces,
      33–40
    characterization of internal surfaces,
      40–41
    complexation models for external
      surfaces, 33–39
    deficiencies of surface complexation
      models, 39–40
  interfacial processes related to external,
      internal surfaces, 41–46
    adsorption, 41
    evaluation of interfacial processes, 45–46
    ion exchange, 42–45
    precipitation, 45
    separation of interfacial processes, 45–46
  kinetics of interfacial processes, 69–79
    disposal of nuclear waste in geological
      formations, 77–79
    kinetics of heterogeneous isotope
      exchange, 75
    migration of water-soluble substances,
      75–77
    steps in interfacial reactions, 70–75
  quantitative treatment of interfacial
      processes, 46–67
    adsorption isotherms, 46–51
    BET isotherm, 50–51
    comparison of law of mass action and
      isotherms, 58–61
    competitive Langmuir adsorption
      isotherm, 49–50
    Gibbs isotherm, 46
    heterogeneous isotope exchange, 66–67
    ion exchange by law of mass action,
      52–58
    ion-exchange isotherms, 64–65
    ion-exchange processes, 51–65
    isotherms, heterogeneous surfaces, 49
    Langmuir adsorption isotherm, 46–50
    multilayer adsorption, 50–51

pitfall of applications of adsorption, 64–65
simultaneous ion-exchange, adsorption processes, 65–66
sorption isotherms, 58–61
structure of interface, 30–32
Interfacial acid–base properties of soils, 212–218
Interfacial processes
geological systems
aluminol sites, 111
atomic percent, 163
Ca-montmorillonite/hydrogen-ion exchange, 124
cation-exchanged montmorillonites, characterization, 103–107
hydrogen ions in dissolution processes of montmorillonite, acidic destruction of montmorillonite, 127–128
hydrogen ions in interfacial processes of montmorillonite, 120–128
lead oxide fine particle, 160–169
montmorillonite as model substance, layer charge of montmorillonite, 92–94
nanolayer, 150–153
palladium ion reduction, 158–160
protonation, aluminol sites, 111
silanol sites, 111
montmorillonite model substance, 91–177
acid–base properties, cation-exchanged montmorillonites, 109–113
calcium-lead cation exchange, selectivity coefficient, 163
cation exchange, outer-sphere complexation, 98–103
cation-exchanged montmorillonites, synthesis, 103–107
cation-exchanged montmorillonites prepared from, 106
cation-exchange processes of, 101
clay-water interactions, 95–96
cobalt-calcium ion exchange, 127
cobalt ions, equilibrium distribution of, 101
complexation agents, effect of, 128–142
crystal structure, 92–94
dissolved quantity of sodium ion as function of sorbed palladium ion, 118
edge charges of montmorillonite, 96–97
EDTA/Ca-montmorillonite, 132–136
EDTA/Ca-montmorillonite/lead ion system, 140–142
EDTA/Ca-montmorillonite/manganese ion system, 132–138
EDTA on montmorillonite, 143–144
effect of chemical modification of bentonite on sorption properties, 172–177

elements of lead-montmorillonite, concentration profiles, 166
Fe ion reduction, 160
formation of edge sites, 107–108
formation of iron oxyhydroxide nanolayer, interlayer space of montmorillonite, 153–158
fresh Mn-bentonite, 150
heterogeneous nucleation, 160–169
hydrogen ion-exchange isotherm, 124
hydrogen ion on cation-exchange processes, 122–127
hydrogen ions in interfacial processes of montmorillonite, 120–128
hydrogen ions in interfacial processes of montmorillonite, acidic destruction of montmorillonite, 127–128
ion adsorption, external surfaces, 113–118
ion reduction, 158–160
ion sorption processes, 114–118
lead enrichments on montmorillonites, 165
lead ions on montmorillonite, 118
manganese ion, montmorillonite, 116
montmorillonite as model substance, 92–98
Mössbauer parameters, bentonites, 156
old Mn-bentonite, 150
organic matter on minerals, 142–149
oxidation of Mn ion, 150–153
palladium ion, montmorillonite, 116–118
permanent charge, edge charges study, 108–109
protonation, silanol, 111
scanning electron microscopic, 159
separation of interfacial processes of montmorillonite, 118–120
silver reduction, 158–160
stability constants on cation composition of interlayer space, 138–140
structure of montmorillonite, 94
surface acid-base properties of montmorillonite, 107–113
thermoanalytical curves, 159
transformations initiated by interfacial processes of montmorillonite, 149–171
valine on montmorillonite, 143–149
valine species, cation-exchanged montmorillonites, 148
zinc ion, montmorillonite, 114–116
zinc ion/calcium-montmorillonite interfacial reactions, 118–119
Interfacial processes related to external, internal surfaces, 41–46
adsorption, 41
evaluation of interfacial processes, 45–46
ion exchange, 42–45

precipitation, 45
separation of interfacial processes, 45–46
Interfacial properties, 193–197
Interfacial reactions, 187–228
    acidic brown forest soil with clay
        illuviation, 214
    amphibolite, 205
    ankerite, 205
    bentonite rocks from Carpathian basin
        external specific surface, 202
        mineral composition, external specific
            surface area, 202
    biotite, 5, 202, 203, 205–206
    brown forest soil with clay illuviation, 214
    calcarous meadow soil, 214
    calcarous sandy soil, 214
    carbonates, 2–3, 9, 16, 45, 108, 206, 207, 234
    cation-exchange capacity, bentonite rocks
        from Carpathian basin, 202
    cesium, bentonites from Sajóbábony deposit,
        migration, 211
    cesium-137 sorption, kinetic parameters, 203
    chemical composition of bentonite
        samples, 190
    chloride, bentonites from Sajóbábony deposit,
        migration, 211
    clayey meadow soil, 214
    clay minerals, 206
    copper, chemical species of cyanide ion, 220
    cristobalite, 205
    cyanide anion, soil, sediment, 218–224
    cyanide ion vs. pH, 219
    deprotonation, silanol, aluminol, 195
    distribution coefficients, ^{137}Cs ion on
        minerals, 206
    dolomite, ankerite, 206
    geological origin, bentonite clay, 187–198
        bentonites of interfacial properties,
            197–198
        geological characteristics, 188–189
        geological origin, 196–197
        interfacial properties, 193–197
        mineral characteristics, 188–189
    group of minerals, 206
    humid soil, 214
    humus sand, 214, 223
    Hungarian rock samples, cation-exchange
        capacity, 205
    hydrogen cyanide vs. pH, 219
    illite, 205
        rectorite, 206
    interfacial acid-base properties of soils,
        212–218
    kalifeldspar, 189, 202, 203, 205–206, 215,
        223
        plagioclase, 206

loamy soil, 214
meadow solonetz, 214
mica plate, suspension of natural clay
    sediment, 212
migration cell, 208
mineral composition of bentonite
    samples, 190
montmorillonite, 4–5, 43, 99, 103–120,
    124–130, 132–138, 140–146, 149–151,
    153–154, 158–161, 166–173, 187–193,
    196–197, 202, 203, 205–206, 215, 223
paligorscite, 205
plagioclase, 205
protonation, silanol, aluminol, 195
quartz, cristobalite, 206
River Tisza samples, 215, 223
soil sample characteristic parameters, 214
sorption of phosphate anion on soils, 225–228
strontium ions, bentonites from Sajóbábony
    deposit, migration, 211
tectosilicates, 206
Tiszalök samples, cyanide, 224
Tiszavasvári samples, cyanide, 224
water-soluble substances in rocks, 198–212
    carrier-free radioactive isotopes, 198–208
    linear model application, 201–208
    migration of ions in bentonite, 208–210
    model predicting migration rate, 198–201
    precipitation in migration, 210–212
zinc, chemical species of cyanide ion, 221
Ion adsorption, external surfaces, 113–118
    dissolved quantity of sodium ion as function
        of sorbed palladium ion, 118
    lead ions on montmorillonite, 118
    ion sorption processes, 114–118
        manganese ion, montmorillonite, 116
        palladium ion, montmorillonite, 116–118
        zinc ion, montmorillonite, 114–116
Ion exchange, 42–45, 51–65 .
    comparison of law of mass action and
        isotherms, 58–61
    ion exchange by law of mass action, 52–58
    ion-exchange isotherms, 64–65
    by law of mass action, 52–58
    model calculation, 64
    pitfall of applications of adsorption, 64–65
    sorption isotherms, 58–61
Ion-exchange isotherms, 64–65
Iron, 23
    potential-pH equilibrium, 18
Iron oxyhydroxide nanolayer, interlayer space of
    montmorillonite, 153–158
Isotherms, heterogeneous surfaces, 49

Kalifeldspar, 189, 202, 203, 205–206, 215, 223
    plagioclase, 206

Kaolinite, 7, 22, 76, 98, 128, 168, 202, 205, 223
Kelvin's capillary condensation equation, 237
Ketone, 12, 224
Kinetics of heterogeneous isotope exchange, 75
Kinetics of interfacial processes, 69–79
    disposal of nuclear waste in geological
        formations, 77–79
    kinetics of heterogeneous isotope
        exchange, 75
    migration of water-soluble substances, 75–77
    steps in interfacial reactions, 70–75

Langmuir adsorption isotherm, 46–50
Langmuir model, 177
Lanthanide-bentonites, structural change upon
    heating, 171
Law of mass action, 58–61
Layered clay minerals, 7–8
Lead enrichments on montmorillonites, 165
Lead-montmorillonite, concentration profiles,
    166
Lead oxide fine particles, montmorillonite,
    160–169
Lepidokrokit, 10
Light lanthanide and ittrium ions, 169–171
Liquid, soil, groundwater solutions, 15–29
    composition of natural water, 14–15
    dissolution, 22–27
    parameters affecting chemical speciation,
        15–22
    precipitation, 22–27
    properties of very dilute solutions, 27–29
Loamy soil, 214

Magnesite, 11
Magnesium ions, 23
Magnetite, 10
Manganese, potential-pH equilibrium, 18
Meadow solonetz, 214
Mercury ion, 98–99
Mica plate, suspension of natural clay
    sediment, 212
Mirabilite, 11
Mixing entropy, 66
Mn-bentonite, 150
Mono and bivalent ions, 62
Mono and trivalent ions, 62
montmorillonite, 127–128
    external surfaces, ion adsorption
        aluminol sites, 111
        atomic percent, 163
        calcium-lead cation exchange, selectivity
            coefficient, 163
        calcium-montmorillonite, 101, 106
        Ca-montmorillonite/hydrogen-ion
            exchange, 124

cation exchange, outer-sphere
    complexation, 98–103
cation-exchanged montmorillonites,
    103–107
clay-water interactions, 95–96
cobalt-calcium ion exchange, 127
cobalt ions, equilibrium distribution of, 101
complexation agents, effect of, 128–142
crystal structure, 92–94
deprotonation, 111
dissolved quantity of sodium ion as
    function of sorbed palladium ion, 118
edge charges of montmorillonite, 96–97
effect of chemical modification of
    bentonite on sorption properties,
    172–177
elements of lead-montmorillonite,
    concentration profiles, 166
formation of iron oxyhydroxide nanolayer,
    interlayer space of montmorillonite,
    153–158
hydrogen ion-exchange isotherm, 124
hydrogen ions in dissolution processes of
    montmorillonite, acidic destruction of
    ion sorption processes, 114–118
    lead ions on montmorillonite, 118
    manganese ion, montmorillonite, 116
    palladium ion, montmorillonite, 116–118
    zinc ion, montmorillonite, 114–116
fresh Mn-bentonite, 150
hydrogen ions in interfacial processes of
    montmorillonite, 120–128
    acidic destruction of montmorillonite,
        127–128
    hydrogen ion on cation-exchange
        processes, 122–127
layer charge of, 92–94
lead enrichments on, 165
as model substance, 92–98
Mössbauer parameters, bentonites, 156
old Mn-bentonite, 150
organic matter on minerals, 142–149
    EDTA on montmorillonite, 143–144
    valine on montmorillonite, 143–149
protonation
    aluminol sites, 111
    silanol sites, 111
separation of interfacial processes of,
    118–120
silver-montmorillonite
    scanning electron microscopic, 159
    thermoanalytical curves, 159
structure of, 94
surface acid-base properties of, 107–113
    acid–base properties, cation-exchanged
        montmorillonites, 109–113

formation of edge sites, 107–108
permanent charge, edge charges study,
      108–109
transformations initiated by interfacial
      processes of, 149–171
   Fe ion reduction, 160
   heterogeneous nucleation, 160–169
   ion reduction, 158–160
   lanthanide ion exchange, 169–171
   lead oxide fine particle, 160–169
   light lanthanide and ittrium ions, 169–171
   nanolayer, 150–153
   oxidation of Mn ion, 150–153
   palladium ion reduction, 158–160
   silver reduction, 158–160
   structural change upon heating
      lanthanide-bentonites, 171
   valine species, cation-exchanged
      montmorillonites, 148
zinc ion/calcium-montmorillonite interfacial
      reactions, 118–119
Morphology study, 236–240
Mössbauer parameters, bentonites, 156
Mössbauer spectrum, 170, 171
Multilayer adsorption, 50–51

Native soda, 11
Natron, 11
Nesosilicates, 5
Nuclear magnetic resonance (NMR) method, 237

Old Mn-bentonite, 150
Opal, 10, 202, 205
Organic matter in soil, 9–12
Organic matter on minerals, 142–149
   EDTA on montmorillonite, 143–144
   valine on montmorillonite, 143–149
Oxides, 9–10, 23

Paligorscite, 202, 205
Parameters affecting chemical speciation, 15–22
Particle size
   classifications of soils by, 13
   external surface, 12–14
   internal surface, 12–14
Peptide, 12, 69
Phenolic OH, 12
Phosphate, 173
phosphate anion, sorption of, 225–228
Phyllosilicates, 4, 5, 92, 203, 206
Plagioclase, 6, 128, 189–190, 192, 202, 203,
      205–206, 215, 223
Positron annihilation techniques, 238–239
Positronium, 239
Potassium ions, 23
Precipitation, 22–27, 45

Properties of very dilute solutions, 27–29
Protonation
   aluminol, 195
   aluminol sites, 111
   silanol sites, 111, 195
Pyrite, 11
Pyrolusite, 10

Quantitative treatment, interfacial processes,
      46–67
   adsorption isotherms, 46–51
      BET isotherm, 50–51
      competitive Langmuir adsorption
         isotherm, 49–50
      Gibbs isotherm, 46
      isotherms, heterogeneous surfaces, 49
      Langmuir adsorption isotherm, 46–50
      multilayer adsorption, 50–51
   heterogeneous isotope exchange, 66–67
   ion-exchange processes, 51–65
      comparison of law of mass action and
         isotherms, 58–61
      ion exchange by law of mass action,
         52–58
      ion-exchange isotherms, 64–65
      pitfall of applications of adsorption,
         64–65
      sorption isotherms, 58–61
   simultaneous ion-exchange, adsorption
      processes, 65–66
Quartz, 6, 10, 25–26, 98, 128, 189–190, 192, 202,
      203, 205–207, 215, 223
   cristobalite, 206
Quinone, 12

Rectorite, 202, 203, 206–207
Redox reactions, 176
River Tisza samples, 215, 223

Sajóbábony bentonites
   geological characteristics, 188–189
   mineral characteristics, 188–189
Separation of interfacial processes of
         montmorillonite, 118–120
Siderite, 11
Silicates, 3–9
Silicon, 23
Silicon oxides, stability diagram, 24
Silver bentonite, 173
Silver-montmorillonite
   scanning electron microscopic, 159
   thermoanalytical curves, 160
Smectite, 8, 40, 77, 92, 205
Sodium chloride, 11
Sodium ions, 23
Solid phase analysis, 233–240

chemical analysis methods, 233–235
mineral composition study, 235–236
morphology study, 236–240
soil organic matter study, 240
Sorption isotherms, 58–61
Steps in interfacial reactions, 70–75
Strengite, 11
Strontium ions, bentonites from Sajóbábony
    deposit, migration, 211
Structure of montmorillonite, 94
Sulfide, 11
Surface acid-base properties, montmorillonite,
    107–113
    acid–base properties, cation-exchanged
        montmorillonites, 109–113
    formation of edge sites, 107–108
    permanent charge, edge charges study, 108–109
Surface complexation models
    characteristics of, 35–38
    external surfaces, 33–39
Surface relaxation strength, 237

Tectosilicates, 2, 6, 203, 206, 207
Transformations initiated by interfacial processes
    of montmorillonite, 149–171
    Fe ion reduction, 160
    heterogeneous nucleation, 160–169
    ion reduction, 158–160

lanthanide ion exchange, 169–171
lead oxide fine particle, 160–169
light lanthanide and ittrium ions, 169–171
nanolayer, 150–153
oxidation of Mn ion, 150–153
palladium ion reduction, 158–160
silver reduction, 158–160
structural change upon heating lanthanide-
    bentonites, 171
Types of silicate, 5–6

Valine species, cation-exchanged
    montmorillonites, 148
Variscite, 11
Vermiculite, 7, 40
Vivianite, 11

Water-soluble substances, rock, 198–212
    carrier-free radioactive isotopes, 198–208
        linear model application, 201–208
        model predicting migration rate, 198–201
    migration of ions in bentonite, 208–210
    precipitation in migration, 210–212
Weathering of silicates, 22

Zinc, chemical species of cyanide ion, 221
Zinc ion/calcium-montmorillonite interfacial
    reactions, 118–119

Taylor & Francis eBooks

www.taylorfrancis.com

A single destination for eBooks from Taylor & Francis with increased functionality and an improved user experience to meet the needs of our customers.

90,000+ eBooks of award-winning academic content in Humanities, Social Science, Science, Technology, Engineering, and Medical written by a global network of editors and authors.

TAYLOR & FRANCIS EBOOKS OFFERS:

A streamlined experience for our library customers

A single point of discovery for all of our eBook content

Improved search and discovery of content at both book and chapter level

REQUEST A FREE TRIAL
support@taylorfrancis.com

 Routledge
Taylor & Francis Group

 CRC Press
Taylor & Francis Group